IUTAM SYMPOSIUM ON MULTISCALE PROBLEMS IN MULTIBODY SYSTEM CONTACTS

IUTAM BOOKSERIES

Volume 1

Aims and Scope of the Series

The IUTAM Bookseries publishes the proceedings of IUTAM symposia under the auspices of the IUTAM Board.

For a list of related mechanics titles, see final pages.

IUTAM Symposium on Multiscale Problems in Multibody System Contacts

Proceedings of the IUTAM Symposium held in Stuttgart, Germany, February 20–23, 2006

Edited by

PETER EBERHARD

University of Stuttgart, Germany

Springer

A C.I.P. Catalogue record for this book is available from the Library of Congress.

ISBN 978-1-4020-5980-3 (HB)
ISBN 978-1-4020-5981-0 (e-book)

Published by Springer,
P.O. Box 17, 3300 AA Dordrecht, The Netherlands.

www.springer.com

Printed on acid-free paper

Contents

Preface

The investigations of multiscale problems in multibody system contacts is a most interesting and timely topic which is the subject of intensive research for more than a decade. Many questions have already been answered and the mechanically sound description and simulation is increasingly being applied to practical engineering problems. The IUTAM Symposium on "Multiscale Problems in Multibody System Contacts" in Stuttgart, Germany which was held February 20-23, 2006 facilitated discussions between researchers active in the field and enabled us to review the current state of the subject, and to identify the important topics which require further effort.

Multiscale problems occur very naturally in contact mechanics. Typically the contact forces and stresses are very high and they are transmitted within a very short period of time. This leads to questions, e.g. how the slow rigid body motion and the fast motion changes can be considered simultaneously or how interface effects couple with wave propagation and the large-scale motion.

The purpose of the symposium was to provide a basis for discussion and exchange of new concepts and ideas between scientists from all over Europe and the world. Emphasis was placed also on sharing algorithms and concepts with young researchers who only recently entered the field of mechanical contacts.

When organizing such an event, one can never know whether one will succeed, whether people will be interested and whether one will get a good blend of different topics and aspects. In the announcement it was mentioned that one of the goals was to bring together scientists from closely related but traditionally distinct fields such as multibody systems contact, molecular dynamics, finite element contact, collision detection and the mathematics of contact. Although it is improper for us to judge whether the Symposium was an overall success, we are happy to report that at least this goal was reached.

We had participants from mechanical engineering, civil engineering, chemical engineering, mathematics, physics, chemistry, from universities and industrial companies. This emphasizes the interdisciplinary nature of the subject, the strong component of basic research and the increasing importance in industrial applications. It can be observed that the once clear boundaries between these fields blur and an exchange of ideas will accelerate the development of mutual benefits.

Some numbers: The meeting attracted 57 fully registered participants from 17 different countries, i.e.,

> Belgium (1), Brazil (1), Canada (1), China (4), Denmark (1), Finland (3), France (1), Germany (31), Italy (1), Lithuania (1), Netherlands (2), Poland (2), Portugal (2), Russia (1), Spain (2), Switzerland (2), UK (1)

as well as several students and PhD-students who attended the talks, too.

After approval by IUTAM the members of the Scientific Committee, Jorge Ambrosio from Portugal, Christoph Glocker from Switzerland, Peter Eberhard from Germany (Chairman), Anders Klarbring from Sweden, Stefan Luding from The Netherlands, Panos Papadoulos from the USA, Xiaoting Rui from the P.R. China, Bill Stronge from the UK and Werner Schiehlen from Germany (IUTAM observer) did the review procedure and selection of the different contributions for the conference talks.

We heard 42 well presented talks, organized in sequential sessions. All contributions lasted 30 minutes including 5 minutes of discussion. Although there were no special plenary talks intended or given during this IUTAM Symposium, I want to mention the talks by Dr. Harting, Prof. Kuhn and Prof. Wohlmuth who were specially invited. Their topics were closely related to the symposium but partially unknown to many of us - particles in flows, fracture and cracks and mortar methods in FEM.

The programme of the symposium was organized in 13 sessions:

- Particle Adhesion
- Impact of Elastic Bodies
- Contact in Applications
- Frictional Contact
- Granular Media I and II
- Gears and Bearings
- Multiscale Aspects
- Fracture
- Complementarity
- FE Contact and Mortar
- Non-Smooth Models
- Contact in Multibody Systems

It may be mentioned that there was not a single 'no-show' at the conference - neither in the talks nor among the registered participants. Besides the presentations and discussions in the scientific program there was time for social interaction during two receptions, an excursion and the conference dinner.

Since many presentations and papers are related to more than one of the session topics, the contributions in this book are organized in alphabetical order with respect to the name of the presenting author.

All my coworkers helped in the preparation of this symposium and I want to thank all of them. From various responses of participants I know that they did very well to make the stay of our guests as comfortable and interesting as possible and they made it possible to enjoy a smoothly running IUTAM Symposium.

Many thanks are deserved by R. Prommersberger/B. Muth (registration, conference bags), A. Eiber (excursion, conference dinner), R. Seifried/F. Fleissner (receptions), K. Sedlaczek/P. Ziegler (computers, presentations), T. Gaugele/C. Breuninger (lecture room, fotos, equipment), C. Henninger/M. Ackermann (coffee breaks, beverages), A. Ast/H. Alkhaldi (www, program booklet), M. Lehner (Proceedings), S. Ebrahimi/B. Mavroudakis/Y. Jiang (misc) as well as by many students who supported our activities (M. Buettner, M. Dippon, M. Hoefert, P. Jakob, T. Kurz, K. Mast, J. Metzger, A. Puzik, S. Sanzenbacher, F. Seybold, C. Simader, J. Tertilt).

Finally it remains to thank all members of the Scientific Commitee, the session chairs, all participants and all authors for their efforts and contributions.

Stuttgart, October 2006 *Peter Eberhard*

Welcome Address by Prof. Wolfgang Ehlers, Vice Rector for Organization at the University of Stuttgart

On behalf of the rectorate of our University, I would like to express a few words of welcome to the organizers of this event and to you, the distinguished participants of this symposium. First of all, I would like to thank Professor Peter Eberhard from the Faculty of Mechanical Engineering, Institute of Engineering and Computational Mechanics, for the fact that he took the initiative to successfully apply for a IUTAM Symposium on Multiscale Problems in Multibody System Contacts. IUTAM Symposia, in general, as you will all know, are outstanding events of high scientific impact, and our university is proud that, again, in the line of several comparable events before, one of our members succeeded in the organization of one of this important scientific events. As far as I am aware, it is the policy of IUTAM to send a representative to such an event in order to supervise and to control the high standard of IUTAM symposia. Today we have two representatives here: Werner Schiehlen and Dick van Campen. It is a great honor for me to welcome especially Professor Dick van Campen from the Technical University of Eindhoven. Dick van Campen is not only the IUTAM representative but he is also the present IUTAM Secretary General in charge, so nice to have you here!

The modern and advanced methods in Continuum Mechanics as well as in Particle Mechanics and Molecular Dynamics are the basic tools for nowadays computational engineering. Especially the computation of particle mechanics problems with its generally extremely high number of particles leads to considerable numerical difficulties. Introducing contact mechanics as an independent field of research yields a new insight both in the field of particle mechanics as well as in the combination of continuum and particle methods. Mechanics, in general, is obviously a big issue not only in research but also in the field of scientific education in mechanical engineering, in civil and environmental engineering and in a variety of further branches that we all have here at our university.

Ladies and gentlemen, let me say some words about the University of Stuttgart. Founded in 1829 as a Technical School, the University of Stuttgart became a Polytechnical School in 1840 and a Polytechnicum in 1876 before it got an Institute of Technology in 1890. Thus, the nowadays University of Stuttgart is in the same line like other well-noted schools in Germany as there are the Technical Universities of Aachen, Berlin, Braunschweig, Darmstadt, Dresden, Hannover, Karlsruhe and München. Today, these schools together with the University of Stuttgart form the so-called club of 9 Technical Universities, an informal association with the goal of transferring the interests of German Institutes of Technology into both policy and society. As you might know, we are just now living in very exciting times, namely, since the Excellence Initiative by the German Federal and State Governments to promote science and research at German universities has been initiated last summer.

In these days, the first of two application periods is closed and 41 of all together 157 initiatives to found clusters of excellence with a financial volume of 6.5 Mio Euro per year have been selected for final application. We are glad to say that we have been successful with our application for a cluster of Simulation Technology. Furthermore, there have been selected 39 of all together 135 initiatives to found International Graduate Schools with a financial volume of 1.5 Mio. Euro per year. Again, we are glad to say that we have been successful with our application for a graduate school for Advanced Manufacturing Engineering. All these initiatives will pass a final ranking in October 2006. If we finally succeed to reach one of the first 15 positions in the frame of the clusters of excellence and one of the first 20 positions of the graduate schools, then we will have a lot of money both for research and scientific education in the next five and possibly in the next ten years. Of course, it is interesting to see what are the goals of such an initiative. The German universities are well-known all over the world, however, their visibility could be improved. Thus, it is one of the goals of the Excellence Initiative to promote some few universities as a result of a very competitive procedure such that these universities, in the sense of universities of excellence, have the chance to glare over the whole scientific community.

In the frame of this procedure, the German universities are undergoing a variety of changes both in their scientific domains as well as in their educational structure. From the scientific point of view, it will be necessary to concentrate on those fields, where the universities have been most successful in the last years. Concerning the University of Stuttgart, it is only natural that we will concentrate on our engineering and natural science branches. From the educational point of view, the German universities are changing in these years from their traditional system of diploma and magister curricula towards the international system of bachelor and master degrees. In addition, it is our goal to found international graduate schools for the promotion of young scientists and we furthermore want to offer junior professorships for the best of them. However, this is a complicated process that will take some years, but we sincerely hope, I should better say, we are sure, to succeed.

Ladies and gentlemen, I would like to take this opportunity of wishing you a successful symposium, a nice and pleasant stay at our university and, of course, at the city of Stuttgart, and every success you want for this event and for your scientific future.

Welcome Address by Prof. Dick van Campen, IUTAM Secretary General, TU Eindhoven

It is a great pleasure and honor for me to deliver not only a welcome address on behalf of IUTAM, but also - as I have just learnt - to officially open this symposium.

First, I will introduce IUTAM to you and then I will say some words about this symposium. IUTAM can be considered to be the world umbrella organization in the field of solid and fluid mechanics with couplings to related sciences. For the purpose of this symposium I put solids ahead of fluids.

IUTAM was officially formed in 1946 and it was preceded by world congresses starting already in 1924. Of course, I am somewhat proud that Delft was the first place where these conferences was held.

IUTAM is an adhering member of the International Council of Science (abbreviated ICSU). At present, IUTAM embraces 49 adhering organizations, representing nations all over the world. The represntatives of these organizations are the main constituents of the General Assembly of IUTAM.

Furthermore at the moment there are 17 affiliated organizations representing subareas of mechanics or special geographical regions. For example the European Mechanics Society (EUROMECH) is an affiliated organization of IUTAM. An example of an affiliated organization of another category is CISM, the International Centre for Mechanical Sciences in Unine, Italy, which may be well known to many of you. CISM keeps the IUTAM archive, but it also organizes, on a regular basis, many interesting courses and very often, once per year or once per two years, there is an IUTAM-CISM summer course.

Other important examples of affiliated organizations are the International Association of Computational Mechanics, the International Association for Vehicle System Dynamics and the International Society for Structural and Multidisciplinary Optimization.

It is interesting to have a look at the organizational scheme of IUTAM, which I will not discuss in detail and which is established on the IUTAM website. I will show the web address in a while and I invite you to study this scheme from your computer.

The IUTAM Symposia and the IUTAM Summer Schools constitute major IUTAM activities. The present IUTAM Symposium here in Stuttgart is one of the Symposia accepted at the 2004 biannual meeting of the General Assembly. Prior to the General Assembly meeting the proposals for Symposia and Summer Schools are reviewed by the IUTAM Symposia Panels.

Another major IUTAM activity is the IUTAM World Congress (abbreviated ICTAM), held once every four years. The Congress Committee, whose members are appointed by the General Assembly, is responsible for the program of these World Congresses. The last ICTAM was held in Warsaw in August 2004.

The motto of all these world congresses is that they are a kind of Olympics in mechanics. In Warsaw there were about 1250 contributions, about 1500 participants, 6 mini-symposia and 60 prenominated sessions. On the occasion of the millennium congress in Chicago a special poster was designed by an artist. This poster is named Meters of Motion. All details are not very clearly visible on the screen here, but depicted are a number of famous historical scientists in mechanics and also some present days well-known figures related to the field. Together with an explanation this picture can be found in the history section on the IUTAM website, whose internet address is shown here. Again, I invite you to study this nice poster in detail from your computer.

The next world congress will be held in Adelaide, Australia, in 2008, and I show you here a part of the currently available homepage of that World Congress. The associated the web address still looks a little bit complex, but maybe it will be replaced by a simpler one afterwards.

The major output of IUTAM consists of the proceedings of the IUTAM Symposia, the proceedings of the World Congresses, the annual reports and the newsletters. I've taken some copies with me of the last annual report and of the newsletter, you can have them for free. If there are not enough copies, please contact me.

Now, let me move on to the present IUTAM symposium. I think that Professor Ehlers has already already given a kind of introduction to the Symposium contents. I just want to add that nowadays multibody dynamics has become a very flourishing area. Moreover, Professor Schiehlen, who is present here, has guided the field over the past decades and has provided major impulses to the florishing state of the field.

As can be deducted from the introduction by Professor Ehlers, the present Symposium is very timely and also very multidisciplinary. Also, as was mentioned already by Professor Eberhard, there are interactions with new disciplines. Because the number of presentations and the number of participants have already been given by Professor Eberhard, there is need to repeat them now.

The chairman of this IUTAM Symposium is Professor Peter Eberhard, whereas Professor Werner Schiehlen acts as the official IUTAM representative. You may well know that there is a tradition here in Stuttgart, with respect to well-organized IUTAM symposia under the responsibility of Professor Schiehlen. Hence, the appointment of Professor Werner Schiehlen as a kind of IUTAM watchdog provided an additional opportunity to have this Symposium organized in the sprit of the Stuttgart tradition. Of cause, this is just a little joke and given the reputation of Professor Peter Eberhard we can be sure that the Symposium will go anyway according to the Stuttgart tradition.

So, I am very glad that I have the opportunity to open this Symposium and to welcome all of you. I want to express my sincere thanks to the chairman, the organizers and the scientific committee for all the hard work done. I am sure that the symposium will be a great success, on behalf of IUTAM I greet all of you and wish you a very pleasant, successful and fruitful Symposium.

List of Contributors

Jorge Ambrósio
Technical University of Lisbon
IDMEC - Instituto Superior Técnico
Av. Rovisco Pais
1049-001 Lisboa, Portugal
jorge@dem.ist.utl.pt

Martin Arnold
Institute of Mathematics
Martin Luther University Halle-Wittenberg
06099 Halle (Saale), Germany
martin.arnold@mathematik.uni-halle.de

Gunther Blankenhorn
Institut für Mechanik
Universität Karlsruhe (TH)
Englerstraße 2
76131 Karlsruhe, Germany
mechanik@ifm.uni-karlsruhe.de

Michael Breidt
Lehrstuhl für Ingenieurinformatik im Bauwesen
Ruhr Universität Bochum
44780 Bochum, Germany
hartus@inf.bi.rub.de

Li-Qun Chen
Department of Mechanics
Shanghai University
Shanghai 20444, China
lqchen@staff.shu.edu.cn

Peter Eberhard
Institute of Engineering and Computational Mechanics
University of Stuttgart
Pfaffenwaldring 9
70569 Stuttgart, Germany
eberhard@itm.uni-stuttgart.de

Saeed Ebrahimi
Institute of Engineering and Computational Mechanics
University of Stuttgart
Pfaffenwaldring 9
70569 Stuttgart, Germany
ebrahimi@itm.uni-stuttgart.de

Alexander Fidlin
LuK GmbH & Co. oHG
Industriestraße 3
77815 Bühl, Germany
alexander.fidlin@luk.de

Florian Fleissner
Institute of Engineering and Computational Mechanics
University of Stuttgart
Pfaffenwaldring 9
70569 Stuttgart, Germany
fleissner@itm.uni-stuttgart.de

Christoph Glocker
IMES - Center of Mechanics
ETH Zurich
8092 Zurich, Switzerland
glocker@imes.mavt.ethz.ch

Veli-Matti Järvenpää
Tampere University of Technology
P.O. Box 589
33101, Tampere, Finland
veli-matti.jarvenpaa@tut.fi

Rimantas Kačianauskas
Laboratory of Numerical Modelling
Vilnius Gediminas Technical University
Saulėtekio al. 11
10223 Vilnius-40, Lithuania
rkac@fm.vtu.lt

Rainer Keppler
Institut für Technische Mechanik
Universität Karlsruhe (TH)
Kaiserstraße 12
76131 Karlsruhe, Germany
keppler@itm.uni-karlsruhe.de

Erno Keskinen
Laboratory of Machine Dynamics
Tampere University of Technology
P.O. Box 589
33101, Tampere, Finland
erno.keskinen@tut.fi

Alexander Konyukhov
Institut für Mechanik
Universität Karlsruhe
Kaiserstraße 12
76131 Karlsruhe, Germany
Konyukhov@ifm.uni-karlsruhe.de

Alexander S. Kuleshov
Department of Mechanics and Mathematics
Moscow State University
Main building of MSU
Leninskie gory
Moscow 119992, Russia
kuleshov@mech.math.msu.su

Sirpa Launis
John Deere Forestry
Lokomonkatu 21, Box 474
33101 Tampere, Finland
LaunisSirpaS@johndeere.com

Caishan Liu
Department of Mechanics & Engineering Science
Peking University
Beijing, 100871, China
liucs@pku.edu.cn

Stefan Luding
Particle Technology, Nanostructural Materials
DelftChemTech
TU Delft
Julianalaan 136
2628 BL Delft, The Netherlands
s.luding@tudelft.nl

Alexander Lünenschloß
Dynamics Group
Dept. of Mechanical Engineering
TU Darmstadt
Hochschulstraße 1
64289 Darmstadt, Germany
luenenschloss@dyn.tu-darmstadt.de

Steffen Mattern
Institut für Mechanik
Universität Karlsruhe (TH)
Englerstraße 2
76131 Karlsruhe, Germany
mechanik@ifm.uni-karlsruhe.de

Sean McNamara
Institut für Computerphysik
Universität Stuttgart
Pfaffenwaldring 27
70569 Stuttgart, Germany
sean@icp.uni-stuttgart.de

Jaap P. Meijaard
School of MMME
The University of Nottingham
University Park
Nottingham NG9 2RD, United Kingdom
jaap.meijaard@nottingham.ac.uk

Michael Möller
IMES - Center of Mechanics
ETH Zurich
8092 Zurich, Switzerland
michael.moeller@ imes.mavt.ethz.ch

Pauli Pedersen
Dept. of Mechanical Engineering, Solid Mechanics
Technical University of Denmark
Nils Koppels Allé, Building 404
2800 Kgs.Lyngby, Denmark
pauli@mek.dtu.dk

Ettore Pennestrì
Università di Roma Tor Vergata
Dipartimento di Ingegneria Meccanica
via del Politecnico
1 00133 Roma, Italy
pennestri@mec.uniroma2.it

Frederico Grases Rauter
SNCF-Direction de l'Innovation et de la Recherche
45 Rue de Londres
75008 Paris, France
frederico.grases@sncf.fr

Xiaoting Rui
Institute of Power Engineering
Nanjing University of Science and Technology
Nanjing, 210094, China

Bernhard Schweizer
Fachbereich Maschinen- und Energietechnik
HTWK Leipzig
Koburger Straße
04416 Markkleeberg, Germany
schweizer@me.htwk-leipzig.de

Karl Schweizerhof
Institut für Mechanik
Universität Karlsruhe
Kaiserstraße 12
76131 Karlsruhe, Germany
Schweizerhof@ ifm.uni-karlsruhe.de

Wolfgang Seemann
Institut für Technische Mechanik
Universität Karlsruhe (TH)
Kaiserstraße 12
76131 Karlsruhe, Germany
seemann@itm.uni-karlsruhe.de

Robert Seifried
Institute of Engineering and Computational Mechanics
University of Stuttgart
Pfaffenwaldring 9
70569 Stuttgart, Germany
seifried@itm.uni-stuttgart.de

Werner Schiehlen
Institute of Engineering and Computational Mechanics
University of Stuttgart
Pfaffenwaldring 9
70569 Stuttgart, Germany
schiehlen@itm.uni-stuttgart.de

Irena Sielamowicz
Civil Engineering Department
Bialystok Technical University
Wiejska 45 E
15-351 Bialystok, Poland
sieliren@pb.bialystok.pl

Wolfgang Stamm
LuK GmbH & Co. oHG
Industriestraße 3
77815 Bühl, Germany
wolfgang.stamm@luk.de

Jürgen Tomas
Mechanical Process Engineering
The Otto-von-Guericke-University
Universitätsplatz 2
39106 Magdeburg, Germany
juergen.tomas@
vst.uni-magdeburg.de

Hans I. Weber
Department of Mechanical Engineering
PUC/Rio
Rua Marques de Sao Vicente 225
22453-900 Rio de Janeiro, Brazil
hans@mec.puc-rio.br

Roland Zander
Lehrstuhl für Angewandte Mechanik
Technische Universität München
Boltzmannstraße 15
85748 Garching, Germany
zander@amm.mw.tum.de

Multi-Rate Time Integration for Large Scale Multibody System Models

Martin Arnold

Martin Luther University Halle–Wittenberg, Institute of Mathematics
06099 Halle (Saale), Germany, martin.arnold@mathematik.uni-halle.de

1 Introduction

In industrial applications of multibody dynamics the standard time integration methods are implicit. They are tailored to small and medium-sized nonlinear differential-algebraic equations of motion that may contain stiff components resulting from stiff force elements [6].

In high-end applications with thousands of degrees of freedom these implicit solvers show a dramatical loss of efficiency. Typical examples are multibody system models of vehicles that move along large elastic structures [13] and the dynamical simulation of combustion engines with chain drives [7].

Large scale problems that are composed of subsystems with different time scales may be handled efficiently by *multi-rate* methods that use different time stepsizes in different subsystems. In the present paper, we consider theoretical and practical aspects of multi-rate time integration in multibody dynamics.

The paper is organized as follows. The class of problems to be solved is introduced in Section 2. For practical reasons, multi-rate methods with weak coupling between the subsystems are favourable, see Section 3. Convergence and stability of these methods are discussed in Sections 4 and 5.

2 Coupled mechanical systems

The class of multi-rate methods that will be considered in Section 3 below is tailored to coupled technical systems with a clear modular structure. We focus on pure mechanical systems and restrict ourselves to the simulation of two coupled subsystems to keep notation simple. However, the multi-rate methods and their stability and convergence analysis may be carried over straightforwardly to systems with three or more coupled subsystems.

The state of the i-th subsystem is described by position coordinates q_i and velocity coordinates $v_i := \dot{q}_i$, $(i = 1, 2)$. The structure of the coupled equations of motion depends on the physical model for the coupling forces.

Peter Eberhard (ed.), Multiscale Problems in Multibody System Contacts, 1–10.

If the subsystems are coupled by applied forces $F_c(t, q_1, \dot{q}_1, q_2, \dot{q}_2)$ the equations of motion are given by

$$M_1(q_1)\ddot{q}_1 = f_1(t, q_1, \dot{q}_1, F_c(t, q_1, \dot{q}_1, q_2, \dot{q}_2)), \tag{1a}$$
$$M_2(q_2)\ddot{q}_2 = f_2(t, q_2, \dot{q}_2, F_c(t, q_1, \dot{q}_1, q_2, \dot{q}_2)) \tag{1b}$$

with M_i denoting the mass matrix of the i-th subsystem and f_i the corresponding vector of forces and momenta, $i = 1, 2$.

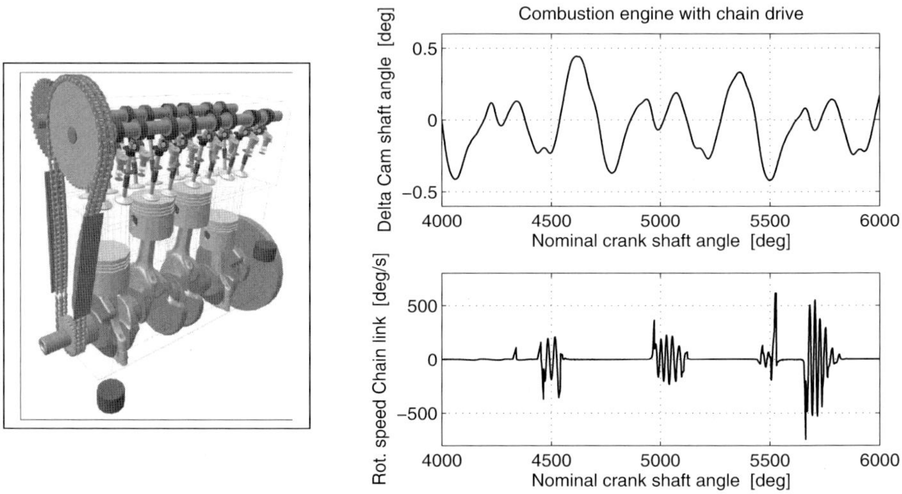

Fig. 1. Dynamical simulation of a combustion engine with chain drive [7].

Fig. 1 shows a combustion engine with chain drive [7] as typical example. Very stiff contact forces between the chain links introduce high frequency oscillations in the subsystem "Chain drive", see the lower plot in Fig. 1 that shows the velocity of one of the chain links. The upper plot illustrates that the coupled system engine / chain drive is a multiscale problem with clearly separated time scales. The deviation Δ_c of the cam shaft angle from its nominal position is a typical state variable of the slow subsystem "Core engine". Both subsystems are coupled by contact forces F_c acting between gear wheels and chain links, see Fig. 1. These forces are modelled by stiff unilateral spring-damper elements.

As an alternative to (1) we consider the coupling of subsystems by n_b holonomic constraints $0 = b(t, q_1, q_2)$ resulting, e.g., from rigid body contact conditions:

$$M_1(q_1)\ddot{q}_1 = f_1(t, q_1, \dot{q}_1) - B_1^\top(t, q_1, q_2)\lambda, \tag{2a}$$
$$M_2(q_2)\ddot{q}_2 = f_2(t, q_2, \dot{q}_2) - B_2^\top(t, q_1, q_2)\lambda, \tag{2b}$$
$$0 = b(t, q_1, q_2). \tag{2c}$$

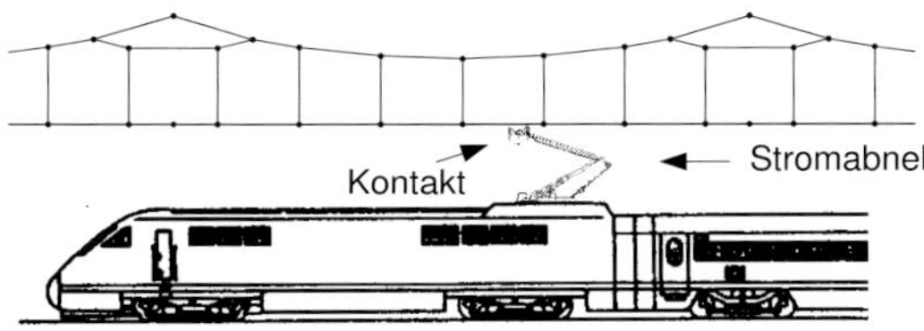

Fig. 2. Dynamical interaction between pantograph and catenary [13].

In (2) the constraints (2c) imply constraint forces $-B_i^\top \lambda$ with $B_i(t, q_1, q_2) := (\partial b / \partial q_i)(t, q_1, q_2)$ in both subsystems $i = 1, 2$. The Lagrangian multipliers $\lambda(t) \in \mathbb{R}^{n_b}$ are implicitly defined by the constraints (2c) that have to be satisfied at all time instances t.

Fig. 2 shows a typical multiscale problem that results in equations of motion (2), see [11, 12, 13]. The coupled system consists of the two subsystems "Pantograph" and "Catenary" that are connected by rigid body contact between pantograph head and overhead line with a rigid body contact condition of form (2c). Pantograph and overhead equipment guarantee the power supply of modern high-speed trains. The high-speed motion of the pantograph along the overhead equipment (subsystem "Catenary") results in nearly undamped high frequency oscillations in the catenary. The catenary is modelled by a combination of beams and strings that are semi-discretized in space by finite differences or finite elements [2, 11].

The changes in the state variables of subsystem "Pantograph" are slower by one order of magnitude. In the practical application, the dynamical simulation of the subsystem "Pantograph" caused nevertheless substantial difficulties since its multibody system model is strongly nonlinear [12].

Both in (1) and in (2) we suppose that there are no *inner* constraints in the subsystems, i.e., the subsystems are described by ODEs $M_i \ddot{q}_i = f_i \left[- B_i^\top \lambda \right]$. This restriction helps to keep the notation compact, but all results remain valid as well in the case of inner constraints, i.e., in the case of coupled DAEs, see [1].

3 Multi-rate methods with weak coupling

The time integration of multiscale problems by standard methods gets inefficient since the time stepsize h for the overall system is restricted by the time scale of the fastest subsystem. *Multi-rate methods* [4] that allow to use different time stepsizes h_i in different subsystems ($i = 1, 2$) offer a large potential for savings of computing time in this case.

The approach is well established in electrical circuit simulation. In this field, the errors that are introduced by the different time stepsizes h_i are often compensated by techniques that rely on frequent data exchange between the subsystems, see, e.g., [5].

In the simulation of large scale multibody systems we follow a different approach that reduces the amount of communication between the subsystems to a minimum (multi-rate methods with *weak coupling*). In this way, the subsystems may not only be solved with different time stepsizes h_i but the solvers itself may be adapted to the special structure of the subsystems. In a co-simulation framework, the subsystems may even be integrated in different simulation packages [8, 9].

The pantograph / catenary system, see Fig. 2, may be considered as a typical example for this approach: In the industrial application, the catenary is simulated in a specially tailored simulation tool by an explicit ODE method with small time stepsize. The time integration of the pantograph subsystem is performed in an industrial multibody system package using an implicit DAE method for constrained mechanical systems. Both simulation tools are coupled via an IPC interface [12].

A multi-rate method with weak coupling proceeds in *macro steps* $T_m \to T_{m+1} = T_m + H$ of macro stepsize H. The data exchange between subsystems is restricted to the discrete *synchronization points* T_m. Inside the macro step, the subsystems $i = 1, 2$ are solved separately with micro stepsizes h_i. Variable stepsizes H, h_1, h_2 could be considered as well but will be omitted here to keep notation simple.

The decoupled time integration of the subsystems makes it necessary to modify the evaluation of the coupling terms $F_c(t, q_1, \dot{q}_1, q_2, \dot{q}_2)$ and $-B_i^\top \lambda$ in (1) and (2) accordingly. If in the time integration of the i-th subsystem the current values $(q_j(t), \dot{q}_j(t))$, $(j \neq i)$ of state variables from other subsystems are not available for a time instance $t \in (T_m, T_{m+1}]$, they may simply be substituted by the corresponding function values $(q_j(T_m), \dot{q}_j(T_m))$ at the previous synchronization point T_m. Applied to (1), this Jacobi iteration like approach requires in each macro step $T_m \to T_{m+1}$ the time integration of the two fully decoupled ODEs

$$M_1(q_1)\ddot{q}_1 = f_1(t, q_1, \dot{q}_1, F_c(t, q_1, \dot{q}_1, q_2^{(m)}, \hat{q}_2^{(m)})) , \tag{3a}$$

$$M_2(q_2)\ddot{q}_2 = f_2(t, q_2, \dot{q}_2, F_c(t, q_1^{(m)}, \hat{q}_1^{(m)}, q_2, \dot{q}_2)) \tag{3b}$$

with *constant* extrapolated function values

$$q_j^{(m)}(t) := q_j(T_m) , \quad \hat{q}_j^{(m)}(t) := \dot{q}_j(T_m) , \quad (t \in [T_m, T_{m+1}]) . \tag{4}$$

Accuracy may be improved by higher order extrapolation steps involving function values of $r \geq 1$ previous macro steps. *Linear* extrapolation is defined by

$$q_j^{(m)}(t) := q_j(T_m) + (t - T_m) \frac{q_j(T_m) - q_j(T_{m-1})}{T_m - T_{m-1}} , \quad (t \in [T_m, T_{m+1}]) \tag{5}$$

and the corresponding formula with q_j being substituted by $\dot{q}_j$ in the right hand side of (5) to define $\hat{q}_j^{(m)}(t)$. Alternatively, the second order structure of (1) may be exploited [7] using

$$q_j^{(m)}(t) := q_j(T_m) + (t - T_m)\dot{q}_j(T_m) + \frac{1}{2}(t - T_m)^2\ddot{q}_j(T_m)\,, \quad \hat{q}_j^{(m)}(t) := \dot{q}_j^{(m)}(t)\,. \tag{6}$$

Method (3) allows straightforwardly a parallel implementation. On a sequential hardware platform, a Gauss–Seidel like method that exploits in the time integration of the second subsystem all data that have already been computed before in the time integration of the first subsystem may be more favourable.

We studied this approach in great detail for coupled DAEs (2), see [1, 10]. In the case of coupled ODEs (1), Stage 1 of macro step $T_m \to T_{m+1}$ is identical to (3a) but in Stage 2, the function values $(q_1(T_{m+1}), \dot{q}_1(T_{m+1}))$ are used to get an improved estimate for $(q_1(t), \dot{q}_1(t))$ by interpolation [7]:

$$M_1(q_1)\ddot{q}_1 = f_1(t, q_1, \dot{q}_1, F_c(t, q_1, \dot{q}_1, q_2^{(m)}, \hat{q}_2^{(m)}))\,, \tag{7a}$$

$$M_2(q_2)\ddot{q}_2 = f_2(t, q_2, \dot{q}_2, F_c(t, \bar{q}_1^{(m)}, \dot{\bar{q}}_1^{(m)}, q_2, \dot{q}_2)) \tag{7b}$$

with

$$\bar{q}_1^{(m)}(t) := q_1(T_m) + (t - T_m)\,\frac{q_1(T_{m+1}) - q_1(T_m)}{T_{m+1} - T_m}\,, \quad (\,t \in [T_m, T_{m+1}]\,)\,. \tag{8}$$

4 Global error of multi-rate methods

Methods (3) and (7) are representatives of the large class of weakly coupled multi-rate methods. Many similar approaches are known from the literature. They all have in common that extrapolation and interpolation introduce additional errors in the overall time integration method and may even lead to numerical instability.

If the i-th subsystem in (3) or (7) is solved by any classical ODE time integration method of order p_i, the global error

$$\epsilon := \max_{0 \le t_n \le T} \left(\|q_{1,n} - q_1(t_n)\| + \|\dot{q}_{1,n} - \dot{q}_1(t_n)\| + \|q_{2,n} - q_2(t_n)\| + \|\dot{q}_{2,n} - \dot{q}_2(t_n)\| \right) \tag{9}$$

is composed of the global errors $\mathcal{O}(h_i^{p_i})$ in the subsystems and the errors $\mathcal{O}(H^{r+1})$ caused by extrapolation and interpolation. Here, r denotes the order of the extrapolation and interpolation methods with $r = 0$ in the case of constant extrapolation (4) and $r = 1$ for linear extrapolation (5).

The classical convergence analysis for ODE time integration methods [3] may be extended straightforwardly to get a global error bound for the multi-rate time integration of (3) and (7) whenever the macro stepsize $H > 0$ is sufficiently small:

$$\epsilon \le C_0(h_1^{p_1} + h_2^{p_2} + H^{r+1})\,, \quad (\,H \in (0, H_0]\,)\,. \tag{10}$$

The constants H_0 and C_0 in (10) are positive and independent of H. Furthermore, the micro stepsizes h_1, h_2 are supposed to be bounded by $H \le H_0$.

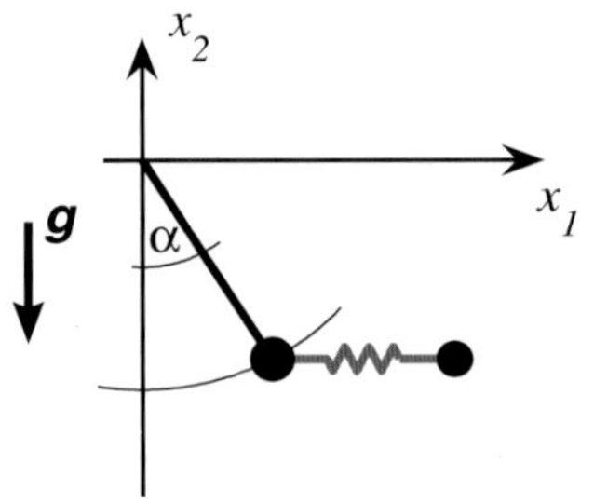

$$
\begin{aligned}
m_{\text{pend}}\, l\ddot{\alpha} &= -m_{\text{pend}}\, g \sin\alpha + \cos\alpha\, F_{\text{c}}(\alpha, \dot{\alpha}, x_1, \dot{x}_1) \\
m_{\text{osc}}\, \ddot{x}_1 &= -F_{\text{c}}(\alpha, \dot{\alpha}, x_1, \dot{x}_1) \\
F_{\text{c}} &= k \times (x_1 - l\sin\alpha) + d \times (\dot{x}_1 - l\dot{\alpha}\cos\alpha)
\end{aligned}
$$

$$
\begin{aligned}
&l = 1.0\,\text{m}\,,\ g = 9.81\,\text{m/s}^2,\ m_{\text{pend}} = m_{\text{osc}} = 1.0\,\text{kg}\,, \\
&\alpha(0) = 5^\circ,\ \ x_1(0) = 0.1\,\text{m} + 1.0\,\text{m} \times \sin\alpha(0)\,, \\
&\dot{\alpha}(0) = 0\,\text{rad/s}\,,\ \ \dot{x}_1(0) = 0\,\text{m/s}\,.
\end{aligned}
$$

Fig. 3. Test problem: Mathematical pendulum coupled to an oscillator.

Practical implications of error bound (10) are illustrated by numerical test results for the multi-rate time integration of (3) for a mathematical pendulum that is coupled to a damped oscillator with a horizontal degree of freedom, see Fig. 3.

Fig. 4 shows ϵ vs. H in the non-stiff case ($k = 10.0\,\text{N/m}$, $d = 1.0\,\text{Ns/m}$). For small macro stepsizes H, the error terms $\mathcal{O}(h_1^{p_1}) + \mathcal{O}(h_2^{p_2})$ resulting from the time integration of the subsystems dominate. Both subsystems are integrated by backward Euler method ($p_1 = p_2 = 1$). Therefore, the minimal global error ϵ in the lower plot of Fig. 4 (with $h_1 = h_2 = 0.05\,\text{ms}$) is smaller by a factor of 2 compared to the upper plot ($h_1 = h_2 = 0.1\,\text{ms}$).

For larger macro stepsizes H, the error term $\mathcal{O}(H^{r+1})$ that is introduced in each macro step by extrapolation from $t \le T_m$ to $t \in (T_m, T_{m+1}]$ dominates the overall global error ϵ. Quadratic ($r = 2$) and linear ($r = 1$) extrapolation are clearly superior to the brute force approach (4).

For comparison the classical backward Euler method was applied to the coupled system (1) with time stepsize $h = H$. In the non-stiff case, the error of multi-rate time integration with macro stepsize H and constant extrapolation ($r = 0$) nearly coincides with the error of the classical backward Euler method with stepsize $h = H$.

For an example for the use of multi-rate techniques in industrial applications we refer to Hippmann et al. [7]. There, the coupled system "Chain drive + Core engine" of Fig. 1 is solved successfully by a multi-rate method for (7) with macro stepsize $H = 1.0\,\mu\text{s}$ and an explicit Runge–Kutta discretization (method of Dormand and Prince, $p_1 = 5$) of subsystem "Chain drive" combined with a BDF discretization ($1 \le p_2 \le 5$) of subsystem "Core engine". Standard ODE stepsize control algorithms are used to select appropriate (variable) micro stepsizes h_1, h_2. In the time integration of (7a), quadratic extrapolation of q_2 according to (6) is applied. For (7b), linear interpolation of q_1 according to (8) proved to be useful.

For a typical benchmark problem with a chain drive containing 216 chain links, see [7], the overall computing time of the multi-rate method (7) is smaller by a factor of 20 compared to a classical BDF discretization of (1).

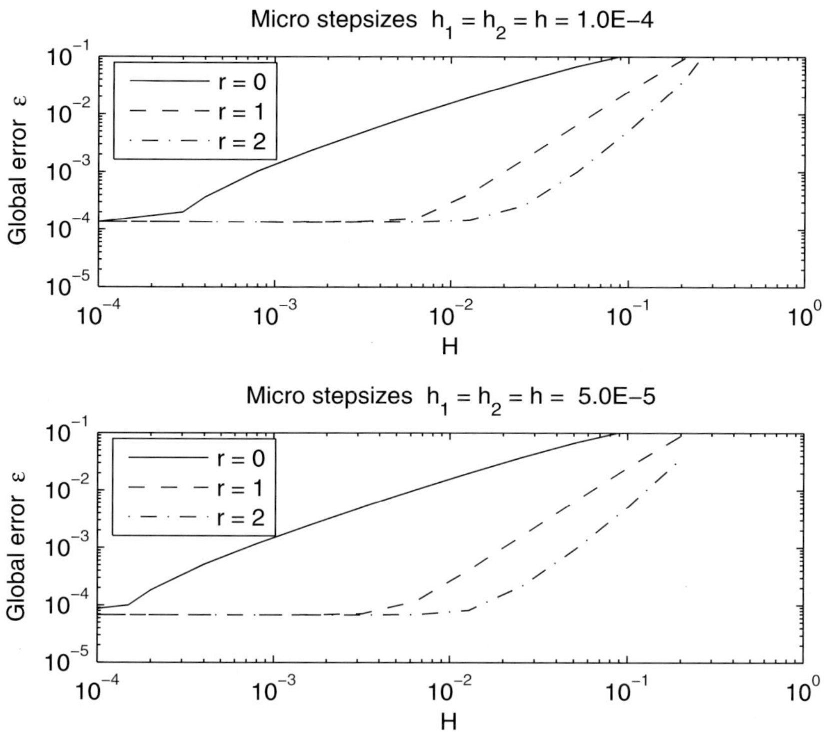

Fig. 4. Error of multi-rate time integration methods: The non-stiff case, $t \in [0\,\mathrm{s}, 4\,\mathrm{s}]$.

5 Stability of multi-rate methods

The very small macro stepsize $H = 1.0\,\mu\mathrm{s}$ in the multi-rate time integration of combustion engines with chain drives is chosen to resolve all dynamical interaction effects up to a frequency of 1 MHz. Furthermore, this small macro stepsize helps to guarantee the stability of the multi-rate method.

For less challenging standard applications in multibody dynamics, macro stepsizes in the range of $H = 1.0\,\mathrm{ms}$ are considered to be appropriate to meet the accuracy requirements and to cover the frequency range of interest. For macro stepsizes of this size, the multi-rate method applied to problems (1) with stiff coupling forces F_c may get unstable because of the use of polynomial extrapolation $q_2^{(m)}(t)$ in (3) and (7) that makes the method in part explicit.

For the example of Fig. 3 with stiff oscillator ($k = 10^6\,\mathrm{N/m}$, $d = 10^2\,\mathrm{Ns/m}$), an exponential instability is observed for all three extrapolation methods in use ($r = 0, 1, 2$) at $H \approx 0.8\,\mathrm{ms}$, see Fig. 5. This instability does not contradict the global error bound (10) that remains valid for all $H \leq H_0 = 0.5\,\mathrm{ms}$.

From the practical viewpoint, the results for constant extrapolation ($r = 0$) are useless because of errors in the size of more than 10%. In a comparison

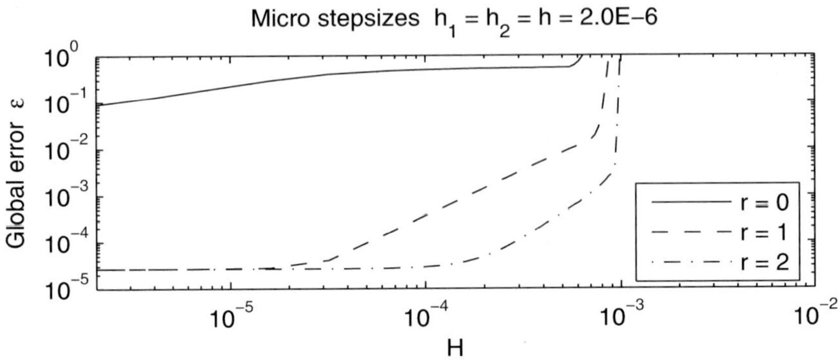

Fig. 5. Error of multi-rate time integration methods: The stiff case, $t \in [0.1\,\mathrm{s}, 0.2\,\mathrm{s}]$.

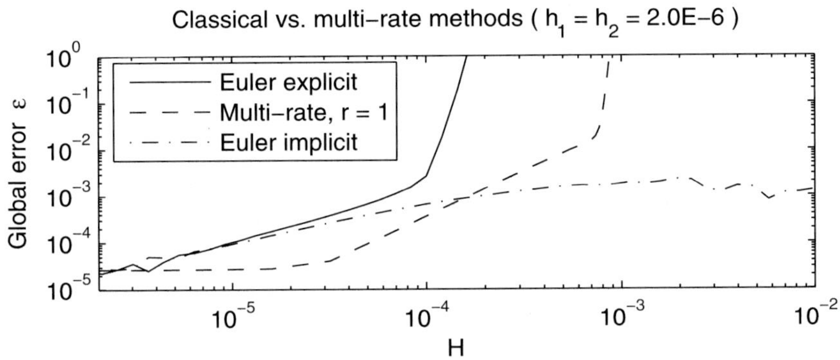

Fig. 6. Instability of time integration methods in the stiff case, $t \in [0.1\,\mathrm{s}, 0.2\,\mathrm{s}]$.

to classical time integration methods we consider therefore linear extrapolation (5) with $r = 1$. Fig. 6 shows results for the classical explicit Euler method with stepsize $h = H$ that gets unstable for $h \geq 0.1\,\mathrm{ms}$ and for the classical backward Euler method that remains stable for any $h > 0$. In the multi-rate method, that uses an unconditionally stable method in both subsystems, the instability for $H > 0.8\,\mathrm{ms}$ results from the linear extrapolation step (5). The critical stepsize in the multi-rate method is, however, substantially larger than for classical explicit methods like explicit Euler method.

In the case of coupled ODEs (1), the instability phenomenon is not yet understood in full detail. The stability analysis for classical ODE time integration methods, see, e.g., [3], may be extended to show that the multi-rate method is stable if all subsystems are integrated by stable time integration methods and the condition

$$H \times \max \left\{ \|\frac{\partial f_1}{\partial q_2}\|, \|\frac{\partial f_1}{\partial \dot{q}_2}\|, \|\frac{\partial f_2}{\partial q_1}\|, \|\frac{\partial f_2}{\partial \dot{q}_1}\| \right\} \leq \tilde{C} \tag{11}$$

is satisfied with a sufficiently small constant $\tilde{C} = \mathcal{O}(1)$. However, the stepsize restriction (11) is known to be too pessimistic in many practical applications.

The stability bound (11) gets more and more restrictive with increasing stiffness of the coupling forces F_c. The coupling by constraint forces, see (2), may be considered as the limit case of "infinite" stiffness. For (2), the stability of multi-rate methods has been studied in great detail [1, 9, 10]. In the DAE case (2), the multi-rate method gets unstable for *any* $H > 0$ unless the contractivity condition

$$\alpha := \max_{t \in (0,T]} \|(B_2 M_2^{-1} B_2^\top)^{-1} (B_1 M_1^{-1} B_1^\top)\| \leq \tilde{\alpha} \tag{12}$$

is satisfied. The constant $\tilde{\alpha}$ depends (slightly) on the extrapolation method in use, but it is always of size $\tilde{\alpha} = \mathcal{O}(1)$, see [1, 10]. The multi-rate method may be stabilized by a modified method for computing $\lambda(t)$ that involves the mass matrices M_1, M_2, the Jacobians B_1, B_2 of the constraints and the Jacobians $B_1^\top$, $B_2^\top$ of the constraint forces, see [1, 10] for a detailed discussion.

Jacobian information can be used as well to stabilize multi-rate methods in the case of stiffly coupled ODEs (1). Let us consider a fictitious micro step $t_n \to t_{n+1} = t_n + h_2$ for the solution of (3a) in macro step $T_m \to T_{m+1}$ with $t_0 := T_m$. Equation (3a) is formally transformed to a first order system that is solved by a simplified linearly implicit Euler step

$$x_{2,n+1} = x_{2,n} + h_2 f_{2,0} + h_2 J_{21}(x_1(t_{n+1}) - x_1(t_n)) + h_2 J_{22}(x_{2,n+1} - x_{2,n}) \tag{13}$$

with $x_1(t) = \big(q_1(t)\,,\, \dot{q}_1(t) \big)^\top$, $x_{2,n} = \big(q_2^{(m)}(t_n)\,,\, \hat{q}_2^{(m)}(t_n) \big)^\top$,

$$f_{2,0} = \big(\dot{q}_2(t)\big|_{t=T_m}\,,\, f_2(t, q_1, \dot{q}_1, F_c)\big|_{t=T_m} \big)^\top,$$

and Jacobians

$$J_{21} = \Big(\frac{\partial f_2}{\partial q_1}\ \frac{\partial f_2}{\partial \dot{q}_1} \Big)\Big|_{t=T_m}, \quad J_{22} = \Big(\frac{\partial f_2}{\partial q_2}\ \frac{\partial f_2}{\partial \dot{q}_2} \Big)\Big|_{t=T_m}. \tag{14}$$

Taking into account that $x_{2,0} = x_2(T_m)$, the recursive application of (13) defines the stabilized extrapolation $x_{2,n}$:

$$(I - h_2 J_{22})(x_{2,n} - x_2(T_m)) = (t_n - T_m) f_{2,0} + h_2 J_{21}(x_1(t_n) - x_1(T_m)). \tag{15}$$

In the same way, Jacobians J_{11} and J_{12} are used to define a stabilized extrapolation $x_{1,n} = (q_1^{(m)}(t_n), \hat{q}_1^{(m)}(t_n))^\top$ for integrating (3b) from T_m to T_{m+1}. Stabilized extrapolation (15) extends the amount of data exchange between the subsystems ($f_{i,0}$, J_{ij}) but as before the data exchange is restricted to the synchronization points T_m (weak coupling).

In the application to the stiff test problem of Fig. 7 the use of Jacobian information in the stabilized extrapolation (15) improves accuracy and stability of the multi-rate method substantially, see Fig. 7. The results are slightly more accurate than the ones for linear extrapolation (5) with $r = 1$. The stabilized multi-rate integration is stable up to $H = 1.2\,\mathrm{ms}$ but further investigations will be necessary to get multi-rate methods with weak coupling that are unconditionally stable in the stiff case.

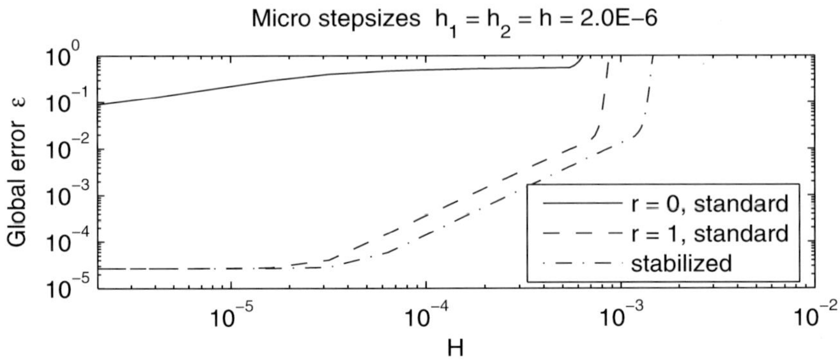

Fig. 7. Stabilization of multi-rate time integration, $t \in [0.1\,\mathrm{s}, 0.2\,\mathrm{s}]$.

References

1. M. Arnold and M. Günther. Preconditioned dynamic iteration for coupled differential-algebraic systems. *BIT Numerical Mathematics*, 41:1–25, 2001.
2. M. Arnold and B. Simeon. Pantograph and catenary dynamics: a benchmark problem and its numerical solution. *Applied Numerical Mathematics*, 34:345–362, 2000.
3. U. Ascher and L.R. Petzold. *Computer Methods for Ordinary Differential Equations and Differential–Algebraic Equations.* SIAM, Philadelphia, 1998.
4. C.W. Gear and R.R. Wells. Multirate linear multistep methods. *BIT*, 24:484–502, 1984.
5. M. Günther and P. Rentrop. Multirate ROW methods and latency of electric circiuts. *Applied Numerical Mathematics*, 13:83–102, 1993.
6. E. Hairer and G. Wanner. *Solving Ordinary Differential Equations. II. Stiff and Differential-Algebraic Problems.* Springer–Verlag, Berlin, 2nd edition, 1996.
7. G. Hippmann, M. Arnold, and M. Schittenhelm. Efficient simulation of bush and roller chain drives. In J.M. Goicolea, J. Cuadrado, and J.C. García Orden, editors, *Proc. of Multibody Dynamics 2005*, Madrid, Spain, 2005.
8. W. Kortüm, W.O. Schiehlen, and M. Arnold. Software tools: From multibody system analysis to vehicle system dynamics. In H. Aref and J.W. Phillips, editors, *Mechanics for a New Millennium*, pages 225–238, Dordrecht, 2001. Kluwer.
9. R. Kübler and W. Schiehlen. Modular simulation in multibody system dynamics. *Multibody System Dynamics*, 4:107–127, 2000.
10. B. Möller. Multirate–Verfahren für gekoppelte differentiell-algebraische Systeme. Master Thesis, TU Munich, Department of Mathematics, 2003.
11. G. Poetsch, J. Evans, R. Meisinger, W. Kortüm, W. Baldauf, A. Veitl, and J. Wallaschek. Pantograph/catenary dynamics and control. *Vehicle System Dynamics*, 28:159–195, 1997.
12. A. Veitl. *Integrierter Entwurf innovativer Stromabnehmer.* Fortschritt-Berichte VDI Reihe 12, Nr. 449. VDI–Verlag, Düsseldorf, 2001.
13. A. Veitl and M. Arnold. Coupled simulation of multibody systems and elastic structures. In J.A.C. Ambrósio and W.O. Schiehlen, editors, *Advances in Computational Multibody Dynamics*, pages 635–644, IDMEC/IST Lisbon, 1999.

Multiscale Analysis of a Cantilever with a Contact Boundary

Li-Qun Chen

Department of Mechanics, Shanghai University, Shanghai 20444, China
Shanghai Institute of Applied Mathematics and Mechanics, Shanghai 200072, China lqchen@staff.shu.edu.cn

Summary. This paper investigates nonlinear vibration in a forced cantilever with a contact boundary. The cantilever is assumed as an Euler-Bernoulli beam, and the contact is specified by the Derjaguin-Müller-Toporov theory. The mathematical model is a linear non-autonomous partial-differential equation with a nonlinear autonomous boundary condition. The method of multiple scales is applied to calculate the steady-state response in principal resonance. The equation of response curves is derived from the solvability condition of eliminating secular terms. Numerical examples are presented to demonstrate the effects of the excitation amplitude, the damping coefficient, and the coefficients related to the contact boundary.

1. Introduction

Vibration of linearly elastic beams with boundary is a classical problem that can be approached via different techniques such as the harmonic balance method [1] and the method of multiple scales [2]. A cantilever with a contact boundary plays a crucial role in the tapping atomic force microscope [3]. Hence many researchers investigated dynamical behavior of the cantilever. Turner et al modeled the contact between the cantilever and the sample as a boundary condition consisting of a spring and a damping, and solved the flexural beam equation to examine the limits and validity of the first-mode approximation [4]. Stark and Heckl considered the periodic impact force induced by the tip-sample Hertzian contact in tapping mode atomic force microscopy and used the modal expansion to discretize the governing equation into a multi-degree-of-freedom system [5]. Lee et al applied the Galerkin method to tapping mode atomic force microscopy, with van der Waals force for noncontact state and Derjaguin-Müller-Toporov force for contact state, and developed a one-degree-of-freedom nonlinear model to compute free, forced, and parametric vibration [6]. Lee et al further used the model to compute responses in monostable and bistable regions as well as bifurcation sets [7]. Stark et al modeled a cantilever with a contact end as a linear time-invariant system with a nonlinear output feedback [8]. Turner applied the method of multiple scales to analyze the

Peter Eberhard (ed.), Multiscale Problems in Multibody System Contacts, 11–21.

cantilever with a nonlinear boundary condition due to the Hertzian contact [9]. Based on the model in [6,7], Yagasaki used the averaging method and an extended version of the subharmonic Melnikov method to analyze vibration of a cantilever with a Derjaguin-Müller-Toprov contact [10]. Abdel-Rahman and Nayfeh further used the subharmonic resonance of the cantilever, obtained by the method of multiple scales, to identify the contact force between the tip and the sample [11].

This paper treats principal resonance of a cantilever with a Derjaguin-Müller-Toporov contact. The problem is similar to those studied in [9,11]. However, [9,11] dealt with a linear autonomous governing equation with a nonlinear non-autonomous boundary condition. The problem reformulated here is a linear non-autonomous governing equation with a nonlinear autonomous boundary condition. The method of multiple scales is applied to derive the steady-state response in the principal resonance. Numerical examples demonstrate the effects of relevant parameters on the amplitude of the steady-state response.

2. Modelling

The governing equation of a cantilever modeled as an Euler-Bernoulli beam is

$$\rho A\ddot{w}(\xi,\tau) + \varepsilon^n C\dot{w}(\xi,\tau) + EIw''''(\xi,\tau) = 0 \tag{1}$$

where $w(\xi,\tau)$ is the deflection at axial coordinate ξ and time τ, measured from the static deformation, ρ is the density, A is the cross-sectional area, C is the viscous damping coefficient, E is the elastic modulus, I is the area moment, ε is a small positive number indicating the fact that the viscous damping term is very small, and n is an integral to be determined.

Consider a small base motion

$$b(\tau) = \varepsilon^m A\sin(\varpi\tau) \tag{2}$$

where A and ϖ are respectively the amplitude and the frequency of the base excitation, and m is an integral. Then the deflection relative to a non-inertial frame attached to the moving base is

$$u(\xi,\tau) = w(\xi,\tau) - b(\tau) \tag{3}$$

Therefore, equation (1) becomes

$$\begin{aligned} &\rho A\ddot{u}(\xi,\tau) + \varepsilon^n C\dot{u}(\xi,\tau) + EIu''''(\xi,\tau) \\ &= \varepsilon^m \rho A\varpi^2 Y\sin(\varpi\tau) - \varepsilon^m\varepsilon^n C\varpi Y\cos(\varpi\tau) \end{aligned} \tag{4}$$

In the non-inertial reference frame moving with the base, the boundary conditions for the end deflection $u^*(L)$ are

$$\begin{gathered} u(0,\tau)=0, u'(0,\tau)=0 \\ u''(L,\tau)=0, EIu'''(L,\tau)-\frac{3EI}{L^3}u^*(L)+F_{\mathrm{DMT}}=0 \end{gathered} \tag{5}$$

where the force F_{DMT} due to the contact between the tip and the sample is given by the Derjaguin-Müller-Toprov model [12,13] as

$$F_{\mathrm{DMT}}=-\frac{4E^*a_0^3}{3R}+\frac{4}{3}E^*\sqrt{R}\delta^{\frac{3}{2}} \tag{6}$$

In the Derjaguin-Müller-Toprov model, a_0 is the contact radius at zero load, R is the tip radius, δ is the deformation, and E^* is the effective elastic modulus. The deformation is expressed by the surface offset z_0, $u^*(L)$ and $u(L,t)$ as

$$\delta=z_0-u^*(L)-u(L,\tau) \tag{7}$$

where $u^*(L)$ is given by

$$3\frac{EI}{L^3}u^*(L)-\frac{4E^*a_0^3}{3R}+\frac{4}{3}E^*\sqrt{R}[z_0-u^*(L)]^{\frac{3}{2}}=0 \tag{8}$$

As the deflection of the beam is very small, assume that $u(\xi,\tau)=\varepsilon v(\xi,\tau)$. Fix n=2 and m=3. Thus equations (4) and (5) can be rewritten as

$$\begin{aligned} &\rho A\ddot{v}(\xi,\tau)+EIv''''(\xi,\tau) \\ &=-\varepsilon^2C\dot{v}(\xi,\tau)+\varepsilon^2\rho A\varpi^2Y\sin(\varpi\tau)-\varepsilon^4C\varpi Y\cos(\varpi\tau) \end{aligned} \tag{9}$$

$$v(0,\tau)=0, v'(0,\tau)=0; v''(L,\tau)=0 \tag{10}$$

$$EI\varepsilon v'''(L,\tau)-3\frac{EI}{L^3}u^*(L)-\frac{4E^*a_0^3}{3R}+\frac{4}{3}E^*\sqrt{R}[z_0-u^*(L)-\varepsilon v(L,\tau)]^{\frac{3}{2}}=0 \tag{11}$$

Substitution of equation (8) and the Taylor expansion into equation (11) yield

$$\begin{aligned} &EIv'''(L,\tau)-\frac{4E^*}{3}\sqrt{R[z_0-u^*(L)]}v(L,\tau) \\ &+\frac{\varepsilon E^*\sqrt{R}v^2(L,\tau)}{6\sqrt{[z_0-u^*(L)]}}+\frac{\varepsilon^2E^*\sqrt{R}v^3(L,\tau)}{12[z_0-u^*(L)]^{\frac{3}{2}}}+o(\varepsilon^3)=0 \end{aligned} \tag{12}$$

Introduce new time variable

$$t=\tau\sqrt{\frac{EI}{\rho A}}L^2, x=\frac{\xi}{L} \tag{13}$$

and still use a dot and a prime to denote the temporal and spatial partial derivatives with respect to the new time and spatial variables. Then equation (9) can be cast into the form

$$\ddot{v}(x,t)+v''''(x,t)=-\varepsilon^2\mu\dot{v}(x,t)+\varepsilon^2F\sin(\Omega t)+o(\varepsilon^3) \tag{14}$$

where

$$\mu = \frac{C}{L^2\sqrt{EA\rho A}}, F = \frac{\rho A\varpi^2 Y}{EIL^4}, \Omega = \frac{\varpi}{L^2}\sqrt{\frac{\rho A}{EI}} \tag{15}$$

Equations (10) and (12) can be rewritten as

$$v(0,t) = 0, v'(0,t) = 0; v''(1,t) = 0 \tag{16}$$

$$v'''(1,t) - \alpha_1 v(1,t) = -\varepsilon\alpha_2 v^2(1,t) - \varepsilon^2\alpha_3 v^3(1,t) + o(\varepsilon^2) = 0 \tag{17}$$

where

$$\begin{aligned}\alpha_1 &= \frac{4E^*}{3EIL^3}\sqrt{R[z_0 - u^*(L)]}\\ \alpha_2 &= \frac{E^*\sqrt{R}}{6EIL^3\sqrt{[z_0 - u^*(L)]}}\\ \alpha_3 &= \frac{E^*\sqrt{R}}{12EIL^3[z_0 - u^*(L)]^{\frac{3}{2}}}\end{aligned} \tag{18}$$

3. Analysis

The method of multiple scales will be employed to solve equation (14) under the boundary conditions (16) and (17). A second order uniform approximation is sought in the form

$$\begin{aligned}&v(x,t;\varepsilon)\\ &= v_0(x,T_0,T_1,T_2) + \varepsilon v_1(x,T_0,T_1,T_2) + \varepsilon^2 v_2(x,T_0,T_1,T_2) + o(\varepsilon^2)\end{aligned} \tag{19}$$

where $T_j = t^j$ $(j = 1,2,3)$ are the different time scales of the problem. Substituting of equations (19) as well as the following relationships

$$\begin{aligned}\frac{\partial}{\partial t} &= D_0 + \varepsilon D_1 + \varepsilon^2 D_2 + o(\varepsilon^2)\\ \frac{\partial^2}{\partial t^2} &= D_0^2 + 2\varepsilon D_1 D_2 + \varepsilon^2(2\varepsilon D_0 D_2 + D_1^2) + o(\varepsilon^2)\end{aligned} \tag{20}$$

for $D_j = \partial/\partial T_j$ $(j = 1,2,3)$ into equations (14), (16) and (17), and equating the coefficients of like powers of ε in the resulting equations finally yield

$$D_0^2 v_0(x,T_0,T_1,T_2) + v_0''''(x,T_0,T_1,T_2) = 0 \tag{21}$$

$$v_0(0,T_0,T_1,T_2) = 0, v_0'(0,T_0,T_1,T_2) = 0; v_0''(1,T_0,T_1,T_2) = 0 \tag{22}$$

$$v_0'''(1,T_0,T_1,T_2) - \alpha_1 v_0(1,T_0,T_1,T_2) = 0 \tag{23}$$

at the order ε^0,

$$D_0^2 v_1(x,t) + v_1''''(x,t) = -2D_0 D_1 v_0 \tag{24}$$

$$v(0,T_0,T_1,T_2)=0, v'(0,T_0,T_1,T_2)=0; v''(1,T_0,T_1,T_2)=0 \tag{25}$$

$$v_1'''(1,T_0,T_1,T_2)-\alpha_1 v_1(1,T_0,T_1,T_2)=-\alpha_2 v_0^2(1,T_0,T_1,T_2) \tag{26}$$

at the order ε^1, and

$$\begin{aligned} &D_0^2 v_2(x,T_0,T_1,T_2)+v_2''''(x,T_0,T_1,T_2) \\ &= -2D_0D_2v_0(x,T_0,T_1,T_2)-\mu D_0 v_0(x,T_0,T_1,T_2)+F\sin(\Omega T_0) \end{aligned} \tag{27}$$

$$v_2(0,T_0,T_1,T_2)=0, v_2'(0,T_0,T_1,T_2)=0; v_2''(1,T_0,T_1,T_2)=0 \tag{28}$$

$$\begin{aligned} &v_2'''(1,T_0,T_1,T_2)-\alpha_1 v_2(1,T_0,T_1,T_2) \\ &= -2\alpha_2 v_0(1,T_0,T_1,T_2)v_1(1,T_0,T_1,T_2)-\alpha_3 v_0^3(1,T_0,T_1,T_2) \end{aligned} \tag{29}$$

at the order ε^2.

Under the boundary conditions (22) and (23), equation (21) has the solution

$$\begin{aligned} &v_0(x,T_0,T_1,T_2) \\ &= \sum_{m=1}^{\infty}[\phi_m(x)A_m(T_1,T_2)\mathrm{e}^{\mathrm{i}\omega_m T_0}+\bar{\phi}_k(x)\bar{A}_m(T_1,T_2)\mathrm{e}^{-\mathrm{i}\omega_m T_0}] \end{aligned} \tag{30}$$

where $\phi_m(x)$ is the mth normalized complex eigenfunction, $A_m(T_1,T_2)$ is the coefficients to be determined later, ω_m is the mth natural frequency, and the over bar denotes complex conjugation. The mth natural frequency can be solved from the transcendental equation

$$\sqrt{\omega^3}(1+\cos\sqrt{\omega}\cosh\sqrt{\omega}+\alpha_1(\sin\sqrt{\omega}\cosh\sqrt{\omega}-\cos\sqrt{\omega}\sinh\sqrt{\omega})=0 \tag{31}$$

as the mth root. The mth normalized complex eigenfunction with the normalization condition $\phi_m(1)=1$ is

$$\begin{aligned} \phi_m(x) = &\frac{(\cos\lambda_m+\cosh\lambda_m)(\sin\lambda_m x-\sinh\lambda_m x)}{2(\sin\lambda_m\cosh\lambda_m-\sinh\lambda_m\cos\lambda_m)} \\ &-\frac{(\sin\lambda_m+\sinh\lambda_m)(\cos\lambda_m x-\cosh\lambda_m x)}{2(\sin\lambda_m\cosh\lambda_m-\sinh\lambda_m\cos\lambda_m)} \end{aligned} \tag{32}$$

where

$$\lambda_m=\sqrt{\omega_m} \tag{33}$$

If the excitation frequency Ω approaches a natural frequency of equation (14), the principal resonance may occur. A detuning parameter σ_j is introduced to quantify the deviation of Ω from ω_j ($j=1,2,<$), and Ω is described by

$$\Omega=\omega_j+\sigma_j\varepsilon^2 \tag{34}$$

To investigate the jth principal parametric resonance, the solution (30) of equation (21) can be expressed as

$$v_0(x,T_0,T_1,T_2)=\phi_j(x)A_j(T_1,T_2)\mathrm{e}^{\mathrm{i}\omega_j T_0}+cc \tag{35}$$

where cc represents the complex conjugate of all preceding terms on the right hand of an equation.

Substitution of equation (35) into equations (24) and (26) respectively leads to

$$D_0^2 v_1(x,t) + v_1''''(x,t) = 2i\omega_j \phi_j(x) D_1 A_j(T_1,T_2)\mathrm{e}^{\mathrm{i}\omega_j T_0} + cc \tag{36}$$

$$\begin{aligned} &v_1'''(1,T_0,T_1,T_2) - \alpha_1 v_1(1,T_0,T_1,T_2) \\ &= -\alpha_2(A_j^2 \mathrm{e}^{2\mathrm{i}\omega_j T_0} + 2A_j\bar{A}_j + \bar{A}_j^2 \mathrm{e}^{-2\mathrm{i}\omega_j T_0}) \end{aligned} \tag{37}$$

Hence the solvability condition for equation (36) demands that

$$D_1 A_j(T_1,T_2) = 0 \tag{38}$$

which means $A_j(T_1,T_2) = A_j(T_2)$ depending only on T_2. Under the boundary conditions (25) and (37), the solution to equation (36) can be expressed as

$$v_1(x,T_0,T_1,T_2) = -\alpha_2[A_j^2\phi_j \mathrm{e}^{2\mathrm{i}\omega_j T_0} + \frac{x^3 - 3x^2}{6+2\alpha_1} A_j\bar{A}_j] + cc \tag{39}$$

where

$$\begin{aligned} \phi_j(x) = {}& \frac{(\cos\kappa_j + \cosh\kappa_j)(\sinh\kappa_j x - \sin\kappa_j x)}{2\kappa_j^3(1+\cos\kappa_j\cosh\kappa_j) + 2\alpha_1(\sin\kappa_j\cosh\kappa_j - \cos\kappa - j\sinh\kappa_j)} \\ &+ \frac{(\sinh\kappa_j + \sin\kappa_j)(\cos\kappa_j x - \cosh\kappa_j x)}{2\kappa_j^3(1+\cos\kappa_j\cosh\kappa_j) + 2\alpha_1(\sin\kappa_j\cosh\kappa_j - \cos\kappa - j\sinh\kappa_j)} \end{aligned} \tag{40}$$

$$\kappa_j = \sqrt{2\omega_j} \tag{41}$$

Substitution of equations (34) and (35) into equations (27) yields

$$\begin{aligned} &D_0^2 v_2(x,T_0,T_1,T_2) + v_2''''(x,T_0,T_1,T_2) \\ &= [-2\mathrm{i}\omega_j D_2 A_j \phi_j(x) - \mathrm{i}\omega_j \mu A_j \phi_j(x) - \frac{\mathrm{i}}{2} F \mathrm{e}^{\mathrm{i}\sigma T_2}]\mathrm{e}^{\mathrm{i}\omega_j T_0} + cc \end{aligned} \tag{42}$$

Substitution of equations (29) and (33) into equation (23) gives

$$\begin{aligned} &v_2''''(1,T_0,T_1,T_2) - \alpha_1 v_2(1,T_0,T_1,T_2) \\ &= [2\alpha_2^2\phi_j(1) - \frac{4\alpha_2^2}{3+\alpha_1} - 3\alpha_3]A_j^2\bar{A}_j \mathrm{e}^{\mathrm{i}\omega_j T_0} + cc + NST \end{aligned} \tag{43}$$

The nonhomogeneous problem defined by equations (42), (28) and (43) has solutions only if a certain solvability condition is satisfied [14]. To seek the condition, a solution to equation (42) is expressed as a term proportional to $\exp(\mathrm{i}\omega_j T_0)$ plus other terms. Substituting the solution into equations (42), (28) and (43) and equating the coefficients of $\exp(\mathrm{i}\omega_j T_0)$ on both sides of each equation result in a nonhomogeneous boundary problem with respect to the spatial variable x only. Application of the procedure for general 4 order ordinary differential equation with nonhomogeneous boundary conditions [14] yield the condition after some algebraic and integration manipulations

$$\begin{aligned}[2\alpha_2^2\phi_j(1) - \frac{4\alpha_2^2}{3+\alpha_1} - 3\alpha_3]A_j^2\bar{A}_j \\ = \mathrm{i}\omega_j(-2D_2A_j - \mu A_j)g(\lambda_j) - \frac{\mathrm{i}}{2}f(\lambda_j)\mathrm{e}^{\mathrm{i}\sigma_j T_2}\end{aligned} \tag{44}$$

where

$$\begin{aligned}g(\lambda) = &\frac{-\lambda\cos(2\lambda) + 6(1+\cos\lambda\cosh\lambda)(\sinh\lambda\cos\lambda - \cosh\lambda\sin\lambda)}{8\lambda(\cosh\lambda\sin\lambda - \cos\lambda\sinh\lambda)^2} \\ &+ \frac{\lambda(\cosh(2\lambda) + 4\sin\lambda\sinh\lambda)}{8\lambda(\cosh\lambda\sin\lambda - \cos\lambda\sinh\lambda)^2}\end{aligned} \tag{45}$$

$$f(\lambda) = \frac{F(\cos\lambda - 1)(\cosh\lambda - 1)}{\lambda(\cos\lambda\sinh\lambda - \sin\lambda\cosh\lambda)} \tag{46}$$

The solution A_j to equation (44) is expressed in the polar form

$$A_j(T_2) = p_j(T_2)\mathrm{e}^{\mathrm{i}q_j(T_2)} \tag{47}$$

Then the real-valued functions p_j and q_j represent respectively the amplitude and the phase angle of the response. Substituting equation (47) into equation (44), separating the resulting equation into real and imaginary parts and solving the derivatives of p_j and q_j respect to T_2 from the resulting equations leads to the

$$\dot{p_j} = -\frac{1}{2}p_j - \frac{f(\lambda_j)\cos\gamma_j}{2\omega_j g(\lambda_j)} \tag{48}$$

$$\dot{q_j} = \sigma_j - \frac{p_j^2}{8\omega_j g(\lambda_j)}[2\alpha_2^2\varphi_j(1) - \frac{4\alpha_2^2}{3+\alpha_1} - 3\alpha_3] + \frac{f(\lambda_j)\sin\gamma_j}{2\omega_j g(\lambda_j)p_j} \tag{49}$$

where

$$\gamma_j(T_2) = \sigma_j T_2 - q_j(T_2) \tag{50}$$

The fixed points (p_{j0}, γ_{j0}) of equations (48) and (49), at which the amplitude p_j and the new phase angle γ_j are constant, are corresponding to the steady-state response. Let p_j and γ_j are constant, and then equations (48) and (49) yield

$$-\frac{1}{2p_{j0}} - \frac{f(\lambda_j)\cos\gamma_{j0}}{2\omega_j g(\lambda_j)} = 0 \tag{51}$$

$$\sigma_j - \frac{p_{j0}^2}{8\omega_j g(\lambda_j)}[2\alpha_2^2\varphi_j(1) - \frac{4\alpha_2^2}{3+\alpha_1} - 3\alpha_3] + \frac{f(\lambda_j)\sin\gamma_{j0}}{2\omega_j g(\lambda_j)p_{j0}} = 0 \tag{52}$$

Eliminating γ_{j0} from equations (51) and (52) gives the relation between the steady-state response amplitude and the detuning parameter for the jth mode

$$\{\frac{p_{j0}^3}{4}[2\alpha_2^2\varphi_j(1) - \frac{4\alpha_2^2}{3+\alpha_1} - 3\alpha_3] - 2\omega_j\sigma_j g(\lambda_j)p_{j0}\}^2 - [\omega_j g(\lambda_j)p_{j0}]^2 \\ = [f(\lambda_j)]^2 \tag{53}$$

The amplitudes of the stead-state periodic motion can be solved from equation (53) for prescribed parameters.

For calculated (p_{j0}, γ_{j0}), equations (19), (35) and (47) yield the approximation periodic motion in the jth principal resonance

$$v(x, t, \varepsilon) = p_{j0}\phi_j(x)\cos(\Omega t - \gamma_{j0}) + O(\varepsilon) \tag{54}$$

when the excitation frequency Ω tends to the jth natural frequency of the linear system.

4. Numerical Examples

Equation (35) defines a response curve. Observing the curve, one can determine the effects of the excitation amplitude F, the damping coefficient μ, and the contact parameters α_1, α_2, and α_3, on the relation between the steady-state amplitude response and the detuning parameter, which reflects the excitation frequency.

Figure 1 demonstrates the effect of the excitation amplitude F. In the calculation, μ=0.05, α_1=2400, α_2=2400, α_3=100, and F=20 (solid lines), 30 (dash lines), 40 (dot lines), 50 (dash-dot lines). The response amplitude increases with F. When the detuning parameter is close to zero, the amplitude becomes rather large, and hence the resonance occurs. The non-linearity bends the curve to the left slightly, which indicates the non-linearity due to contact is soft one. In addition, the resonance response amplitude in the first mode is much larger than that in the second mode. It can be inferred that the amplitude in higher order mode are very small, and hence only the first two modes are treated in the present investigation.

Figure 2 illustrates the effect of the damping coefficient μ. For the first principal resonance, F=2, α_1=600, α_2=150, α_3=25, and μ=0.1 (solid), 0.05 (dash), 0.01 (dot). For the second principal resonance F=50, α_1=1800, α_2=450, α_3=75, and μ=0.15 (solid), 0.1 (dash), 0.05 (dot). The response amplitude increases with μ. Besides, the smaller the damping, the more the curves turn to the left. Hence the increasing damping weakens the effect of nonlinearity.

Figure 3 depicts the effect of the square term coefficient α_2. For the first principal resonance, F=2, μ=0.008, α_1=600, α_3=25, and α_2=0 (solid), 150 (dash), 300 (dot). For the second principal resonance F=50, μ=0.05, α_1=1800, α_3=75, and α_2=0 (solid), 250 (dash), 450 (dot). The larger α_2 leads the curves bending more to the left. For rather large α_2, when σ is slightly larger than

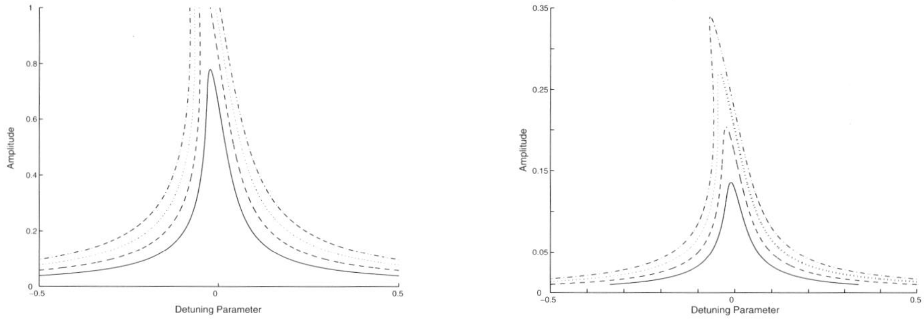

(a) the first principal resonance

(b) the second principal resonance

Fig. 1. The effect of the excitation amplitude

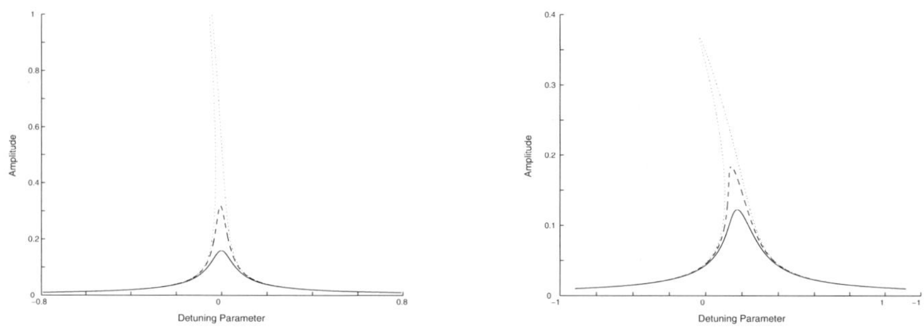

(a) the first principal resonance

(b) the second principal resonance

Fig. 2. The effect of the damping coefficient

zero, there exist more than one value of a_n for a given σ, and a narrow multi-valued region forms. The multivaluedness is a consequence of the nonlinearity.

Figure 4 shows the effect of the cubic term coefficient α_3. For the first principal resonance, F=2, μ=0.008, α_1=600, α_2=150, and α_3=0 (solid), 50 (dash), 150 (dot). For the second principal resonance F=50, μ=0.05, α_1=1800, α_2=450, and α_3=0 (solid), 300 (dash), 500 (dot). The larger α_3 also leads the curves bending more to the left.

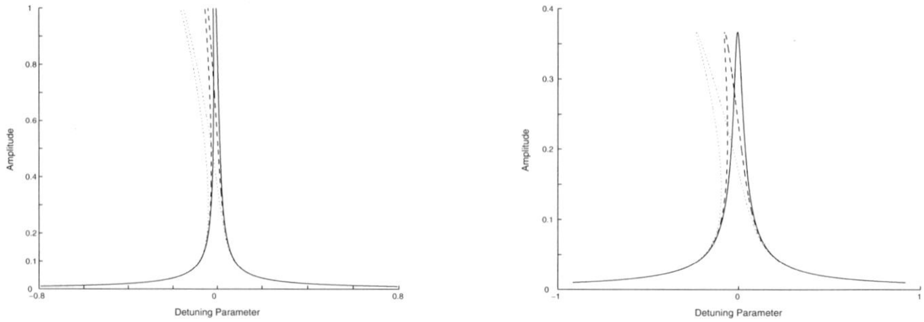

(a) the first principal resonance

(b) the second principal resonance

Fig. 3. The effect of the square term coefficient

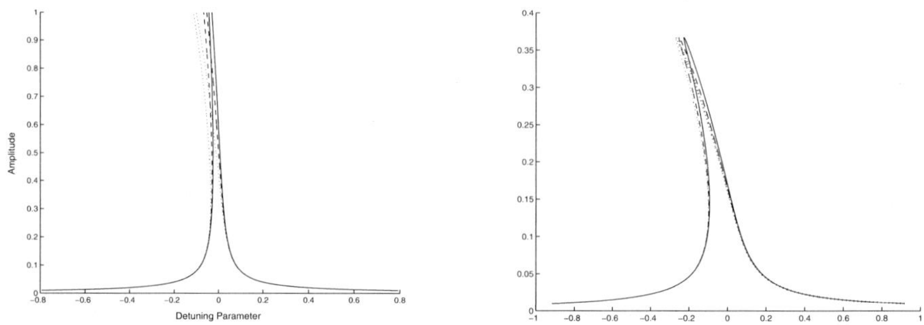

(a) the first principal resonance

(b) the second principal resonance

Fig. 4. The effect of the cubic term coefficient

5. Conclusions

This paper is devoted to principal resonance of a cantilever with a contact end. The cantilever is modeled as an Euler-Bernoulli beam. The contact is described by the Derjaguin-Müller-Toprov model. The method of multiple scales is applied to obtain the relation between the steady-state amplitude response and the detuning parameter. Numerical calculations are performed to explore the effects of related parameters. The large response occurs in the first mode. The larger excitation amplitude and the smaller damping coefficient lead to the larger response amplitude. As the nonlinear coefficients increase, the response curves bend more to the right.

Acknowledgements:

This work was supported by the National Natural Science Foundation of China (Project No. 10472060), Natural Science Foundation of Shanghai Municipality (Project No. 04ZR14058) and Shanghai Leading Academic Discipline Project (Project No.Y0103). The author is indebted to Mr. Qing-Quan Hu , Mr. Xiao-Jun Li and Mr. Ji-Guang Zhang for the assistance.

References

[1] Szemplinska-Stupnicka, W., *The Behavior of Nonlinear Vibrating Systems*, Kluwer, Dordrecht, 1990.

[2] Nayfeh, A. H., *Nonlinear Interactions: Analytical, Computational, and Experimental Methods*, John Wiley & Sons, New York, 2000.

[3] García, R. and Pérez, R., 'Dynamic atomic force microscopy methods', *Surface Science Reports* **47**, 2002, 179-301.

[4] Turner J. A., Hirsekorn, S., Rabe, U., and Arnold, W., 'High-frequency response of atomic-force microscope cantilevers', *Journal of Applied Physics* **82**, 1997, 966-979.

[5] Stark, R. W. and Heckl, W. M., 'Fourier transformed atomic force microscopy: tapping mode atomic force microscopy beyond the Hookian approximation', *Surface Science* **457**, 2000, 219-228.

[6] Lee, S. I., Howell, S., Raman, A., and Reifenberger, R., 'Nonlinear dynamics of microcantilevers in tapping mode atomic force microscopy: comparison between theory and experiment', *Physical Review B* **66**, 2002, Article No. 115409.

[7] Lee, S. I., Howell, S., Raman, A., and Reifenberger, R., 'Nonlinear dynamic perspectives on dynamic force microscopy', *Ultramicroscopy* **97**, 2003, 185-198.

[8] Stark, R. W., Schitter, G., Stark, M., Guckenberger, R., and Stemmer, A., 'State-space model of freely vibrating and surface-coupled cantilever dynamics in atomic force microscopy', *Physical Review B* **69**, 2004, Article No. 085421.

[9] Turner, J. A., 'Non-linear vibrations of a beam with cantilever-Hertzian contact boundary conditions', *Journal of Sound and Vibration* **275**, 2004, 177-191.

[10] Yagasaki, K., 'Nonlinear dynamics of vibrating microcantilevers in tapping-mode atomic force microscopy', *Physical Review B* 70, 2004, Article No. 245419.

[11] Abdel-Rahman, E. M. and Nayfeh, A. H., 'Contact force identification using the subharmonic resonance of a contact-mode atomic force microscopy', *Nanotechnology* **16**, 2005, 199-207.

[12] Derjaguin, B. V., Müller V. M. and Toporov Y. P., 'Effect of contact deformations on the adhesion of particles', *Journal of Colloid Interface Science* **53**, 1975, 314-326.

[13] Cappella, B. and Dietler G. 'Force-distance curve by atomic force microscopy', *Surface Science Reports* **34**, 1999, 1-104.

[14] Nayfeh, A. H., *Introduction to Perturbation Techniques*, John Wiley & Sons, New York, 1981.

Frictional Impact of Planar Deformable Bodies

Saeed Ebrahimi and Peter Eberhard

Institute of Engineering and Computational Mechanics, University of Stuttgart
Pfaffenwaldring 9, 70569 Stuttgart, Germany
[ebrahimi, eberhard]@itm.uni-stuttgart.de

Summary. This paper extends the formulation of frictionless impact analysis of planar deformable bodies presented in [6] to frictional impact. In the approach presented in that paper, a linear complementarity problem (LCP) on position level based on the Signorini conditions for the impact problem of continua was formulated. For this purpose, the normal gap distance between possible contact points was represented in terms of normal impact forces. Now, in order to consider the friction, this formulation may be appended to the formulation of tangential contact forces developed for continual contact in [4, 5]. The key issue behind this approach arises from this fact that in the case of deformable bodies, the behavior of impact in the tangential direction is similar to that of the continual contact for a short period of time. However, it is obvious that this assumption is valid only for impact analysis of deformable bodies and in the case of rigid bodies impact analysis, these two events must be distinguished.

1 Introduction

There are many examples of mechanical systems in which considering impact with friction is an essential and inevitable task and plays a fundamental role in the function of the system. Therefore, the development of numerical procedures for solving such problems accurately and efficiently has become the subject of many researches.

There are extensive researches addressed in the literature about investigation of frictional contact/impact of elastic bodies in the theory of elasticity consisting of formulations based on the FEM, BEM and those derived from analytical and mathematical procedures for static and quasi-static cases, see e.g. [20]. However, in the case of frictional impact of deformable bodies in multibody systems, almost all utilized approaches use the same formulation as approaches used for rigid body impact problems. In other words, in such situations a "rigid body like" impact model may be used.

In this framework the works which use the coefficient of restitution as a measure of energy loss during impact may be mentioned [19, 22, 23]. In

Peter Eberhard (ed.), Multiscale Problems in Multibody System Contacts, 23–32.

the work of Wasfy and Noor [19] the conservation of momentum and restitution equations were used as local velocity constraints to determine the "post-impact" velocities of the impacting nodes in the impact problem between flexible multibody systems. In this approach two coefficient of restitutions have been used, one for normal direction and the other in the tangential direction for considering friction.

In a second approach, the impact duration is finite and a constitutive law describes the force-deformation relationship while the bodies are in contact. Different types of constitutive laws have been provided by many researchers. In some of these contributions, compliant contact/impact models using a penalty-based formulation have been used for this purpose, see e.g. [2, 7, 10, 12, 18]. In [18] an approach of a compliant contact model is explained to overcome the numerical difficulties arising from introducing friction.

In some other approaches, the attention is put on the investigation of the transient dynamic response of deformable bodies subjected to impact using a so-called constraint addition-deletion technique, see e.g. [21]. In fact, this technique is based on the Lagrange multipliers method which has been widely used for contact analysis in multibody systems.

In some other contributions the investigation of frictional contact/impact problems is based on time-stepping techniques [1, 18]. In [1] a time-stepping algorithm has been presented for frictional contact problems of deformable bodies. This approach leads to positive energy dissipation and the integration of the friction law is based on a properly defined slip function. In [15, 8] a bi-potential method using a first order integration has been utilized where the frictional effects are taken into account by physical energy dissipation.

Another important approach which may be used formulates the contact/impact problems as a Linear Complementarity Problem (LCP). In this framework, the approaches which originate from LCPs but use another mathematical formulation may also be mentioned, e.g. [13]. These approaches have been well established and frequently used in the contex of dealing with rigid bodies, see e.g. [9, 11, 14, 16, 17]. However, for the case of dealing with deformable bodies there is a very limited number of publications which uses this approach.

In the work of Duriez et al. [3], the Signorini contact law together with Coulomb's friction law has been formulated as a computer haptics basis. In doing so, a Gauss-Seidel like algorithm served to solve a linear complementarity problem with a guaranteed convergence. However, the treatment of contact/impact problems in computer graphic techniques has a significant difference to those of mechanical engineering. In the former, the interest is put more on the real time animation of systems consisting of colliding objects by following some geometrical techniques while in the latter, the goal is to obtain the contact/impact forces accurately and to investigate the response of the system after collision.

The approach presented in this paper belongs to the mechanically oriented category and formulates the frictional impact problem of planar deformable bodies as an LCP. In the next section, the methodology to reach a formulation for investigation of such problems will be described.

2 Frictional Impact Analysis

2.1 Kinematics of Impact in Normal Direction

The formulation of impact in normal direction may be started by obtaining a relation between gap distances of possible contact pairs and generalized coordinates of bodies. To do this, one can write the relation of the lth normal gap distance $g^l_{N_{ij}}$ between two possible colliding points p_i and p_j located on bodies i and j, respectively, see [6]

$$g^l_{N_{ij}} = (\mathbf{W}^l_{g_{ij}})^T \cdot \mathbf{q}_{ij} + \bar{\mathbf{w}}^l_{ij} \ , \tag{1}$$

where the vector $\mathbf{q}_{ij}$ contains the generalized coordinates of both bodies. For consideration of all k contact pairs between bodies i and j, the normal gap distance $\mathbf{g}_{N_{ij}}$ can be rearranged in the form

$$\mathbf{g}_{N_{ij}} = \begin{bmatrix} g^1_{N_{ij}} \\ \vdots \\ g^k_{N_{ij}} \end{bmatrix} = \begin{bmatrix} \mathbf{W}^1_g \dots \mathbf{W}^k_g \end{bmatrix}^T_{ij} \cdot \mathbf{q}_{ij} + \begin{bmatrix} \bar{\mathbf{w}}^1 \\ \vdots \\ \bar{\mathbf{w}}^k \end{bmatrix}_{ij} = \mathbf{W}^T_{g_{ij}} \cdot \mathbf{q}_{ij} + \bar{\mathbf{w}}_{ij} . \tag{2}$$

Based on this relation, the lth component of $\mathbf{g}_{N_{ij}}$ is a non-active contact pair when $g^l_{N_{ij}} > 0$ and becomes active when $g^l_{N_{ij}} = 0$. In the latter case, for preventing any penetration normal impact forces must be applied at the corresponding contact points. Following the same notation, a general relationship for all possible contact pairs between n_b bodies in the multibody system may be obtained

$$\mathbf{g}_N = \mathbf{W}^T_g \cdot \mathbf{q} + \bar{\mathbf{w}} \tag{3}$$

where the vector $\mathbf{q}$ denotes the system generalized coordinates considering all bodies. A detailed derivation of these relations can be found in [6].

2.2 Kinematics of Impact in Tangential Direction

In contrast to the formulation of normal impact in which the kinematics of impact is considered on position level, the kinematics of tangential impact has to be considered on acceleration level. Considering again the lth contact pair of bodies i and j, the relative acceleration in the tangential direction is written as, see [4]

$$\ddot{g}^l_{T_{ij}} = (\mathbf{W}^l_{T_{ij}})^T \cdot \ddot{\mathbf{q}}_{ij} + (\mathbf{w}^l_{T_{ij}})^T \cdot \dot{\mathbf{q}}_{ij} \ , \tag{4}$$

where

$$\mathbf{W}^l_{Tij} = \begin{bmatrix} (\mathbf{t}^l_i \cdot \mathbf{L}^l_i)^T \\ -(\mathbf{t}^l_i \cdot \mathbf{L}^l_j)^T \end{bmatrix}, \qquad \mathbf{w}^l_{Tij} = \begin{bmatrix} (\dot{\mathbf{t}}^l_i \cdot \mathbf{L}^l_i + \mathbf{t}^l_i \cdot \dot{\mathbf{L}}^l_i)^T \\ -(\dot{\mathbf{t}}^l_i \cdot \mathbf{L}^l_j + \mathbf{t}^l_i \cdot \dot{\mathbf{L}}^l_j)^T \end{bmatrix}.$$

Here, $\mathbf{L}$ is a matrix which projects the generalized velocity vector $\dot{\mathbf{q}}$ onto the velocities in the global coordinate system and $\mathbf{t}$ is the tangent vector at the contact point. This equation which holds for contact point l between bodies i and j can be used to obtain the matrix form of the relative tangential accelerations $\ddot{\mathbf{g}}_T$ for all n_c contact points between n_b bodies

$$\ddot{\mathbf{g}}_T = \mathbf{W}_T^T \cdot \ddot{\mathbf{q}} + \mathbf{w}_T^T \cdot \dot{\mathbf{q}} \,, \tag{5}$$

where the vectors $\dot{\mathbf{q}}$ and $\ddot{\mathbf{q}}$ are the generalized velocities and accelerations of the system consisting of all n_b bodies.

2.3 Equations of Motion

Considering a system of deformable bodies, one can write its nonlinear equations of motion as

$$\mathbf{M} \cdot \ddot{\mathbf{q}} + \mathbf{C} \cdot \dot{\mathbf{q}} + \mathbf{K} \cdot \mathbf{q} = \mathbf{F}_{ext} + \mathbf{F}_v + \mathbf{F}_c \,. \tag{6}$$

Here $\mathbf{M}$ is the system mass matrix, $\mathbf{C}$ and $\mathbf{K}$ are the damping and stiffness matrices arising from the elastic coordinates, $\mathbf{F}_{ext}$, $\mathbf{F}_v$ and $\mathbf{F}_c$ are the nonlinear vectors of generalized external forces, Coriolis forces and contact forces, respectively. Equation (6) may be rewritten in the simpler form

$$\mathbf{M} \cdot \ddot{\mathbf{q}} = \mathbf{F} + \mathbf{F}_c \,, \tag{7}$$

where the vector $\mathbf{F}$ contains all force vectors except inertial and contact forces. The contact force vector $\mathbf{F}_c$ for n_c contact points can be supposed to be the summation of normal and tangential forces which is written in terms of two different vectors $\boldsymbol{\lambda}_N$ and $\boldsymbol{\lambda}_T$

$$\mathbf{F}_c = \mathbf{W}_N \cdot \boldsymbol{\lambda}_N + \mathbf{W}_T \cdot \boldsymbol{\lambda}_T \,. \tag{8}$$

By separating the tangential contact forces into the two different cases of sliding forces $\mathbf{W}_G \cdot \boldsymbol{\mu}_G \cdot \boldsymbol{\lambda}_N$ and sticking forces $\mathbf{W}_H \cdot \boldsymbol{\lambda}_H$ and supposing $\boldsymbol{\lambda}_H = \boldsymbol{\lambda}_H^{(+)} - \boldsymbol{\lambda}_H^{(-)}$ for sticking contact forces, one can rewrite Eq. (7)

$$\mathbf{M} \cdot \ddot{\mathbf{q}} = \mathbf{F} + \underbrace{\left(\mathbf{W}_N + \mathbf{W}_G \cdot \boldsymbol{\mu}_G \mid \mathbf{W}_H \mid -\mathbf{W}_H \right)}_{\mathbf{W}_{NH}} \cdot \underbrace{\begin{pmatrix} \boldsymbol{\lambda}_N \\ \boldsymbol{\lambda}_H^{(+)} \\ \boldsymbol{\lambda}_H^{(-)} \end{pmatrix}}_{\boldsymbol{\lambda}} = \mathbf{F} + \mathbf{W}_{NH} \cdot \boldsymbol{\lambda} \,, \tag{9}$$

where $\mathbf{W}_G$ and $\mathbf{W}_H$ are matrices extracted from the matrix $\mathbf{W}_T$ which correspond to the sliding and sticking parts, respectively, and $\boldsymbol{\mu}_G$ is a diagonal matrix of friction coefficients.

2.4 Generalized Coordinates versus Impact Forces

In Section 2.1 the normal gap distances vector has been obtained in terms of the generalized coordinates vector. However, the problem in the evaluation of $\mathbf{g}_N$ is that the generalized coordinates vector $\mathbf{q}$ is a function of impact forces which are not known previously. Our task here is to utilize the numerical 4th order Runge-Kutta approach and reach an analytical formulation of the systems generalized coordinates vector $\mathbf{q}_{i+1}$ of step $i+1$ as a function of impact forces $\boldsymbol{\lambda}_i$ including normal and tangential forces.

Assume that the integration process is at step i and the generalized coordinates and generalized velocities of the system are known from the previous integration step. Since the numerical 4th order Runge-Kutta approach requires four evaluations of the equations of motion in order to proceed from step i to $i+1$, the same strategy may be followed manually to obtain an analytical formulation of the vector $\mathbf{q}_{i+1}$ as a function of $\boldsymbol{\lambda}_i$. This can be done by successive evaluations and substitutions of the vector $\mathbf{q}$ for several steps of the 4th order Runge-Kutta. Following such a strategy yields this required formula

$$\mathbf{q}_{i+1} = \underbrace{\frac{h^2}{6}(\mathbf{M}_1^{-1}\cdot\mathbf{W}_{NH_1} + \mathbf{M}_2^{-1}\cdot\mathbf{W}_{NH_2} + \mathbf{M}_3^{-1}\cdot\mathbf{W}_{NH_3})}_{\mathbf{W}_{q_i}}\cdot\boldsymbol{\lambda}_i + \tag{10}$$

$$\underbrace{\mathbf{q}_i + h\dot{\mathbf{q}}_i + \frac{h^2}{6}(\mathbf{M}_1^{-1}\cdot\mathbf{F}_1 + \mathbf{M}_2^{-1}\cdot\mathbf{F}_2 + \mathbf{M}_3^{-1}\cdot\mathbf{F}_3)}_{\mathbf{w}_{q_i}} ,$$

where h is the integration step size and $\mathbf{M}_k^{-1}, \mathbf{W}_{NH_k}$ and $\mathbf{F}_k$ $(k = 1,2,3)$ arise from Eq. (9) but are evaluated at the first three steps of the 4th order Runge-Kutta method.

2.5 Frictional Impact Formulation as an LCP

The kinematics of impact has been investigated by finding a relation between the normal gap distances vector $\mathbf{g}_N$ of contact pairs and the generalized coordinates for normal direction and a relation between the relative tangential accelerations $\ddot{\mathbf{g}}_T$ of contact pairs and the generalized accelerations. On the other hand, the generalized coordinates and accelerations have been shown in terms of impact forces. This information may be used to form a linear complementarity problem consisting of two subproblems, the first one between $\mathbf{g}_N$ and $\boldsymbol{\lambda}$ for normal impact and the second one between $\ddot{\mathbf{g}}_T$ and $\boldsymbol{\lambda}$ for tangential impact.

The required procedure to do so may be started by considering Eq. (5) for sticking contacts illustrated by index H in Eq. (9)

$$\underbrace{\begin{bmatrix} \ddot{\mathbf{g}}_H \\ -\ddot{\mathbf{g}}_H \end{bmatrix}}_{\ddot{\mathbf{g}}_{HG}} = \underbrace{\begin{bmatrix} \mathbf{W}_H^T \\ -\mathbf{W}_H^T \end{bmatrix}}_{\mathbf{W}_{HG}^T} \cdot\ddot{\mathbf{q}} + \underbrace{\begin{bmatrix} \mathbf{w}_H^T\cdot\dot{\mathbf{q}} \\ -\mathbf{w}_H^T\cdot\dot{\mathbf{q}} \end{bmatrix}}_{\mathbf{w}_{HG}} . \tag{11}$$

However, it is clear that the sliding contacts have already been taken into account from the Coulomb's friction law. This equation together with Eq. (3) can be written in the following matrix form

$$\begin{bmatrix} \mathbf{g}_N \\ \ddot{\mathbf{g}}_{HG} \end{bmatrix} = \begin{bmatrix} \mathbf{W}_g^T & \mathbf{0} \\ \mathbf{0} & \mathbf{W}_{HG}^T \end{bmatrix} \cdot \begin{bmatrix} \mathbf{q} \\ \ddot{\mathbf{q}} \end{bmatrix} + \begin{bmatrix} \bar{\mathbf{w}} \\ \mathbf{w}_{HG} \end{bmatrix} . \quad (12)$$

Now, the generalized coordinates and accelerations may be written together with Eqs. (10) and (9), respectively,

$$\begin{bmatrix} \mathbf{q} \\ \ddot{\mathbf{q}} \end{bmatrix} = \begin{bmatrix} \mathbf{W}_q \\ \mathbf{M}^{-1} \cdot \mathbf{W}_{NH} \end{bmatrix} \cdot \boldsymbol{\lambda} + \begin{bmatrix} \mathbf{w}_q \\ \mathbf{M}^{-1} \cdot \mathbf{F} \end{bmatrix} , \quad (13)$$

and after substitution in Eq. (12) one obtains

$$\underbrace{\begin{bmatrix} \mathbf{g}_N \\ \ddot{\mathbf{g}}_{HG} \end{bmatrix}}_{\mathbf{g}_M} = \underbrace{\begin{bmatrix} \mathbf{W}_g^T \cdot \mathbf{W}_q \\ \mathbf{W}_{HG}^T \cdot \mathbf{M}^{-1} \cdot \mathbf{W}_{NH} \end{bmatrix}}_{\mathbf{W}^T} \cdot \boldsymbol{\lambda} + \underbrace{\begin{bmatrix} \mathbf{W}_g^T \cdot \mathbf{w}_q + \bar{\mathbf{w}} \\ \mathbf{W}_{HG}^T \cdot \mathbf{M}^{-1} \cdot \mathbf{F} + \mathbf{w}_{HG} \end{bmatrix}}_{\mathbf{w}} . \quad (14)$$

The final form of the complementarity equations based on our notation and according to the above mentioned points is

$$\begin{gathered} \begin{bmatrix} \mathbf{g} \\ \boldsymbol{\lambda}_{H_0} \end{bmatrix} = \begin{bmatrix} \mathbf{W}^T & \mathbf{I}^T \\ \mathbf{N}_H - \mathbf{I} & \mathbf{0} \end{bmatrix} \cdot \begin{bmatrix} \boldsymbol{\lambda} \\ \mathbf{z} \end{bmatrix} + \begin{bmatrix} \mathbf{w} \\ \mathbf{0} \end{bmatrix} , \\ \begin{bmatrix} \mathbf{g} \\ \boldsymbol{\lambda}_{H_0} \end{bmatrix} \geq \mathbf{0} , \quad \begin{bmatrix} \boldsymbol{\lambda} \\ \mathbf{z} \end{bmatrix} \geq \mathbf{0} , \quad \begin{bmatrix} \mathbf{g} \\ \boldsymbol{\lambda}_{H_0} \end{bmatrix} \cdot \begin{bmatrix} \boldsymbol{\lambda} \\ \mathbf{z} \end{bmatrix} = 0 . \end{gathered} \quad (15)$$

Here, the parameters $\boldsymbol{\lambda}_{H_0}$, $\mathbf{N}_H$, $\mathbf{I}$ and $\mathbf{z}$ in this equation are chosen in the same way as in [17] and have the same meaning. Only the vector $\mathbf{g}$ has different interpretation and can be represented as

$$\mathbf{g} = \begin{pmatrix} \mathbf{g}_N \\ \ddot{\mathbf{g}}_{HG}^{+} \\ \ddot{\mathbf{g}}_{HG}^{-} \end{pmatrix} , \quad (16)$$

where $\ddot{\mathbf{g}}_{HG}^{+}$ and $\ddot{\mathbf{g}}_{HG}^{-}$ correspond to $\ddot{\mathbf{g}}_{H}^{+}$ and $\ddot{\mathbf{g}}_{H}^{-}$ from [17], respectively. These parameters are used in order to formulate the complementarity form of the equations in such a way to handle switching between sliding and sticking cases of contact.

2.6 Implemented Algorithm

In order to solve the frictional impact of planar deformable bodies, one has to follow the described algorithm till the end of simulation for the presented approach based on the 4th order explicit Runge-Kutta formulation.

start
- initialize the systems generalized coordinates and velocities, $\mathbf{q} = \mathbf{q}_0, \dot{\mathbf{q}} = \dot{\mathbf{q}}_0$
- set $\boldsymbol{\lambda}_0 = \mathbf{0}, Tol = 1e-6, k_{max}$ =maximum number of iterations
- initialize the step counter $i = 1$
- find active normal contact pairs $g_{N_l} \leq 0$
- find active sticking contact pairs $|\dot{g}_{T_l}| \leq Tol$

while $(t < t_{end})$
  - initialize the impact force from the previous time step $\boldsymbol{\lambda}_i = \boldsymbol{\lambda}_{i-1}$
  - initialize the iteration counter $k = 1$ and error $err = 1.0$

  while $(err > Tol, k < k_{max})$
    - calculate new $\mathbf{q}_i$ and $\dot{\mathbf{q}}_i$ utilizing a suitable integration approach
    - build the required quantities for normal as well as sliding contacts
    - build the required quantities for sticking contacts
    - generate and solve the LCP of Eq. (15) for unknowns $\mathbf{g}_{i+1}^{k+1}$ and $\boldsymbol{\lambda}_i^{k+1}$
    - calculate the iteration error
      $err = |(\|\boldsymbol{\lambda}_{N_i}^{k+1}\| - \|\boldsymbol{\lambda}_{N_i}^{k}\|)| + |(\|\boldsymbol{\lambda}_{H_i}^{k+1}\| - \|\boldsymbol{\lambda}_{H_i}^{k}\|)|$
    - initialize the impact force for the next iteration $\boldsymbol{\lambda}_i^{k+1} = \boldsymbol{\lambda}_i^{k}$
    - increase the iteration counter $k = k + 1$

  end while
  - increase the step counter $i = i + 1$ and proceed to the next time step $t = t + h$

end while

end

It is important to emphasize that the 4th order Runge-Kutta method is just used to obtain an analytical formulation between $\mathbf{g}_N$ and $\boldsymbol{\lambda}$ and not for the main integration processes to proceed to the next time step. In other words, the approach presented here consists of two integration parts. One is the main integration to proceed from step i to $i+1$ and another one is responsible for calculating the normal gap distances vector as a function of impact forces. Maybe it is also important to point out that the required LCPs for normal and tangential directions must be considered only for their corresponding active contact pairs.

3 Numerical Results

We consider here an example of ten identical elastic rectangular blocks which topple down consecutively, see Fig. (1). At first, all blocks are at rest and have been positioned on the rigid foundation with equal distances. Then, the first block on the left hand side follows a rotational motion and hits the second block. This event will repeat sequentially for the other blocks and after a while the last block on the right hand side will topple down over the rigid foundation. Hence, one can investigate the effect of impact with friction between planar elastic blocks.

For the simulation of this test example, the first block has a rotational velocity of $\omega = 4rad/s$ and following values for the material and geometry of each elastic block have been used

material: $E = 2500\ N/m$, $\nu = 0.3$, $\mu = 0.25$, $M = 4\ kg$,

geometry: $H = 1\ m$, $L = 0.3\ m$,

where we denote by E the Young's modulus, ν the Poisson's ratio, μ the friction coefficient, M the mass, H the height and L the length. The animation of this example is illustrated in Fig. (1).

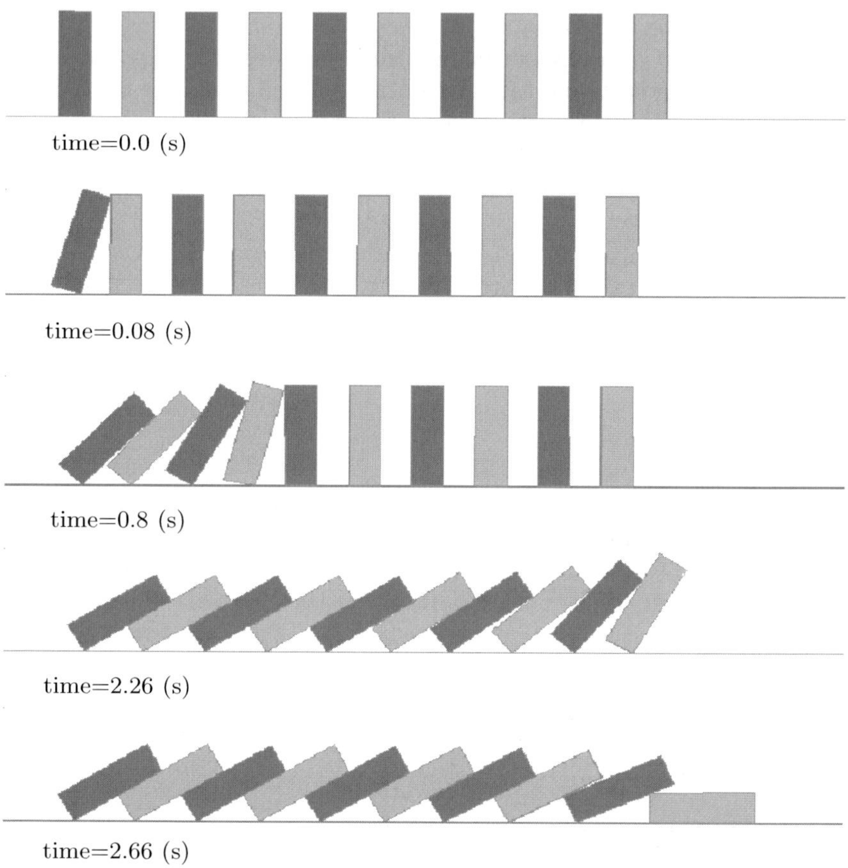

Fig. 1. Animation of the impact between ten identical elastic rectangular blocks

4 Conclusion

A linear complementarity problem formulation for frictional impact analysis of planar deformable bodies was presented. Formulating the normal impact was carried out through finding a relation between the normal gap distances vector and impact forces utilizing the 4th order explicit Runge-Kutta approach. Then, this formulation was appended to the formulation of tangential contact forces developed for continual contact in [4, 5]. The key issue behind this approach arises from the fact that in the case of deformable bodies the behavior of impact in the tangential direction is similar to the one of the continual contact for a short period of time. In this approach, consideration of a impact law in normal direction is done by imposing non-penetrability conditions through the complementarity relation between normal gap distances and normal impact forces.

The presented approach was validated by simulating an example of impacting elastic rectangular blocks falling over each other. The results obtained from this approach show that the softer the material of impacting bodies, the better the agreement with reference solutions may be achieved. This situation is quite reasonable since this approach uses no coefficient of restitution as the measure of energy loss which is considered through the damping in the material law.

References

1. Armero F, Petöcz E (1999) A new dissipative time-stepping algorithm for frictional contact problems: formulation and analysis. Computer Methods in Applied Mechanics and Engineering 179:151–178
2. Bauchau OA (1999) On the modeling of friction and rolling in flexible multibody systems. Multibody systems dynamics 3:209–239
3. Duriez C, Andriot C, Kheddar A (2004) Signorini's contact model for deformable objects in haptic simulations. In: IEEE/RSJ International Conference on Intelligent Robots and Systems (IROS), Sendai, Japan
4. Eberhard P, Ebrahimi S (2005) On the use of linear complementarity problems for contact of planar flexible bodies. In: Goicolea JM, Cuadrado J, Garcia Orden JC (eds) Proceedings of the ECCOMAS Thematic Conference on Multibody Dynamics, Spain
5. Ebrahimi S, Eberhard P (2005) Contact of planar deformable bodies using a linear complementarity formulation. PAMM Proceedings in Applied Mathematics and Mechanics 5:197–198
6. Ebrahimi S, Eberhard P (2006) A linear complementarity formulation on position level for frictionless impact of planar deformable bodies. Submitted to the "Proceedings of the 2nd ECMI Workshop on Numerical Methods in Multibody Dynamics"
7. Ebrahimi S, Hippmann G, Eberhard P (2005) Extension of the polygonal contact model for flexible multibody systems. International Journal of Applied Mathematics and Mechanics 1:33–50

8. Feng ZQ, Feng Z, Domaszewski M (2002) Some computational aspects for analysis of low and high velocity impact of deformable bodies. International Journal of Nonlinear Mechanics 37:1029–1036
9. Kim J, Kwak BM (1996) Dynamic analysis of two-dimensional frictional contact by linear complementarity problem formulation. International Journal of Solids and Structures 30:4605–4624
10. Khulief YA, Shabana AA (1987) A continuous force model for the impact analysis of flexible multibody systems. Mechanism and Machine Theory 22:213–224
11. Kwak BM (1991) Complementarity problem formulation of three-dimensional frictional contact. Journal of Applied Mechanics 58:134–140
12. Lankarani HM, Nikravesh PE (1994) Continuous contact force models for impact analysis in multibody systems. Nonlinear Dynamics 5:193–207
13. Leine RI, Glocker C (2003) A set-valued force law for spatial Coulomb-Contensou friction. European Journal of Mechanics A/Solids 22:193–216
14. Lötstedt P (1981) Coulomb friction in two-dimensional rigid body systems. ZAMM Zeitschrift für Angewandte Mathematik und Mechanik 61:605–615
15. Magnain B, Feng ZQ, Cros JM (2005) Numerical investigation of contact/impact problems between deformable bodies. In: EURODYN Sixth European Conference on Structural Dynamics, Paris, France
16. Pang JS, Trinkle J (1996) Complementarity formulations and existence of solutions of dynamic multi-rigid-body contact problems with Coulomb friction. Mathematical Programming 73:199–226
17. Pfeiffer F, Glocker C (1996) Multibody Dynamics with Unilateral Contacts. John Wiley & Sons, New York
18. Song P, Pang JS, Kumar V (2004) A semi-implicit time-stepping model for frictional compliant contact problem. International Journal for Numerical Methods in Engineering 60:2231–2261
19. Wasfy TM, Noor AK (1997) Computational procedure for simulating the contact/impact responce in flexible multibody systems. Computer Methods in Applied Mechanics and Engineering 147:153–166
20. Wriggers P (2002) Computational Contact Mechanics. John Wiley & Sons, Chichester
21. Wu SC, Haug EJ (1990) A substructure technique for dynamics of flexible mechanical systems with contact-impact. Journal of Mechanical Design 112:390–398
22. Yigit AS, Ulsoy AG, Scott RA (1990) Dynamics of a radially rotating beam with impact, part1: theoretical and computational model. Vibration and Acoustics 112:65–70
23. Zakhariev EV (2001) A numerical method for multibody system frictional impact simulation. Paper No. DETC2001/VIB-21367, In: Proc of ASME 2001 DETC, Pittsburgh PA

Parallel Load Balanced Particle Simulation with Hierarchical Particle Grouping Strategies

Florian Fleissner and Peter Eberhard

Institute of Engineering and Computational Mechanics, University of Stuttgart, Pfaffenwaldring 9, 70569 Stuttgart, Germany
[fleissner, eberhard]@itm.uni-stuttgart.de

Summary. We introduce a new approach for control based load-balanced particle simulation, applying a recursive domain decomposition scheme that enables a minimization of communication expense and efficient hierarchical parallel neighborhood search, especially optimized for multiuser clusters of workstations with fluctuating processor loads.

1 Introduction

The simulation of granular media is a field that is currently under investigation in many research groups. The complex nature of the simulated medium - large quantities of freely moving particles without fixed neighborhood relations - requires grid-free methods following the Lagrangian approach. They enable simulations of phenomena which are beyond the reach of classical continuum simulation methods such as the Finite-Element-Method, the Finite Volume Method and the Finite Difference Method. One of the methods, the Discrete Element Method (DEM) [1] is widely known for the simplicity of its governing equations as it avoids any coupled systems of equations of motion by modeling all bodies in the simulation as free bodies that only interact through applied forces. A large variety of particle geometries as well as force laws have been investigated with this method, reaching from simple spherical particles with frictionless Hertz contacts [5] over hyperellipsoids to nonconvex polyhedra with frictional contacts [6]. Due to high computational expense and memory requirements, larger particle systems can only be simulated by parallel programs. In this paper we describe a parallel particle simulation method optimized for the use on clusters of workstations, the most widespread parallel computation environment. In Section 2 we give an overview about the requirements of parallel particle simulations. The dynamic domain decomposition, that we use for work distribution is described in Section 3, followed by a description of our strategies for optimizing inter-process communication in Section 4. A description of our parallel hierarchical neighborhood search

Peter Eberhard (ed.), Multiscale Problems in Multibody System Contacts, 33–44.

algorithm is given in Section 5. In Section 6 we describe how load balancing between processors can be achieved solving a classical control problem.

2 Requirements of Parallel Simulations

The discrete element method is a comparably simple method with respect to its basic algorithm. It only consists of three major steps, i.e. collision detection, computation of interaction forces and trajectory integration. However, an efficient scalable parallel implementation of this method is challenging, especially for systems with fluctuating loads. These systems require load balancing schemes that reduce the processors' idle time to a minimum. Such programs are usually difficult to implement and hard to debug. Despite of all pitfalls an efficient load-balanced implementation amortizes the time for development due to considerably faster program execution.

Another important aspect of parallel programs is their scalability. Naively parallel programs are expected to run n times faster when executed on n processors in parallel. Unfortunately in reality this behavior can only be achieved for a certain number of processors depending on the size of the simulated problem and it is anyway more difficult to obtain a good scaling behaviour for programs with unstructured data such as DEM-programs. It is obvious that a simulation with n particles assigned to n different processors cannot be n times faster than on one processor due to unavoidable, time consuming network data transfers. However, depending on the quality of the serial program that is compared to the parallel program to obtain a speedup estimation, sometimes even super-linear speedup can be observed. On one hand this can be due to the use of algorithms of nonlinear, higher order complexity that benefit from a kind of "divide and conquer" effect as well as due to a better cache performance of the parallel program that arises from the fact that every parallel processor processes a smaller amount of data than the single processor of the serial program, resulting in a higher probability for data to be present in cache memory. Simple read/write memory access tests on a 3.2 Ghz Pentium 4 with 1 MB L2-cache memory showed that processing cached data (strided access) can be up to eighty times faster than data that is not present in cache memory (random access).

3 Recursive Domain Decomposition

For the distribution of the computation load of a discrete element simulation on different processors, the decision has to be made how to divide simulation work and simulation data. For simulations with only short ranged particle interactions such as collision or adhesional forces, a spatial subdivision is most obvious. In this case the simulation domain is partitioned into subdomains which are assigned to different processors. The particles are hosted by the processor in whose subdomain they reside. Different strategies exists for this decomposition. Amongst those are static and dynamically varying decompositions. Dynamic decompositions allow for adaptive changes of the work distribution during the simulation. These changes can become necessary when the number of objects assigned to the different processors exceeds a certain level of imbalance causing some processors to be severely overloaded. This state has to be avoided by any means as the slowest processor determines the overall simulation speed. Also concurrent processes that require the processor to share its resources affect the program runtime. Both effects manifest as differences of the time that the different processors need to finish their per timestep tasks. The most natural way to cope with these imbalances is to reduce the size of the subdomains of those processors that are overloaded. The domain that is cut off from the overloaded processors' subdomains together with the contained particles is then appended to other, less loaded processors' subdomains.

Zhakhovskii et al. [12] introduced a decomposition scheme that divides the simulation domain into Voronoi regions whose size and shape is varied during the simulation to alleviate the imbalances in processor runtimes. Even though the Voronoi-decomposition is an elegant way of subdivision, it comes with the drawback of the need to deal with the complicated nature of 3D-Voronoi-regions which have to be updated gradually, as well as it complicates the use of neighborhood search with axis aligned bounding boxes. Deng [2] proposed a method that applies arbitrary quadrilaterals with fixed topology for 2D simulations. Approaches based on cubical subdomains were described by Nyland [7] and Hayashi [4].

All methods we found do not take into account the influence of the shape of the subdomains on the amount of necessary inter-process communication, which scales with the subdomain surfaces. However, the amount of data transfered during parallel simulations is crucial for the performance of the computation. Therefore, we propose a new decomposition scheme inspired by the *Orthogonal Recursive Bisection Decomposition (ORB)* introduced by Warren [11] who used quadtrees for decomposition. By using binary trees we gain the choice in which spatial dimension to divide a subdomain, and thus can always choose the optimal dimension as described in Section 4.

As a high percentage of parallel particle simulation time is spend for collision detection, our choice of the decomposition approach was motivated by the requirement for a decomposition scheme that is well suited for the application of the neighborhood search method we already used in our serial implementations. For neighborhood search we apply a flexible and efficient hybrid method that is composed of the *Incremental Sort and Update Algorithm* [10] and a method that introduces secondary rank arrays for a speedup of the bounding-box-list-lookup [9]. The latter is a good choice for simulations of particle systems with greatly varying particles sizes and shapes and is characterized by its robustness against ill shaped bounding boxes. Using axis-aligned bounding boxes as bounding volumina for neighborhood search, it is just natural to apply also axis aligned boxes as subdomain geometry.

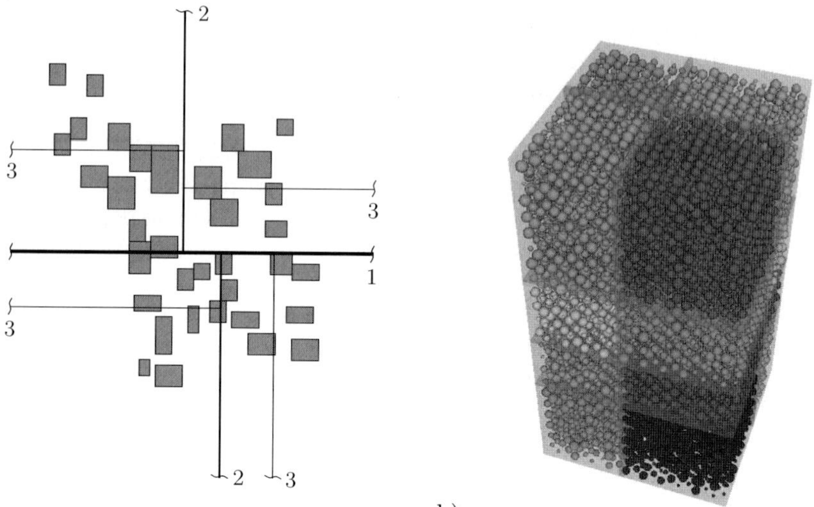

a) b)

Fig. 1. Recursive binary domain decomposition: a) After a first division (1), the division process is repeated recursively (2 and 3) until the number of objects in the leaf domains of the emerging binary decomposition tree is below a given threshold. b) Assignment of objects to 16 processors (colored) by recursive binary domain decomposition in 3D.

For enabling dynamic adaption of the decomposition, we introduce a recursive decomposition pattern with the topology of a binary tree. Following this pattern, the simulation domain is recursively divided in half, resulting in pairs of twin subdomains on every level of the emerging decomposition tree. The leaf subdomains of this decomposition tree are assigned to parallel processor pairs, see Fig. 1.

The same pattern is used on every processor to divide the objects into groups. In this way we benefit of increased cache performance due to improved spatial and temporal locality of memory access [3]. Subgrouping of particles also enables us to divide collision detection and contact force calculation into several subprocesses that can be carried out in arbitrary order, a fact we take advantage of in order to reduce processor idle time, see Section 5.

4 Optimization of Inter-Process Communication

Minimizing the amount of inter-process communication is crucial for good performance of parallel programs. The problem-specific requirements for communication have to be analyzed with respect to different aspects, such as the amount of data to be transfered and the communication pattern.

The costs for transfering data between processors, especially via ethernet are usually much higher than the costs for performing any computations on this data. As communication expense scales linearily with the amount of objects that are overlapping subdomain boundaries and thus with the subdomain surfaces, we dynamically change the dimensions in which subdomains are divided into twin subdomains by the recursive approach in such a way that always yields the minimal possible subdomain surface.

The intersection surface of the subdomains that result from a domain subdivision, see Fig. 2, is calculated from the edges a_i of the parent domain as

$$S_i = a_j a_k, \quad i, j, k \in [1, 2, 3], \quad i \neq j \neq k. \tag{1}$$

The optimal dimension of subdivision is the one that yields the minimal intersection surface

$$S_{min} = \min\{S_1, S_2, S_3\}. \tag{2}$$

If the fraction between the minimal possible intersection surface and the surface of the current dimension of subdivision

$$\frac{S_{min}}{S_{current}} \tag{3}$$

falls below a given threshold tol_{rel}, the dimension of subdivision is switched to the dimension with the minimal intersection surface.

We apply the same algorithm to obtain an optimal initial tree setup.

The mode of data transfer has also an important impact on the process performance. Message passing standards such as PVM and MPI feature point to point and collective communication. The latter is specially useful for use with structured data and fixed communication order, but is of minor interest for our purpose. Point to point communication can be distinguished between blocking and non-blocking communication. While the blocking communication requires the two communicating processors to maintain a data transfer connection that stops program execution until the data is completely transfered, non-blocking communication allows one or both processors to continue the program execution while extra threads performs the data transfer. Especially for computations on unstructured data non-blocking communication is vital, as often the order of communication is not known in advance. Here, blocking communication would in many cases cause the communicating processors to spend to much time waiting on each other and also would potentially enable deadlock situations. For our implementations we use LAM MPI [8] that features full non-blocking communication.

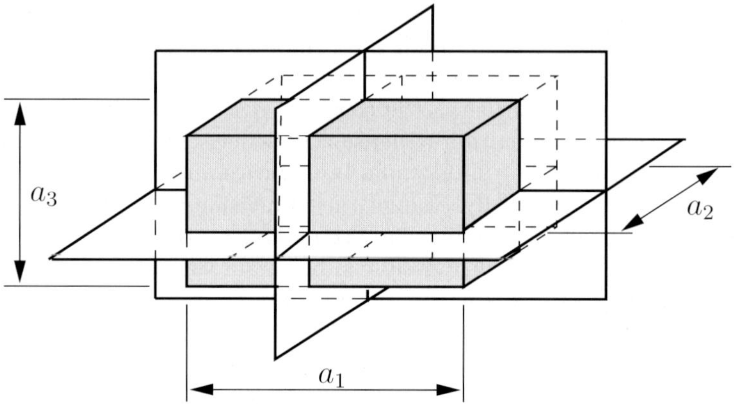

Fig. 2. On every tree level the dimension of subdivision is chosen in such a way that the resulting subdomains have a minimum surface.

5 Hierarchical Neighborhood Search

For improving the performance of the neighborhood search that requires a large amount of the total simulation time, we apply a hierarchical approach. Neighborhood search is performed on three levels, the level of process bounding boxes, the level of object group bounding boxes and the level of object bounding boxes, see Fig. 3.

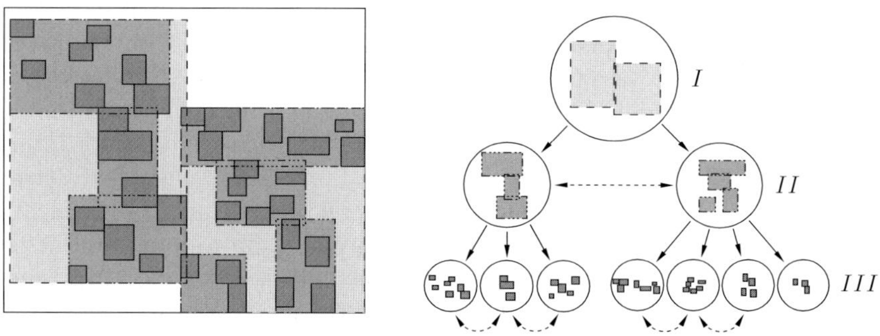

Fig. 3. Hierarchical neighborhood search.

In every timestep an overlap detection between the process bounding boxes, i.e. boxes that enclose all objects assigned to a certain process, is carried out by a manager process that determines which processes have to communicate for collision detection. The manager process assigns master and slave roles to the worker processes for the remote collision detection process.

This role assignment is carried out in a way that every processor has approximately the same master and slave tasks, which is a basis for an even work distribution. At the beginning of every timestep the master processes determine which of their objects are overlapping with the slave processes' bounding boxes. If overlaps are detected that did not occur in the last time step, ghost copies of these objects are transfered to the slave process which are then used for the actual overlap detection. If the overlap has already been present in the last timestep, it is sufficient to transfer the objects states to the slave process. The actual collision detection between bounding boxes of local and ghost objects of remote processors is then carried out by the slave processes, which send back the accumulated interaction forces of the colliding objects to the master processes. After having send the ghost objects or object states to the slave process, the worker processes perform the local collision detection tasks until the slaves reply, i.e. accumulated forces and torques of the ghost objects return from the slave processes. As mentioned above, all processors divide their objects into object groups. Collision detection can thus be divided into subprocesses that may be performed in arbitrary order. Two kinds of collision detection tasks are carried out. Collision detection between objects inside the same group and between objects of different groups whose bounding boxes overlap. This procedure is carried out in a loop over all collision detection tasks. At the beginning of every loop cycle the processors check for messages of other processors. These messages can either be send by master processes, containing ghost objects or object states for remote collision detection or they can be send by slave processes returning collision results. These requests have priority from the local tasks. Only in the case that no remote collision detection tasks are received, the processors perform their remaining local collision detection tasks as "idle tasks". By applying this scheme we are able to reduce the processor waiting time significantly.

6 Dynamic Load Balancing

One of the main goals of our work is the development of an efficient load balancing approach. The spatial binary tree decomposition described in Section 3 enables a simple, robust and efficient control based load balance scheme. A reduction of runtime imbalances between two twin processors is achieved by shifting the separating boundary between the two twin processors' subdomains. A shift of the boundary in the direction of the overloaded processor causes some of its objects to move to the less loaded processor, thus shifting some of the work load to the less loaded processor. This procedure is carried out as long as an imbalance exists. In Fig. 4 snapshots of a load-balanced parallel simulation of 10000 spherical particles are depicted. The load balancing task represents a classical control problem that can easily be accomplished by use of an I-controler that controls the position of the division boundary

Fig. 4. Snapshots of a parallel simulation of 10000 spherical particles collapsing under gravity with dynamically adapted domain decomposition due to load balancing.

based on the mean timestep runtime differences. We define the computation performance of a subprocessor i in time step n as

$$p_n^i = \frac{t_{total}^i - t_{idle}^i}{t_{total}^i}, \tag{4}$$

with the total process time t_{total}^i and the idle process time t_{idle}^i that elapses while the processor is waiting for messages.

As time measure for the calculation of the processor performance we take the process time that elapses while a processor performs its local as well as remote collision detection tasks. We do not count the time spend for the load balancing overhead, neighter the time needed for actual data transfer as it is sensitive to the current network load, which we cannot influence. Following the recursive approach we apply the same method on all higher levels of the decomposition tree to determine the mean computation performance of all processors in the two subjacent branches of node i by the equation

$$p_n^i = 2\frac{p_n^{2i}p_n^{2i+1}}{p_n^{2i} + p_n^{2i+1}} . \tag{5}$$

With the computation performance values p_n^1 and p_n^2 of two twin processors, the new position x_{n+1} of the division boundary after load balancing step n can be calculated as

$$x_{n+1} = x_n - k_I\, a\, \frac{p_n^1 - p_n^2}{p_n^1 + p_n^2} . \tag{6}$$

The control parameter k_I of the I-controler is scaled by the length a of the parent subdomain in the current dimension of subdivision to make it independent of the subdomain size.

By this means we achieve a good balance between the runtimes of all processors within few timesteps, as results of a simulation with 4 processors, depicted in Fig. 5, show. After some initial disturbances the process time per time step stabilizes whereas the number of objects that are assigned to the processors reach an almost steady state. Adequate control parameters can be chosen empirically or calculated by methods from control theory. We found that it is not necessary two rebalance after every timestep, but rather after every tenth or hundredth timestep, depending on the calculation time spend for each time step and thus on the problem size. For different reasons it is important to have a sufficiently long time periode elapsing between to rebalancing operations. Firstly, longer time periodes, i.e. more elapsed timesteps, smooth sudden peaks in the system load. Such peaks can occur when concurrent processes allocate large quantities of memory causing the memory system to swap to hard disc - a state that even on modern operation systems can severely affect the performance of other processes. In such cases too short measures would cause large controler fluctuations. Secondly, the accuracy of time measurments on most of todays workstations is low - often in the order of milliseconds, compared to nanosecods for processor cycles - which requires relatively long measurments periods.

The term performance in the context of parallel implementations usually refers to the scalability of an algorithm or an implementation. Even though we regarded the load balancing behaviour as paramount for our investigations, series of computations of the system depicted in Fig. 5 with ten thousand, hundred thousand and one million particles yield promising results on a cluster of Pentium 4 machines that is used by multiple users. The speedup and parallel efficiency plots (speedup divided by number of nodes) in Fig. 6 show that the program scales linearly up to a certain number of nodes before reaching the saturation point, i.e. the maximum of the speedup curve. Although simulations on other architectures have not been carried out yet, it can be expected that running the program exclusive on some nodes of a supercomputer will yield even better results.

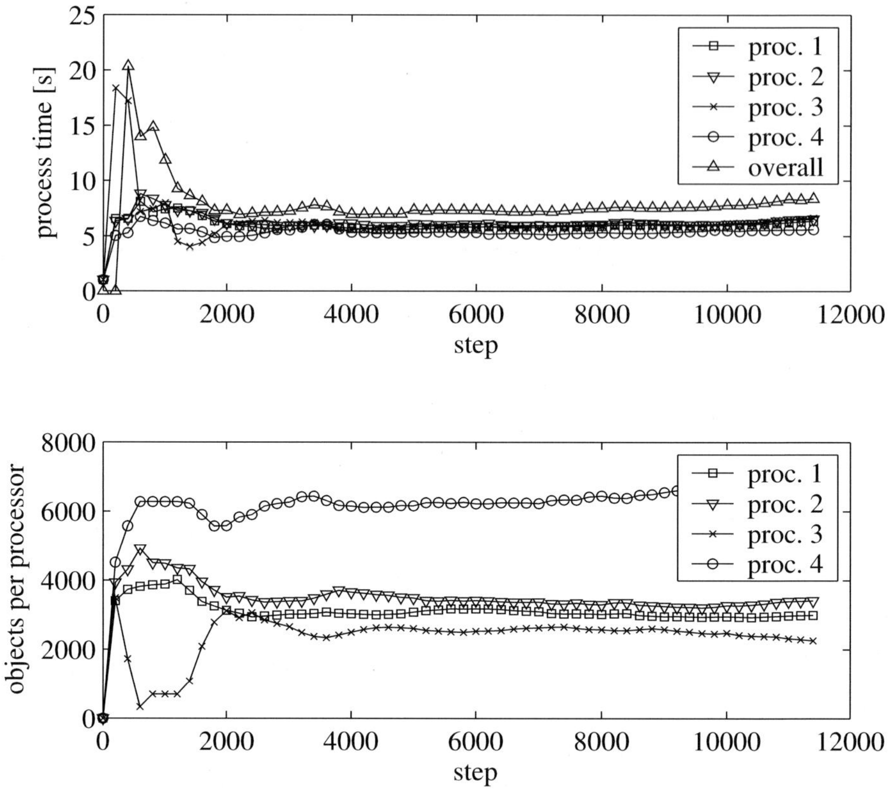

Fig. 5. Load balancing performance for a simulation on 4 processors. After some initial disturbances the number of objects assigned to the processors as well as the per step wall time reaches a steady state.

7 Conclusions

We have introduced a new method that allows for efficient parallel particle simulations on clusters of workstations with fluctuating work load due to many concurrent processes. It applies an adaptive recursive domain decomposition scheme that efficiently reduces imbalances due to changes in the particle distribution during the simulation. Moreover, it significantly reduces imbalances that result from fluctuating loads due to concurrent processes or permanent inhomogenities such as different processor architectures. To achieve this, we apply an $O(n)$ hierarchical bounding-volume-based neighborhood search that is robust with respect to varying particle geometry and size. Apart from grouping of objects for their distribution to different processors, we recursively divide the objects on every processor into cache-optimal subgroups that yield an increase in spatial and temporal locality in the collision detection stage as

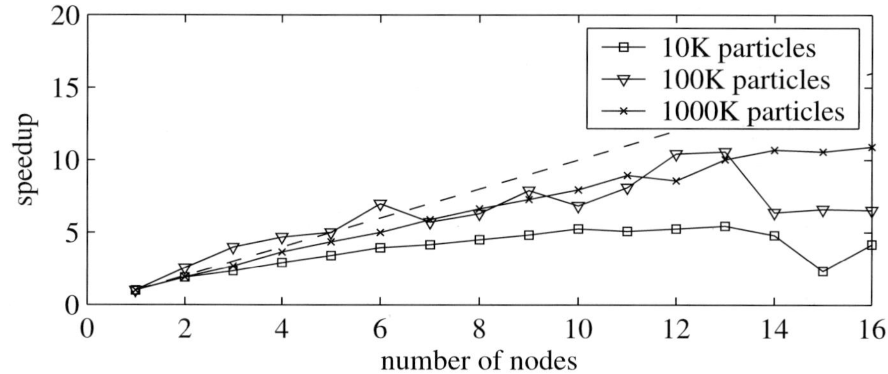

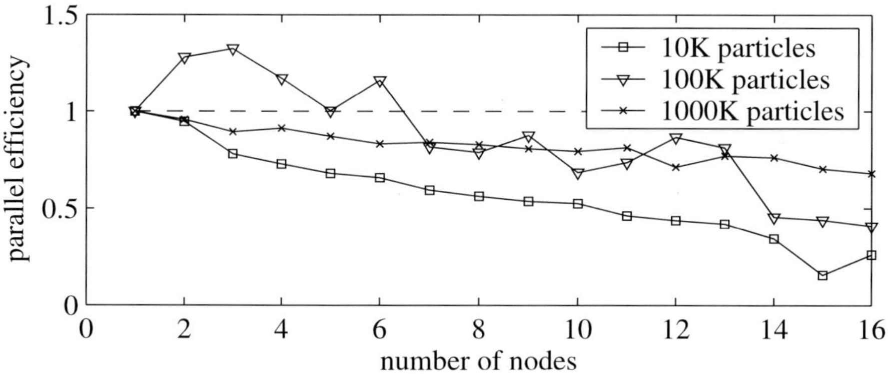

Fig. 6. Parallel speedup and efficiency of simulations of the collapsing objects example, see Fig. 5. Results of simulations series with ten thousand, hundred thousand and one million particles are depicted.

well as it enables splitting of collision detection tasks and thus significantly reduces process waiting times.

References

1. P. Cundall and O. Strack. A discrete numerical model for granular assemblies. *Geotechnique*, 29(1):47–65, 1979.
2. Y. Deng, R. F. Peierls, and C. Riveraz. An adaptive load balancing method for parallel molecular dynamics simulations. *Journal of Computational Physics*, 161:250–263, 2000.
3. F. Fleißner and P. Eberhard. Dynamical particle simulation with parallel cache-aware domain decomposition strategies. In *Proc. Appl. Math. Mech., 5, 657-658*, 2005.

4. R. Hayashi and S. Horiguchi. Relationships between efficiency of dynamic load balancing and particle concentration for parallel molecular dynamics simulation. In *The Fourth International Conference on High-Performance Computing in the Asia-Pacific Region-Volume 2*, page 976, 2000.
5. B. Muth and P. Eberhard. Collision detection for many colliding bodies. *In: Schiehlen, W. and Valasek, M. (Ed.): Virtual Nonlinear Multibody Systems, NATO ASI*, 1:138–143, 2002.
6. B. Muth, M.K. Müller, P. Eberhard, and S. Luding. Contacts between many bodies. *Machine Dynamics Problems*, 28(1):101–114, 2004.
7. L. Nyland, J. Prins, R.H. Yun, J. Hermans, H.C. Kum, and L. Wang. Achieving scalable parallel molecular dynamics using dynamic spatial domain decomposition techniques. *Journal of Parallel and Distributed Computing*, 47(2):125–138, 1997.
8. Open Systems Lab, Indiana University. *LAM/MPI User's Guide Version 7.1.1*, 2004.
9. E. Perkins and J.R. Williams. A fast contact detection algorithm insensitive to object sizes. *Engineering Computations*, 18(1/2):48–61, 2001.
10. A. Schinner. Fast algorithms for the simulation of polygonal particles. *Granular Matter*, 2:35–43, 1999.
11. M.S. Warren and J.K. Salmon. A parallel hashed oct-tree n-body algorithm. In *Supercomputing '93*, pages 12–21, 1993.
12. V. Zhakhovskii, K. Nishihara, Y. Fukuda, and S. Shimojo. A new dynamical domain decomposition method for parallel molecular dynamics simulation on grid. Annual progress report, Institute of Laser Engineering, Osaka University, 2004.

Analogous Non-Smooth Models of Mechanical and Electrical Systems

Michael Möller and Christoph Glocker

IMES - Center of Mechanics, ETH Zurich, 8092 Zurich, Switzerland
michael.moeller@imes.mavt.ethz.ch, glocker@imes.mavt.ethz.ch

Summary. The non-smooth modeling of mechanical and electrical systems allows for ideal unilateral contacts, sprag clutches and dry friction in mechanical systems and for ideal diodes and switches in electrical systems. The formulation of non-smooth electrical models is demonstrated by the example of the DC-DC buck converter using the flux approach. The non-smooth electrical elements are described with set-valued branch relations in analogy with set-valued force laws in mechanics. With the set-valued branch relations, the dynamics of the circuit are described as measure differential inclusions. The measure differential inclusions obtained for the DC-DC buck converter are related to an analogous mechanical system. For the numerical solution, the measure differential inclusions are formulated as a measure complementarity system and discretised with a difference scheme, known in mechanics as time-stepping. For every time-step a linear complementarity problem is obtained.

1 Introduction

The well developed formulations and methods used for non-smooth mechanical systems [5, 8, 9] can be adopted for electrical systems, by extending the classical electromechanical analogy to non-smooth systems. There are basically three approaches for the description of electrical systems, called the charge approach, the flux approach and mixed approaches [3]. The charge approach for electrical systems uses the charges and associated currents as variables while the voltages are balanced. In the flux approach the fluxes with associated voltages form the variables while balancing the currents. In mechanics usually the positions and their associated velocities are used as variables and the forces are balanced. The classical analogy links these approaches in mechanics and electronics. In Table 1 the corresponding variables and linear elements are shown for each approach. The duality between voltage u and current $\imath$ of an electrical system mirror the duality in mechanics between velocity v and force f. Table 1 is therefore completed with a column for the momentum approach, which is dual to the position approach in the same way as the

Peter Eberhard (ed.), Multiscale Problems in Multibody System Contacts, 45–54.

	mechanics position app.	mechanics momentum app.	electronics charge approach	electronics flux approach
local variables	position r velocity v force f	momentum p force f velocity v	charge g current $\imath$ voltage u	flux φ voltage u current $\imath$
inertia	mass m $-f = m\dot{v}$	stiffness k $-v = \frac{1}{k}\dot{f}$	inductivity L $-u = L\dot{\imath}$	capacity C $-\imath = C\dot{u}$
dissipation	damping d $-f = dv$	damping d $-v = \frac{1}{d}f$	resistance R $-u = R\imath$	resistance R $-\imath = \frac{1}{R}u$
energy storage	stiffness k $-f = kr$	mass m $-v = \frac{1}{m}p$	capacity C $-u = \frac{1}{C}g$	inductivity L $-\imath = \frac{1}{L}\varphi$

Table 1. Corresponding variables and elements in mechanics and electronics.

charge approach is dual to the flux approach. In [6], the idealized modeling of switches and diodes in the charge approach is introduced and linked to non-smooth mechanics. In this paper the non-smooth modeling of ideal switches and diodes in the flux approach will be demonstrated by the example of the DC-DC buck converter.

2 DC-DC Buck Converter

The DC-DC buck converter as shown in Fig. 1 is a circuit that allows for the

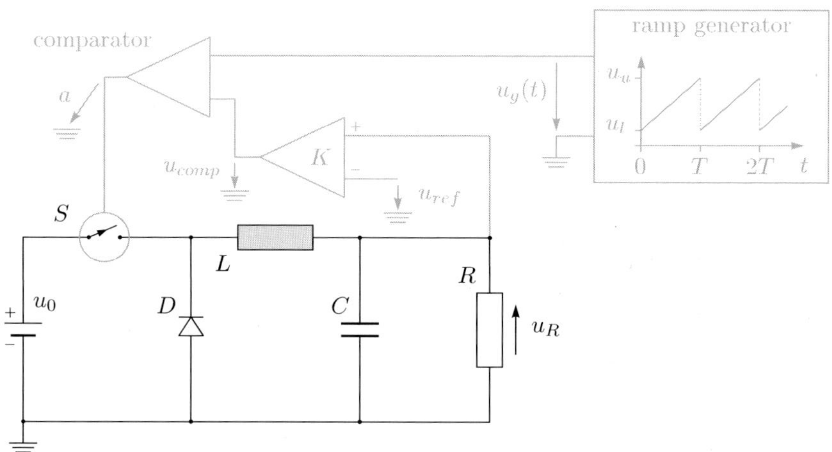

Fig. 1. The DC-DC buck converter.

efficient conversion of DC electrical power from the voltage u_0 supplied by the voltage source to a lower voltage u_R at the resistive load R. Besides the classical elements R, C, L and the voltage source u_0, the circuit consists of an ideal diode D and a unilateral switch S. The part of Fig. 1, which is drawn in grey, shows the switch control of the DC-DC buck converter, which controls the voltage at the load R by operating the unilateral switch S.

2.1 The Switch Control of the DC-DC Buck Converter

The switch control of the DC-DC buck converter consist of an amplifier with gain K, a comparator and a ramp generator with period T, lower voltage u_l and upper voltage u_u (cf. Fig. 1). The output voltage a of the switch control is used to operate the unilateral switch S of the buck converter. With the output voltage u_{comp} of the amplifier

$$u_{comp} = -K(u_R + u_{ref}) \tag{1}$$

and the explicitly time-dependent output voltage $u_g(t)$ of the ramp generator

$$u_g(t) = u_l + \left(\frac{t}{T} - k\right)(u_u - u_l) \ ; \ kT \leq t < (k+1)T \ ; \ k = 0, 1, 2, ... \tag{2}$$

the voltage a at the comparator is set for modeling reasons as

$$a = \begin{cases} 0, & u_{comp} - u_g(t) \leq 0, \\ +\infty, & u_{comp} - u_g(t) > 0. \end{cases} \tag{3}$$

This relation can be simplified by eliminating u_{comp} yielding the following rule for the switch control

$$a = \begin{cases} 0, & -u_R(t) \leq h(t), \\ +\infty, & -u_R(t) > h(t), \end{cases} \qquad h(t) := u_{ref} + \frac{1}{K} u_g(t). \tag{4}$$

For $a = 0$ the switch is closed and behaves as an ideal diode, while the switch is perfectly isolating for $a \to +\infty$.

2.2 The Extended DC-DC Buck Converter

The extended DC-DC buck converter has two additional capacitors C^* and C° compared to the original DC-DC buck converter, in order to obtain a non-singular matrix $\boldsymbol{M}$ of capacitances (cf. Fig. 2). This extension of the circuit is done to ease the analogy to mechanical systems, while the circuit of the original DC-DC buck converter can be obtained by setting the two additional capacities C^* and C° to zero. In analogy with the force impulsion measure $\mathrm{d}\mathcal{F}$ in mechanics, a current impulsion measure $\mathrm{d}\mathcal{I}$ is introduced for the flux approach. The force impulsion measure $\mathrm{d}\mathcal{F}$ consists of the Lebesgue-measurable forces

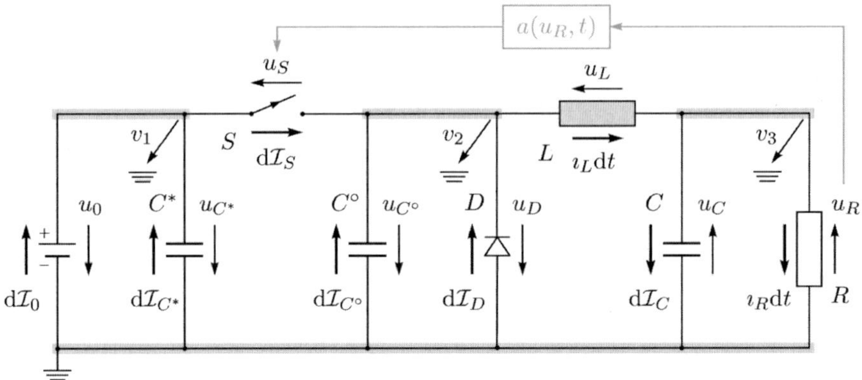

Fig. 2. Electrical model of the extended DC-DC buck converter.

f and the purely atomic impulsive forces F, while $d\mathcal{I}$ consists of Lebesgue-measurable currents $\imath$ and purely atomic impulsive currents I.

$$d\mathcal{F} = f\,dt + F\,d\eta, \quad d\mathcal{I} = \imath\,dt + I\,d\eta. \tag{5}$$

In the flux approach the voltages u are assumed to be functions of bounded variation. Higher order discontinuities are out of the scope of this paper. A general method of classifying such discontinuities may be found in [1].

To describe the dynamics of the circuit, generalized voltages $\boldsymbol{v}$ and associated generalized fluxes $\boldsymbol{q}$ are introduced in analogy with generalized velocities $\boldsymbol{u}$ and generalized coordinates $\boldsymbol{q}$ in mechanics. The circuit of the extended DC-DC buck converter has four nodes, which have been shaded grey in Fig. 2. The voltages v_i of each node with respect to the ground node are chosen as generalized voltages. The vector of nodal voltages $\boldsymbol{v}$ and the associated vector of nodal fluxes $\boldsymbol{q}$ become

$$\boldsymbol{q} := \left(q_1, q_2, q_3\right)^{\mathrm{T}}, \quad \boldsymbol{v} := \left(v_1, v_2, v_3\right)^{\mathrm{T}}, \quad \dot{\boldsymbol{q}} = \boldsymbol{v}\ dt\text{-almost everywhere}, \tag{6}$$

for the extended DC-DC buck converter. All branch voltages u_i satisfying Kirchhoff's voltage law can be expressed as a linear combination of the nodal voltages v_i, defining the nodal transformation of the circuit

$$\begin{aligned}
u_0 &= \boldsymbol{w}_0^{\mathrm{T}}\boldsymbol{v} = v_1 && \Rightarrow \boldsymbol{w}_0^{\mathrm{T}} = (1,0,0)\\
u_{C^*} &= \boldsymbol{w}_{C^*}^{\mathrm{T}}\boldsymbol{v} = v_1 && \Rightarrow \boldsymbol{w}_{C^*}^{\mathrm{T}} = (1,0,0)\\
u_S &= \boldsymbol{w}_S^{\mathrm{T}}\boldsymbol{v} = v_2 - v_1 && \Rightarrow \boldsymbol{w}_S^{\mathrm{T}} = (-1,1,0)\\
u_D &= \boldsymbol{w}_D^{\mathrm{T}}\boldsymbol{v} = v_2 && \Rightarrow \boldsymbol{w}_D^{\mathrm{T}} = (0,1,0)\\
u_C &= \boldsymbol{w}_C^{\mathrm{T}}\boldsymbol{v} = -v_3 && \Rightarrow \boldsymbol{w}_C^{\mathrm{T}} = (0,0,-1)\\
u_L &= \boldsymbol{w}_L^{\mathrm{T}}\boldsymbol{v} = v_3 - v_2 && \Rightarrow \boldsymbol{w}_L^{\mathrm{T}} = (0,-1,1)\\
u_{C^\circ} &= \boldsymbol{w}_{C^\circ}^{\mathrm{T}}\boldsymbol{v} = v_2 && \Rightarrow \boldsymbol{w}_{C^\circ}^{\mathrm{T}} = (0,1,0)\\
u_R &= \boldsymbol{w}_R^{\mathrm{T}}\boldsymbol{v} = -v_3 && \Rightarrow \boldsymbol{w}_R^{\mathrm{T}} = (0,0,-1).
\end{aligned} \tag{7}$$

The nodal transformation (7) holds also in integrated form for the branch fluxes $\varphi_i = \boldsymbol{w}_i^{\mathrm{T}}\boldsymbol{q}$, as well as for the associated virtual flux displacements $\delta\varphi_i = \boldsymbol{w}_i^{\mathrm{T}}\delta\boldsymbol{q}$. Kirchhoff's current law is evaluated in terms of a virtual work approach by demanding that the virtual action $\frac{\mathrm{d}\,\delta}{\mathrm{d}\,W}$ has to vanish for arbitrary virtual branch flux displacements $\delta\varphi_i^{adm}$ that are admissible with the constraints imposed by the topology of the circuit

$$\forall\ \delta\varphi_i^{adm}:\quad 0 = \mathrm{d}\delta W = \sum_i \mathrm{d}\mathcal{I}_i \delta\varphi_i^{adm}. \tag{8}$$

The sum is taken over all elements of the system $i \in \{0, C^*, S, D, C, L, C^\circ, R\}$. With the admissible virtual branch flux displacements $\delta\varphi_i^{adm} = \boldsymbol{w}_i^{\mathrm{T}}\delta\boldsymbol{q}$, obtained by transforming arbitrary virtual nodal flux displacements $\delta\boldsymbol{q}$, the equation (8) can be further simplified,

$$\forall\ \delta\boldsymbol{q}:\quad 0 = \mathrm{d}\delta W = \sum_i \mathrm{d}\mathcal{I}_i \boldsymbol{w}_i^{\mathrm{T}}\delta\boldsymbol{q} = \delta\boldsymbol{q}^{\mathrm{T}}\sum_i \boldsymbol{w}_i \mathrm{d}\mathcal{I}_i, \tag{9}$$

yielding the equilibrium conditions at the nodes

$$\sum_i \boldsymbol{w}_i \mathrm{d}\mathcal{I}_i = 0. \tag{10}$$

The branch relations of the capacitors, the resistor and the inductor are single-valued

$$\begin{aligned} \mathrm{d}\mathcal{I}_{C^*} = -C^*\mathrm{d}u_{C^*},\quad \mathrm{d}\mathcal{I}_C = -C\mathrm{d}u_C,\quad \mathrm{d}\mathcal{I}_{C^\circ} = -C^\circ\mathrm{d}u_{C^\circ}, \\ \imath_R\mathrm{d}t = -\frac{1}{R}u_R\mathrm{d}t,\quad \imath_L\mathrm{d}t = -\frac{1}{L}\varphi_L\mathrm{d}t. \end{aligned} \tag{11}$$

The branch relations for the capacitors are formulated on the level of measures to include impulsive currents in analogy with impulsive forces in mechanics. The branch relations of the diode, the switch and the voltage source are set-valued

$$\begin{aligned} -\mathrm{d}\mathcal{I}_D &\in \mathrm{Upr}(u_D^+), \\ -\mathrm{d}\mathcal{I}_S &\in \mathrm{Upr}(a^+ + u_S^+), \\ -\mathrm{d}\mathcal{I}_0 &\in \mathbb{R},\quad u_0 = u_0(t), \end{aligned} \tag{12}$$

where Upr denotes the unilateral primitive [6]. An ideal diode is an element through which the current may flow only in the positive direction. To prevent the current from flowing in the negative direction, an ideal diode can provide an unbounded voltage at zero current. This characteristic can be expressed with the inclusion $-\imath \in \mathrm{Upr}(u)$ and is analogous to an unilateral kinematic constraint (sprag clutch) in mechanics $-f \in \mathrm{Upr}(v)$. The relation given in (12) for the diode and depicted in Fig. 3 is obtained after completion with an impact law. The unilateral switch is modeled as a series connection of

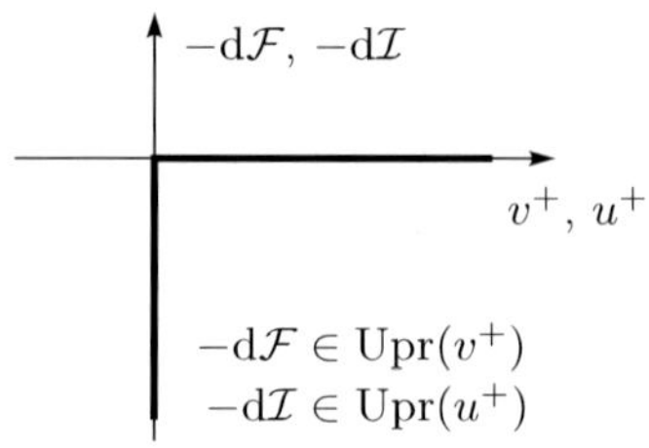

Fig. 3. Characteristic of sprag clutches and diodes.

a spark gap with break-through voltage a and a diode (cf. [6]). The unilateral switch is analogous to a series connection of a sprag clutch and a kinematic excitation with relative velocity a in the position-flux analogy. The unilateral switch can be operated using the break-through voltage, while it is closed for $a = 0$ and open for $a \to \infty$. The characteristic given in (12) for the unilateral switch contains an impact law as well and is depicted in Fig. 4. The third set-valued element in (12) is the voltage source, which is analogous to a bilateral kinematic constraint in mechanics.

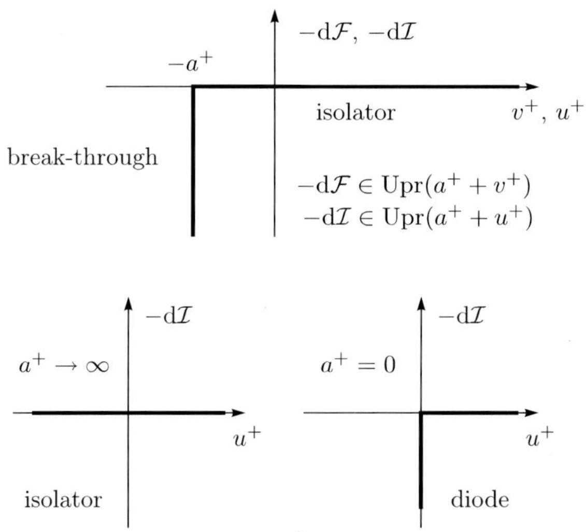

Fig. 4. Characteristic of moving sprag clutches and unilateral switches.

After inserting the single-valued branch relations (11) into (10), replacing all branch variables using the nodal transformation (7) and defining the matrices

$$\begin{aligned} \boldsymbol{M} &:= \boldsymbol{w}_{C^*}\boldsymbol{w}_{C^*}^{\mathrm{T}} C^* + \boldsymbol{w}_{C^\circ}\boldsymbol{w}_{C^\circ}^{\mathrm{T}} C^\circ + \boldsymbol{w}_{C}\boldsymbol{w}_{C}^{\mathrm{T}} C, \\ \boldsymbol{D} &:= \boldsymbol{w}_{R}\boldsymbol{w}_{R}^{\mathrm{T}}\frac{1}{R}, \quad \boldsymbol{K} := \boldsymbol{w}_{L}\boldsymbol{w}_{L}^{\mathrm{T}}\frac{1}{L}, \end{aligned} \tag{13}$$

one obtains the measure differential inclusions

$$\begin{gathered}\boldsymbol{M}\mathrm{d}\boldsymbol{v} + \boldsymbol{D}\boldsymbol{v}\mathrm{d}t + \boldsymbol{K}\boldsymbol{q}\mathrm{d}t - \boldsymbol{w}_D\mathrm{d}\mathcal{I}_D - \boldsymbol{w}_S\mathrm{d}\mathcal{I}_S - \boldsymbol{w}_0\mathrm{d}\mathcal{I}_0 = \boldsymbol{0}, \quad \dot{\boldsymbol{q}} = \boldsymbol{v} \text{ d}t\text{-a.e.} \\ \begin{aligned} u_D &= \boldsymbol{w}_D^{\mathrm{T}}\boldsymbol{v}, & -\mathrm{d}\mathcal{I}_D &\in \mathrm{Upr}(u_D^+), \\ u_S &= \boldsymbol{w}_S^{\mathrm{T}}\boldsymbol{v}, & -\mathrm{d}\mathcal{I}_S &\in \mathrm{Upr}(a^+ + u_S^+), \\ u_0 &= \boldsymbol{w}_0^{\mathrm{T}}\boldsymbol{v}, & -\mathrm{d}\mathcal{I}_0 &\in \mathbb{R}, \quad u_0 = u_0(t), \end{aligned} \quad a = \begin{cases} 0, & v_3(t) \le h(t), \\ +\infty, & v_3(t) > h(t). \end{cases}\end{gathered} \tag{14}$$

describing the dynamics of the DC-DC buck converter. The mechanical model associated with the extended DC-DC buck converter can be obtained from (14) using the position-flux analogy. The mechanical model is illustrated in Fig. 5. The model consists of three masses C^*, C° and C corresponding to the three capacitors of the extended DC-DC buck converter. Since the position-flux analogy is used, the electrical circuit and the mechanical model have the same topology. The sprag clutch acting between the environment and the mass C° is analogous to the diode and allows the mass C° only move to the right. The masses C^* and C° are interacting by the serial connection of a kinematic excitation with relative velocity $-a$ and a sprag clutch, constituting the analog to the unilateral switch. The switch control of the DC-DC buck converter measures the velocity u_R and provides the relative velocity $-a$.

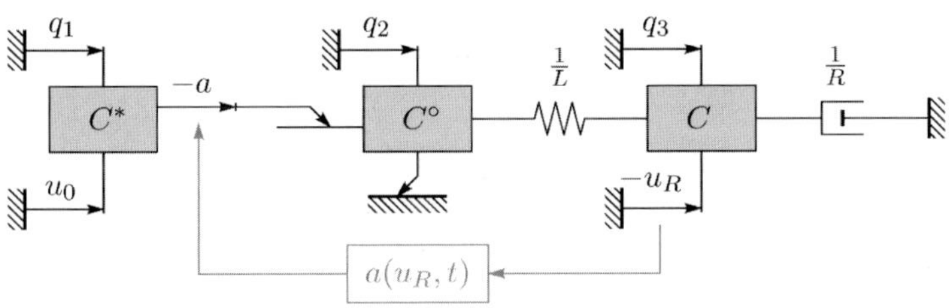

Fig. 5. Mechanical model associated with the extended DC-DC buck converter.

2.3 Numerical Integration

A time-stepping method is used to solve the measure differential inclusions (14) describing the non-smooth dynamics of the circuit for the unknown nodal voltages $\boldsymbol{v}(t)$ and the associated nodal fluxes $\boldsymbol{q}(t)$. Time-stepping methods discretise directly the inclusion (14) over a time step Δt. The problem of solving the measure differential inclusion (14) numerically, is formulated as follows: For the system (14), with given initial nodal charges $\boldsymbol{q}_A$ and initial nodal voltages $\boldsymbol{v}_A$,

$$\boldsymbol{q}_A := \boldsymbol{q}(t_A), \quad \boldsymbol{v}_A := \boldsymbol{v}(t_A), \tag{15}$$

at the initial time t_A, find nodal charges $\boldsymbol{q}_E$ and nodal voltages $\boldsymbol{v}_E$,

$$\boldsymbol{q}_E := \boldsymbol{q}(t_E), \quad \boldsymbol{v}_E := \boldsymbol{v}(t_E), \tag{16}$$

at the time t_E which approximate the exact solution. The time t_E is the end of a chosen time interval $[t_A, t_E]$ with length

$$\Delta t := t_E - t_A. \tag{17}$$

The resulting algorithms are very robust and easy to implement, but have a limited accuracy, see e.g. [2, 7] for some versions of time stepping algorithms. There are different possibilities to treat smooth set valued elements - in this case the voltage source - for numerical integration. Beside state reduction and replacement with two unilateral constraints, the third possibility is to append the current of the voltage source branch to the vector of unknown voltages after discretisation. This approach, also known in electronics as modified nodal analysis, is used in the following for the extended DC-DC buck converter. In order to simplify the expressions, the notations

$$\mathrm{d}\boldsymbol{\mathcal{I}} := \begin{pmatrix} \mathrm{d}\mathcal{I}_D \\ \mathrm{d}\mathcal{I}_S \end{pmatrix}, \quad \boldsymbol{W} := \begin{pmatrix} \boldsymbol{w}_D & \boldsymbol{w}_S \end{pmatrix} \tag{18}$$

are introduced. Using these notations the equality of measures in (14) may be written as

$$\begin{aligned} &\boldsymbol{M}\mathrm{d}\boldsymbol{v} + \boldsymbol{D}\boldsymbol{v}\mathrm{d}t + \boldsymbol{K}\boldsymbol{q}\mathrm{d}t - \boldsymbol{W}\mathrm{d}\boldsymbol{\mathcal{I}} - \boldsymbol{w}_0\mathrm{d}\mathcal{I}_0 = \boldsymbol{0}, \\ &\boldsymbol{w}_0^{\mathrm{T}}\boldsymbol{v} - u_0 = 0, \end{aligned} \tag{19}$$

where the constraint of the voltage source has been added as an additional equation. The notations

$$\boldsymbol{\gamma} := \boldsymbol{W}^{\mathrm{T}}\boldsymbol{v} + \hat{\boldsymbol{w}}, \quad \hat{\boldsymbol{w}} := \begin{pmatrix} 0 \\ a \end{pmatrix} \tag{20}$$

are introduced to formulate the measure inclusions of the ideal diode and the unilateral switch as complementarity conditions

$$\boldsymbol{0} \preceq \boldsymbol{\gamma}^{+} \perp \mathrm{d}\boldsymbol{\mathcal{I}} \succeq \boldsymbol{0}, \tag{21}$$

allowing to set up the linear complementarity problem after discretisation. For non-smooth mechanical systems usually Moreau's midpoint rule is used to discretise the measure differential inclusions. This rule consists of a trapezoidal rule for the positions and an Euler step for the velocities. For the discretisation of the DC-DC buck converter, an implicit Euler scheme is used, which does not require a regular capacitor matrix $\boldsymbol{M}$. The integral of the equality of measures (19) over the time step Δt is approximated as

$$\begin{aligned} &\boldsymbol{M}(\boldsymbol{v}_E - \boldsymbol{v}_A) + \boldsymbol{D}\boldsymbol{v}_E\Delta t + \boldsymbol{K}\boldsymbol{q}_E\Delta t - \boldsymbol{W}\Delta\boldsymbol{\mathcal{I}} - \boldsymbol{w}_0\Delta\mathcal{I}_0 = \boldsymbol{0}, \\ &\boldsymbol{w}_0^{\mathrm{T}}\boldsymbol{v}_E - u_0 = 0, \end{aligned} \tag{22}$$

using end-point terms by applying an implicit Euler scheme. The relation between the nodal voltages $\boldsymbol{v}$ and the nodal fluxes $\boldsymbol{q}$ can be approximated using one implicit Euler step

$$\boldsymbol{q}_E = \boldsymbol{q}_A + \boldsymbol{v}_E \Delta t. \tag{23}$$

The complementarity conditions (21) are expressed in the discretised form

$$\mathbf{0} \preceq \boldsymbol{\gamma}_E \perp \Delta\boldsymbol{\mathcal{I}} \succeq \mathbf{0}, \tag{24}$$

where the vector of local variables at the end-time t_E,

$$\boldsymbol{\gamma}_E = \boldsymbol{W}^{\mathrm{T}} \boldsymbol{v}_E + \hat{\boldsymbol{w}}_A, \tag{25}$$

is formed using the vector $\hat{\boldsymbol{w}}_A$ at the beginning t_A of the time step. This is done in order to avoid a nonlinear dependence on the unknown nodal voltages $\boldsymbol{v}_E$, which would lead to a nonlinear complementarity problem. By using the vector $\hat{\boldsymbol{w}}_A$ instead of $\hat{\boldsymbol{w}}_E$, a small time-delay of Δt is inserted into the switch control feedback of the DC-DC buck converter, which seems reasonable from the modeling point of view as well.

Elimination of the end-point nodal fluxes $\boldsymbol{q}_E$ from the equations (22) with the help of equation (23) yields

$$\begin{aligned} &(\boldsymbol{M} + \boldsymbol{D}\Delta t + \boldsymbol{K}\Delta t^2)\boldsymbol{v}_E - \boldsymbol{w}_0 \Delta\mathcal{I}_0 - \boldsymbol{M}\boldsymbol{v}_A + \boldsymbol{K}\boldsymbol{q}_A \Delta t - \boldsymbol{W}\Delta\boldsymbol{\mathcal{I}} = \mathbf{0}, \\ &-\boldsymbol{w}_0^{\mathrm{T}} \boldsymbol{v}_E + u_0 = 0, \end{aligned} \tag{26}$$

where the terms have already been regrouped for the unknown variables $\boldsymbol{v}_E$ and $\Delta\mathcal{I}_0$. With the definition of the vectors and matrices

$$\begin{aligned} \boldsymbol{\nu} &:= \begin{pmatrix} \boldsymbol{v}_E \\ \Delta\mathcal{I}_0 \end{pmatrix}, \quad \hat{\boldsymbol{M}} := \begin{pmatrix} \boldsymbol{M} + \boldsymbol{D}\Delta t + \boldsymbol{K}\Delta t^2 & -\boldsymbol{w}_0 \\ -\boldsymbol{w}_0^{\mathrm{T}} & 0 \end{pmatrix}, \\ \hat{\boldsymbol{h}} &:= \begin{pmatrix} \boldsymbol{M}\boldsymbol{v}_A - \boldsymbol{K}\boldsymbol{q}_A \Delta t \\ -u_0 \end{pmatrix}, \quad \hat{\boldsymbol{W}} := \begin{pmatrix} \boldsymbol{W} \\ \mathbf{0} \end{pmatrix}, \end{aligned} \tag{27}$$

the notation can be simplified yielding the mixed linear complementarity problem

$$\hat{\boldsymbol{M}}\boldsymbol{\nu} - \hat{\boldsymbol{h}} - \hat{\boldsymbol{W}}\Delta\boldsymbol{\mathcal{I}} = \mathbf{0}, \quad \boldsymbol{\gamma}_E = \hat{\boldsymbol{W}}^{\mathrm{T}}\boldsymbol{\nu} + \hat{\boldsymbol{w}}_A, \quad \mathbf{0} \preceq \boldsymbol{\gamma}_E \perp \Delta\boldsymbol{\mathcal{I}} \succeq \mathbf{0}. \tag{28}$$

If the matrix $\hat{\boldsymbol{M}}$ is regular then the vector $\boldsymbol{\nu}$ can be eliminated from (28) resulting in the linear complementarity problem

$$\underbrace{\boldsymbol{\gamma}_E}_{\boldsymbol{y}} = \underbrace{\hat{\boldsymbol{W}}^{\mathrm{T}} \hat{\boldsymbol{M}}^{-1} \hat{\boldsymbol{W}}}_{\boldsymbol{A}} \underbrace{\Delta\boldsymbol{\mathcal{I}}}_{\boldsymbol{x}} + \underbrace{\hat{\boldsymbol{W}}^{\mathrm{T}} \hat{\boldsymbol{M}}^{-1} \hat{\boldsymbol{h}} + \hat{\boldsymbol{w}}_A}_{\boldsymbol{b}}, \quad \underbrace{\mathbf{0} \preceq \boldsymbol{\gamma}_E \perp \Delta\boldsymbol{\mathcal{I}} \succeq \mathbf{0}}_{\mathbf{0} \preceq \boldsymbol{y} \perp \boldsymbol{x} \succeq \mathbf{0}} \tag{29}$$

in standard form. It has to be noted, that the matrix $\hat{\boldsymbol{M}}$ is regular not only for the extended DC-DC buck converter, but for the original version as well. After solving the linear complementarity problem (29) for the vectors $\boldsymbol{\gamma}_E$ and $\Delta\boldsymbol{\mathcal{I}}$, the vector $\boldsymbol{\nu}$ can be calculated from the first equation in (28), yielding the end-point nodal voltages $\boldsymbol{v}_E$. The nodal fluxes $\boldsymbol{q}_E$ can then be calculated with the help of (23). The numerical results obtained for the original DC-DC buck converter in a chaotic parameter regime, as published in [4, 6], are shown in Fig. 6 and agree with those given in the publications.

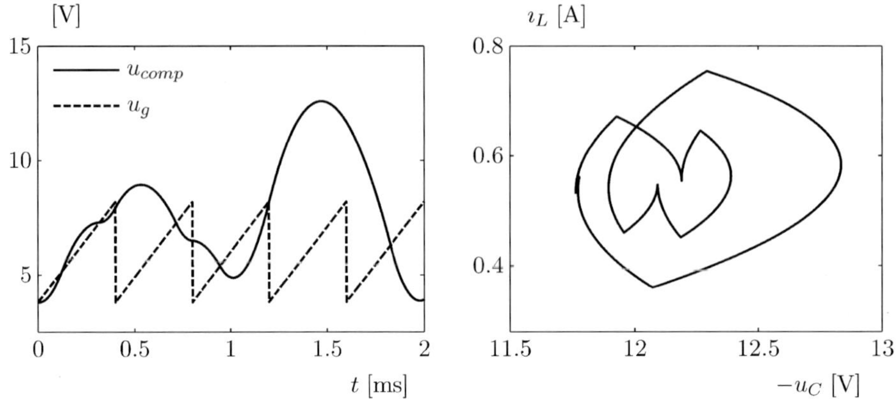

Fig. 6. Phase space plot and comparator voltages of the DC-DC buck converter.

3 Conclusion

Using the example of the DC-DC buck converter the formulation of the measure differential inclusions, their relation to an analogous mechanical model and their solution using the time-stepping method has been shown for the flux approach. Only a small set of non-smooth elements have been described in this paper to show the basic procedure, but the formulations used are not limited to this set of elements only.

References

1. Acary, V., Brogliato, B. Numerical time integration of higher order dynamical systems with state constraints. In *ENOC-2005*, 2005.
2. Anitescu, M., Potra, F.A., Stewart, D.E. Time-stepping for three-dimensional rigid body dynamics. *Comp. Meth. Appl. Mech. Eng.*, 177(3):183–197, 1999.
3. Enge, O., Maißer, P. Modelling Electromechanical Systems with Electrical Switching Components Using the Linear Complementarity Problem. *Multibody System Dynamics*, 13(4):421–445, 2005.
4. Fosas, E., Olivar, G. Study of Chaos in the Buck Converter. *IEEE Transactions on Circuits and Systems-I: Fundamental Theory and Appl.*, 43(1):13–25, 1996.
5. Glocker, Ch. *Set-Valued Force Laws: Dynamics of Non-Smooth Systems*, volume 1 of *Lecture Notes in Applied Mechanics*. Springer, Berlin, 2001.
6. Glocker, Ch. Models of non-smooth switches in electrical systems. *International Journal of Circuit Theory and Applications*, 33:205–234, 2005.
7. Jean, M. The non-smooth contact dynamics method. *Computer Methods in Applied Mechanics and Engineering*, 177:235–257, 1999.
8. Moreau, J.J. *Unilateral contact and dry friction in finite freedom dynamics*, volume 302 of *CISM Courses and Lectures*. Springer Verlag, Wien, 1988.
9. Pfeiffer F., Glocker, Ch. *Multibody Dynamics with Unilateral Contacts*. Wiley, New York, 1996.

Numerical Modeling of Paper Machine Roll Contact with Regenerative Out-Of-Roundness Excitation

Veli-Matti Järvenpää and Lihong Yuan

Tampere University of Technology, Finland
veli-matti.jarvenpaa@tut.fi

Summary. The softened rolling contact of two paper machine rolls is considered. A regenerative excitation source can exists in the system due to the finite relaxation time of the roll cover polymer materials. The modeling of the rolls is described by multibody substructuring methodology and the polymer cover layer between the rolls in contact is modeled by contact spring reaction forces calculated according to the stiffness of the layer. The regenerative excitation is introduced to the system as a time delay term and it is calculated from the penetration history of the cover layer. The solution of the system equations is obtained by numerical time integration by utilizing the method of steps time delay equation solution procedure. Some numerical results are illustrated.

1 Introduction

The rolling contact of two paper machine rolls is one of important processing methods used in the paper manufacturing. The paper web is driven into the contact by the rolling motion of the rolls and the web is processed by the contact to improve its surface texture and interior structure. Typically, the rolls are manufactured from steel, but the contact interaction is softened by using polymer covers on the rolls. Perhaps most common design is to have one roll with a steel surface and another roll with a polymer cover. Many parameters are used to control this rolling contact processes. The main parameter is the contact pressure, or the line load, of the rolls. A constant line load distribution is desired, because all surface of paper web should be processed equally. This must be obtained by using good roll designs. The surfaces of the rolls are manufactured slightly as barrel-shaped to compensate the roll deflections in the contact. This method is referred as the roll crowing and the grinding of the roll surfaces is applied according to pre-calculated crowning functions. Also the polymer cover stiffness must be chosen correctly according to the roll deflection characteristics.

It is observed that the polymer covers can produce complex vibration phenomena, which cannot be fully explained by classical dynamical analyses. These vibrations originate hyperelastic and viscoelastic material behavior. Because the polymer cover is constantly re-entering to the contact due to the rolling motion of the rolls, these non-linear material characteristics become active. If the rotation period of the rolls is shorter than the viscoelastic relaxation time of the cover, it is possible,

Peter Eberhard (ed.), Multiscale Problems in Multibody System Contacts, 55–64.

that an unstable self-excited or regenerative vibration source is generated into the system. More generally, this can be understood as a time delay excitation source and this means, that the history of the polymer cover deformations contributes to the dynamics of the system. The delay time is the rotation period of the roll with the cover. The stability of a vibration system with a time delay depend on the size of the time delay as well as the parameters of the system, in this case, especially, the stiffness of the polymer cover.

The aim of this presentation is to describe the numerical solution of a roll contact dynamics with the time delay effect. The model consists of the two paper machine rolls and the polymer cover layer and the time domain responses are requested. The modeling is divided into two parts. The metallic rolls are modeled by using multibody substructuring methodology [1] [2]. The equations of motion of the rolls are defined by using local coordinate systems, which perform large spin rotational motion in the inertial system. The dynamics of the rolls are described in the local coordinate systems by sets of modal vectors. The roll geometries are modeled by continuum finite element meshes. The polymer cover layer on one of the rolls is not included in the finite element models. Instead, the polymer cover is introduced to the system as spring reaction forces coupling the equations of the rolls. The contact is considered as a line contact and the contact reaction forces are calculated node to node basis according to the distance of the roll contact lines and the stiffness of the polymer cover. The contact line motion on the roll surfaces is obtained by nodal interpolation in the circumferential direction of the rolls. The time delay effect of the cover deformations is included as follows. The deformation history of the cover is recorded as a penetration function and this is introduced to the system equations as a delayed feedback source. The time delay is non-constant depending on the roll rotation time. To obtain the correct deformation history at the right times according to the stepping of the time integration solution procedure, a time domain interpolation of the penetration history is used.

2 System equations

2.1 Rolls

The following assumptions are used in the multibody modeling. Firstly, only one large global rotation about the longitudinal axis, the spin motion, is considered. The other two rotations remain small, because the roll bearings are mounted at roll ends and the rolls are driven in parallel rolling motion by their individual drives. The layout of the rolls is illustrated in the Fig. 1. The two other rotations as well as the reference translations are neglected due to this mounting. This simplifies the coordinate transformation significantly. Secondly, the lumped mass formulation is used and the finite element nodal masses are considered as point masses or mass particles. Linear continuum element meshes are used for determining the modal vectors of the rolls. Also the element types used do not have rotational degrees of freedom. Finally, the deformations of the bodies are described by sets of modal vectors. The modes with bearing boundary conditions are used (Fig. 2).

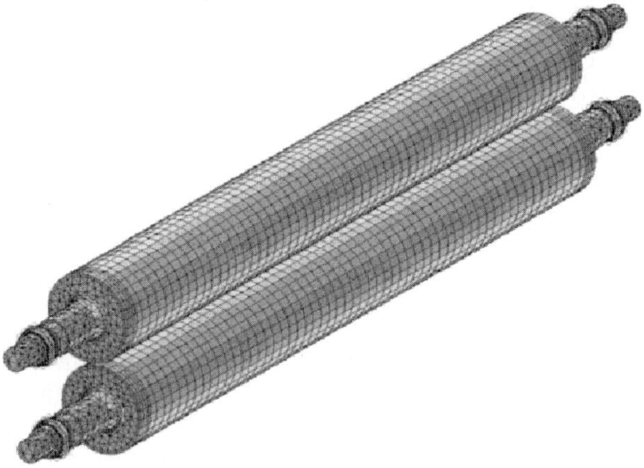

Fig. 1. The roll layout and the finite element meshes.

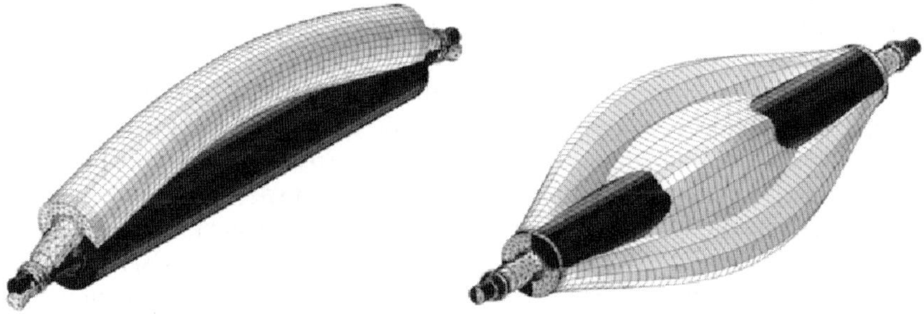

Fig. 2. The 1st bending mode (left) and 1st shell mode (right) of a roll.

The equations of motion of a roll are [3] [4]

$$\begin{bmatrix} (\mathbf{x}_G + \mathbf{\Omega\eta})^T \mathbf{M}_{xy} (\mathbf{x}_G + \mathbf{\Omega\eta}) & (\mathbf{x}_G + \mathbf{\Omega\eta})^T {\mathbf{M}_S}^T \mathbf{\Omega} \\ \mathbf{\Omega}^T \mathbf{M}_S (\mathbf{x}_G + \mathbf{\Omega\eta}) & \mathbf{\Omega}^T \mathbf{M\Omega} \end{bmatrix} \begin{bmatrix} \ddot{\theta} \\ \ddot{\mathbf{\eta}} \end{bmatrix}$$
$$+ \begin{bmatrix} 2(\mathbf{x}_G + \mathbf{\Omega\eta})^T \mathbf{M}_{xy} \mathbf{\Omega}\dot{\mathbf{\eta}} & \mathbf{0} \\ \mathbf{0} & 2\dot{\theta}\mathbf{\Omega}^T \mathbf{M}_S \mathbf{\Omega} \end{bmatrix} \begin{bmatrix} \dot{\theta} \\ \dot{\mathbf{\eta}} \end{bmatrix} \tag{1}$$
$$+ \begin{bmatrix} 0 & \mathbf{0} \\ \mathbf{0} & \mathbf{\Omega}^T \mathbf{K\Omega} \end{bmatrix} \begin{bmatrix} \theta \\ \mathbf{\eta} \end{bmatrix} - \begin{bmatrix} 0 \\ \dot{\theta}^2 \mathbf{\Omega}^T \mathbf{M}_{xy} (\mathbf{x}_G + \mathbf{\Omega\eta}) \end{bmatrix} = \begin{bmatrix} \mathbf{f}^T \dfrac{\partial \mathbf{R}^T}{\partial \theta} (\mathbf{x}_G + \mathbf{\Omega\eta}) \\ \mathbf{\Omega}^T \mathbf{Rf} \end{bmatrix}$$

where

θ	is the rotation angle of the roll,
$\boldsymbol{\eta}$	the vector of modal coordinates,
$\boldsymbol{\Omega}$	the modal matrix,
$\mathbf{x}_G$	the vector of initial x and y nodal coordinates,
$\mathbf{M}$	the lumped mass matrix,
$\mathbf{M}_{xy}$	the mass matrix with nodal mass in x and y directions only,
$\mathbf{M}_s$	skew symmetric mass matrix,
$\mathbf{K}$	the stiffness matrix,
$\mathbf{f}$	the applied force vector and
$\mathbf{R}$	the rotation matrix in xy plane.

The local xyz coordinate system of each roll is oriented that z axis is coaxial to the rotation axis and the xy plane is parallel to the cross-section of the roll.

2.2 Roll contact

The key characteristic of the roll contact is that the polymer cover is much softer than the metallic rolls. This makes the contact spring force modeling most suitable for this type of a contact. The principle of the contact modeling method is illustrated in Fig. 3. The contact is described as a line contact by using the stiffness distribution function $k_n(z)$ between the rolls [4]. This function represents the stiffness of the polymer cover layer of the upper roll. This approach is similar as the elastic foundation model [5]. However, the main difference is that nonlinear contact force of the foundation model depends on the contact width, which here is only considered in the pre-calculation of the stiffness distribution. It also turns out that the pre-calculated stiffness distribution is almost constant due to the proper design of the rolls. The contact line is rolling over element boundaries from element to element, according to the rotations of the rolls. The crowning of the rolls is introduced to the contact by the enforced displacement function $cr(z)$, which is pre-defined. In the roll installation, the crowning is applied on the surfaces of both rolls and the function $cr(z)$ describes the total value of the crowning from both rolls.

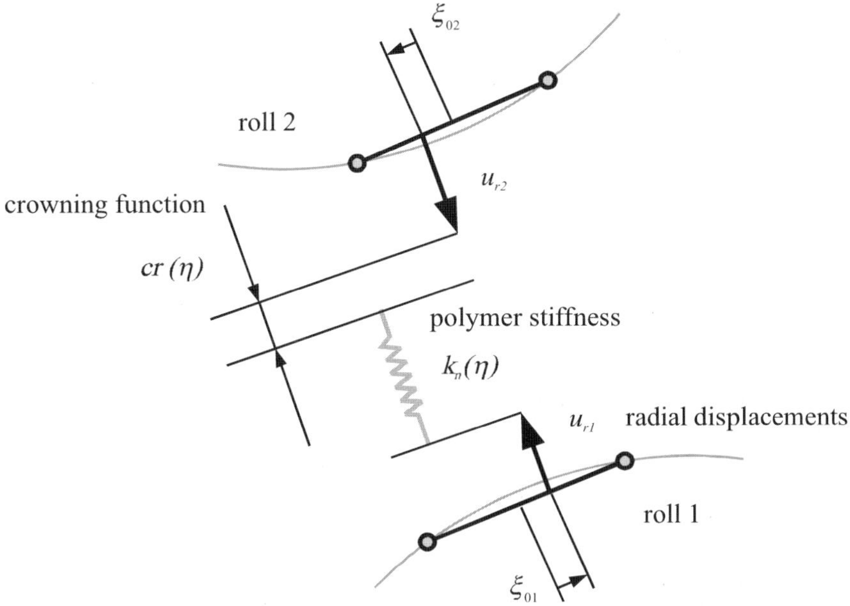

Fig. 3. The contact line load between opposing elements. The stiffness of the polymer coating and the crowning of the rolls are functions of z-axis of rolls.

3 Excitation due to time delay

The contact spring force with the time delay term takes the form [6]

$$N = k_n\left[\varepsilon(t) - \gamma_e \varepsilon(t-T)\right] + c_n \dot{\varepsilon}(t), \tag{2}$$

where k_n is the contact stiffness and c_n is the damping coefficient for a Kelvin-like polymer roll covering material [7]. The contact penetration ε splits into three different components

$$\varepsilon(t) = x_1 - x_2 + z_{pw}(t) \tag{3}$$

where the first two parts x_1 and x_2 represent the relative roll motion at the contact line and the third one $z_{pw}(t)$ is the paper thickness variation, which is not considered here. As assumed here, the polymer cover material may behave in a more complicated way than a purely passive spring-damper system, as it may not return to its initial shape before a finite time interval. So it may enter the nip after one roll revolution without having completely recovered to its original shape. The decay factor is

$$\gamma_e = e^{-T/\tau_{relaxatation}} \tag{4}$$

which contains the ratio of the revolution time T and the material specific relaxation time, which is

$$\tau_{relaxation} = c_n / k_n \tag{5}$$

Depending on this ratio, roll cover dynamics will be affected by the memory of past deformation so that when returning to the contact, the previous cover deformation is not completely vanished.

4 Numerical solution method

The time integration method used to the system equations here is described in [8]. The method belongs to the Newmark time integrator family, but it is actually more commonly known as the average constant acceleration method, due to the selection of the Newmark coefficients. The method is an implicit time integration method based on a predictor-corrector approach and it is suitable for solving non-linear ordinary differential equations because the Newton-Rhapson iteration procedure is included in it.

The method of steps is a conservative method which can be used for the time domain solution of delay equations. Due to the contribution of time delay τ to the vibration of a system, the time domain responses include delay characteristics and the solution is more complicated than for normal ordinary differential equations. There is no simple initial state, but an initial history period. In idea the method of steps procedure is to specify an initial function on some interval of length τ, here the interval $[t_0 - \tau\ ,\ t_0]$ [9]. This so-called initial period of the delay equation is solved without delay effects and it is not necessary that it should satisfy the delay equation. Then one needs to seek a continuous function, which satisfies the delay equation for $t \geq t_0$. This procedure can be continued as far as requested (Fig. 4).

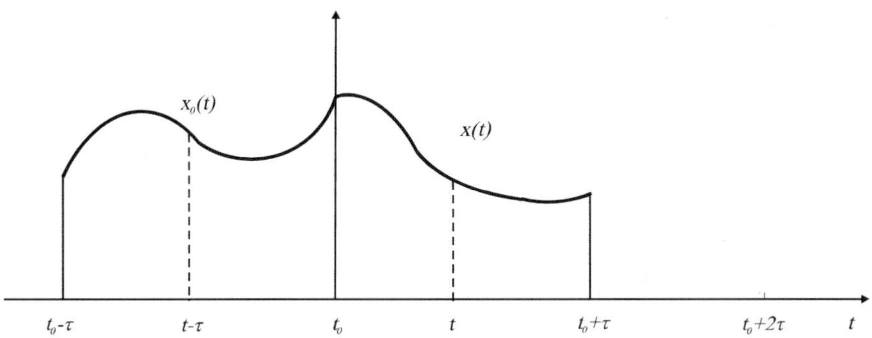

Fig. 4. The method of steps procedure

The problem with the system equations of the roll contact is that the time delay, the rotation time of the polymer covered roll T, is non-constant. During the integration, the delay value and the penetration history at $t-T$ are always needed to complete the following time step. The exact values are not directly available from the data of previous time increments and must be estimated by using some appropriate methods. This is carried out by a two-step polynomial interpolation procedure. In the first step, the unknown value of the time delay is determined from the inverse curve of the rotation angle θ by subtracting 2π from its current value and then interpolating the delayed time $t-T$. In the second step the unknown value of the penetration function $\varepsilon(t-T)$ is interpolated according to the value of the delay time. The cubic interpolation is used. The accuracy of this process, of course, depends on the quality of the interpolation. Generally, this interpolation makes this time domain solution slow because it is repeatedly carried out during the iterations.

5 Numerical results

Numerical simulation results are presented next. The aim is to study how different delay times (i.e. different running speeds) influence to the time domain responses of the system. The result of interest is the averaged contact line load response. The initial history for the time delay determination must be created first. The rolls are accelerated to their desired speeds and the roll contact is closed. Relatively long initial simulation of 100 seconds is made to get rid of any transient effects and to obtain as smooth steady state running condition as possible. The desired line load is set to 14.7 kN/m and moderate Rayleigh damping is used for the rolls. No damping is introduced to the roll contact.

The key feature of the regenerative roll contact vibration is the excitation of the roll beating vibration mode. This mode means that the upper and lower rolls have bending vibrations with opposite phases and the rolls are beating each other. The natural frequency of this mode in the numerical model is about 109.5 Hz. Because this is the most natural way how the rolls should vibrate under regenerative contact excitation this mode will dominate the responses. Three different cases are considered with the different running speeds of 389 m/min, 396 m/min and 403 m/min, which equal to the roll rotation frequencies of 3.77 Hz, 3.84 Hz and 3.91 Hz, respectively. The time domain responses of the contact line load are illustrated in following figures. The first and third cases in Fig. 5 and Fig. 7 show relatively stable contact force responses but the second case in Fig. 6 is clearly unstable immediately. This result is expected because in the time delay systems in general have multiple stability boundaries in the frequency domain. As expected, the highest frequency component in the responses is the beating natural frequency. The responses also show some envelope patterning. The time period of an envelope in the curves is the delay time, which is another typical characteristic of a delay system. More explanation on the stability behavior can be obtained if the rotation frequencies are compared to the beating mode natural frequency. The division of the natural frequency by the rotation frequencies produces ratios of 28, 28.5 and 29, respectively. This means that one resonance and two anti-resonances are encountered. The consequence of this resonances behavior is that the polymer cover layer will have equivalent number of circumferential out-of-roundness waves on its surface profile during the excitation [6].

One big drawback of the modeling presented here is that one expects that clear limit cycles should appear in the responses. The reason why this not the case simply is that linear stiffness is used in contact line load calculation. The contact model should have nonlinear contact force calculation based on the material model of the polymer cover layer. For example, the elastic foundation approach could be used to produce such limit cycle results. Another important remark is that the damping has strong influence to the stability. This makes the estimation of the damping coefficients in the real applications important.

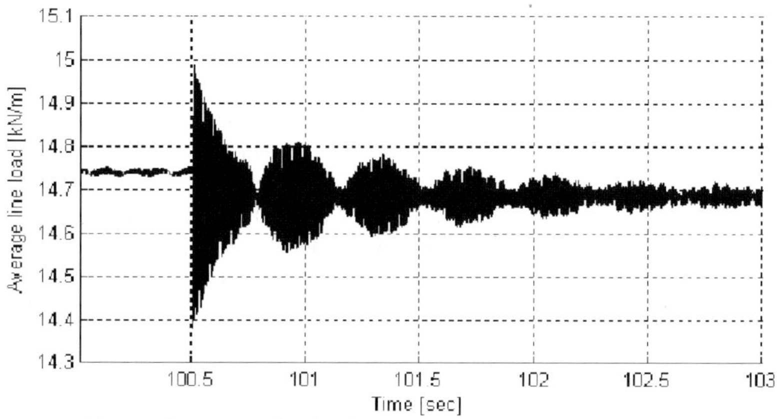

Fig. 5. The contact line load response. Running speed is 389 m/min.

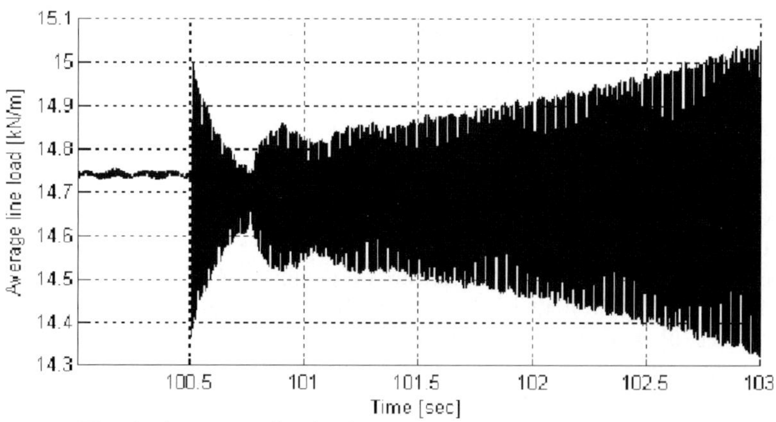

Fig. 6. The contact line load response. Running speed is 396 m/min.

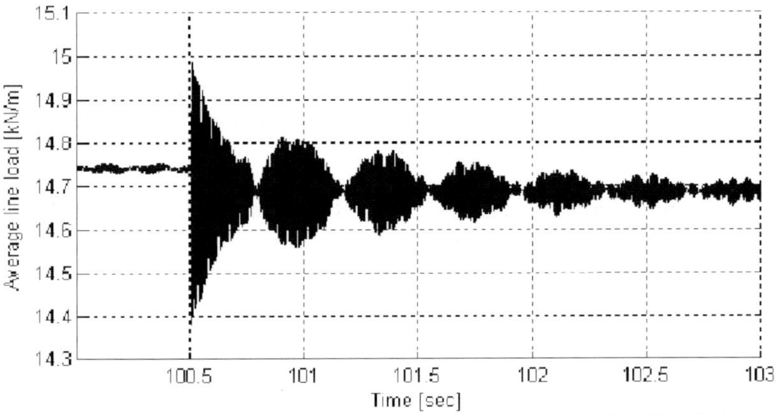

Fig. 7. The contact line load response. Running speed is 403 m/min.

6 Conclusions

The multibody modeling of roll contact with regenerative excitation is presented. The modeling of the rolls, the roll contact and the time delay term is described and the numerical solution procedure is presented. Very basic dynamical characteristics of the roll system are detected in the numerical simulation results and the dynamical behavior can be explained by some simple reasoning based on physics of the system. I can be concluded that the model presented here is suitable for more comprehensive dynamical studies of the roll contact vibrations. Some modifications, however, should be introduced. The contact force calculation should be updated and also the numerical performance of the solution procedure should be improved.

7 Acknowledgments

The authors would like to express their gratitude to The Academy of Finland, which has provided funding for this project.

References

1. Cardona A.; Geradin M., Modelling of superelements in mechanism analysis, International Journal for Numerical Methods in Engineering, v 32, n 8, Dec, 1991
2. Shabana A., Substructure Synthesis Methods for Dynamic Analysis of Multi-body Systems, Computers and Structures, v 20, n 4, 1985

3. Järvenpää V.M., Keskinen E., Dynamical modelling and simulation of calendering roll contact, Fifth Euromech Nonlinear Dynamics Conference, Enoc-2005, August 7-12, 2005, Eindhoven, Netherlands
4. Järvenpää V.M., Numerical Modeling of a Roll Mechanism for Multi-Objective Dynamical Analyses, Doctoral thesis, Tampere University of Technology, Publication 569, Tampere, 2005
5. Johnson K., Contact Mechanics, Cambridge University Press, 2003
6. Yuan L., Analysis of Delay Phenomena in Roll Dynamics, Doctoral Thesis, Tampere University of Technology, 2002
7. Sperling L., Introduction to Physical Polymer Science, Wiley-Interscience, New York, 1985
8. Géradin M., Rixen D., Mechanical vibrations, John Wiley & Sons Ltd., 1997
9. Driver R., Ordinary and Delay Differential Equations, Springer-Verlag, New York, 1977

Discrete Element Method in Simulation of Granular Materials

Rimantas Kačianauskas[1], Robertas Balevičius[1], Darius Markauskas[1], and Algirdas Maknickas[2]

[1] Laboratory of Numerical Modelling, Vilnius Gediminas Technical University, Saulėtekio al. 11, 10223 Vilnius-40, Lithuania rkac@fm.vtu.lt

[2] Laboratory of Parallel Computations, Vilnius Gediminas Technical University, Saulėtekio al. 11, 10223 Vilnius-40, Lithuania

1 Introduction

The discrete (or distinct) element method (DEM) has been recently recognized as efficient numerical tool for solving many scientific and technological problems in various fields of engineering. The method started in the 70-ies with its first application to simulate the dynamic behaviour of granular material in the work of Cundall and Strack [1]. Unlike the continuum approach, the DEM presents particulate material as an assemblage of discrete elements. It is based on the Lagrangian approach, according to which particles of granular material are treated as contacting bodies, while the dynamical parameters (i. e. position, velocity, orientation, etc.) of each body are tracked during the simulation. Some variations on the theme of DEM and granular materials may be found in [2]–[5].

2 Concept and methodology

The granular material considered presents a space filled with particles termed discrete elements here. It is regarded as an assemblage of a finite number N of discrete elements with the predefined geometry and given material properties. A composition of the particle assemblage is time-dependent because individual particles change their position due to free rigid body motion or interactions with neighbouring particles or walls.

Three translations and three independent rotations of each particle i in time t are described by the Newton's second law and expressed in terms of forces and torques acting at the centre of gravity of the particle i as follows:

$$m_i \frac{\mathrm{d}^2 \boldsymbol{x_i}}{\mathrm{d}t^2} = \boldsymbol{F}_i, \; I_i \frac{\mathrm{d}^2 \boldsymbol{\Theta}_i}{\mathrm{d}t^2} = \boldsymbol{T}_i \tag{1}$$

Peter Eberhard (ed.), Multiscale Problems in Multibody System Contacts, 65–74.

where m_i, and I_i are the mass and the inertia moment, while vectors $\boldsymbol{x}_i$ and $\boldsymbol{\Theta}_i$ stand for the position of the centre of gravity and for the orientation of particle i, respectively. Vectors $\boldsymbol{F}_i$ and $\boldsymbol{T}_i$ present the sum of external, contact and environmental (in our case, gravity) forces and torques, which act on the particle i.
The time-driven (TD) discrete element method originally proposed in [1] is explored to simulate the behavior of granular material. It is based on a Lagrangian frame of reference. The TD method is based on the evaluation of contact forces and torgues occuring between particle i and the neighboring particle j:

$$\boldsymbol{F}_i = \boldsymbol{F}_{i,contact} = \sum_{j=1,j\neq i}^{N} \boldsymbol{F}_{ij},$$

$$\boldsymbol{T}_i = \boldsymbol{T}_{i,contact} = \sum_{j=1,j\neq i}^{N} \boldsymbol{T}_{ij} = \sum_{j=1,j\neq i}^{N} \boldsymbol{d}_{ij} \times \boldsymbol{F}_{ij}, \tag{2}$$

where $\boldsymbol{d}_{ij}$ is particle geometry dependent vector pointing from the particle center to contact center.
Methodology of calculation of forces Eq. 2 depends on particle geometry and mechanical properties, a constitutive model of particle interaction and the surrounding media. The simplest commonly applied three-dimensional spherical particles are considered here, however, the ellipsoids, polyhedrons, or composite particles may also be used. Several constitutive models of contact forces may be found throughout references [2][6]. By assuming visco-elastic non-cohesive frictional particles, the contact model considers a combination of elasticity, damping and friction force effects. Actually, the contact between two material particles is modelled by a spring and dashpot in both the normal and tangential directions and an additional slider in tangential direction. Hence, the inter-particle force vector $\boldsymbol{F}_i$ describing contact between the particles i and j may be expressed in terms of the normal and the tangential components $\boldsymbol{F}_{n,ij}$ and $\boldsymbol{F}_{t,ij}$, respectively. The normal component $\boldsymbol{F}_{n,ij}$ presenting, actually, a repulsion force comprises elastic and viscous ingredients. The tangential component $\boldsymbol{F}_{t,ij}$ reflects static or dynamic frictional behaviour. The static force describes friction prior to gross sliding and comprises elastic and viscous ingredients, while the dynamic force describes friction after gross sliding and is expressed by the Coulomb's law. Contact behaviour is characterized by microscopic integrated parameters such as interaction stiffness $k_{n,ij}$ and $k_{t,ij}$ and damping coefficients γ_n and γ_t, which may be explicitly expressed in terms of the interacting particle data. Inter-particle friction is defined by internal friction coefficient μ which may different whether particle-particle or particle-wall contact is considered.
It seems that the equations of particle motion are rather simple, however, their numerical solution presents a difficult task. The solution involves the performance of various computational procedures termed hereafter a computational

technique. Most of them differ considerably from the existing continuum discretisation methods. Special attention should be paid to seting up the initial and boundary conditions, contact searching and time integration of equations of motion (Eq. 1).
In contrast to the continuum, where the initial conditions could be set up by prescribing the appropriate values or restrictions of the state variables, DEM requires an independent simulation step comprising artificial generation of the physically adjustable initial particles state by using Eq. 1. Boundary conditions may be implemented via specific particles or computational procedures. In most cases, problem-oriented technique has to be employed for setting up the initial and boundary conditions.
Another computationally expensive procedure is time integration of equations of motion. Due to desired maximal simplicity, either the one-step explicit (Taylor expansion series, fourth-order Runge-Kutta, velocity Verlet) or the predictor-corrector (fourth-order Hermite, fifth-order Gear) schemes, basically preferring one of corrector iteration are applicable in DEM [2], [8].
The above discussed DEM concept was implemented into the software code DEMMAT [9]. The code employs a zoning concept for contact searching [7] and the 5th order Gear predictor-corrector time integration scheme. A detailed description of DEM technique applied may also be found in [2] and [10].

3 Illustration of discharge in the flat-bottomed hopper

Granular flow in hoppers is an important operation used in industry for storing the materials. Macroscopic continuum models of granular flow based on theoretical considerations and experimental evidence may be found in the famous book of Nedderman [11]. Application of the three-dimensional DEM models (see, for example, Clearly and Sawley [12], Zhu and Yu [13], Balevičius et al [5]) allow us take into account microscopic properties, to handle new flow regimes and particle models and even to formulate optimization problems. The main focus of the current presentation is on investigation of the unsteady flow in the 3D flat-bottomed hopper.
The geometry of the hopper is presented in Fig. 1. The characteristic dimension of the outlet D is assumed to be related to the maximal diameter d of the particle as $D = 8.6d$. Assuming that $d = 0.06$ m, the main geometrical parameters of the hoppers are defined as: $D = 0.6$ m, $L = 1.4$ m, $H = 1.7$ m and $b = 0.3$ m.
The granular material is represented by an assembly of $N = 1980$ particles. The values of the particle radii R_i ranging from 0.03 to 0.035 m are defined randomly with uniform distribution. Total mass M of the material is fixed and is equal to $M = 143.7$ kg. The data on the visco-elastic particle are given in Table 1. Inter-particle and particle-wall friction is defined by the same friction coefficient $\mu = 0.6$.

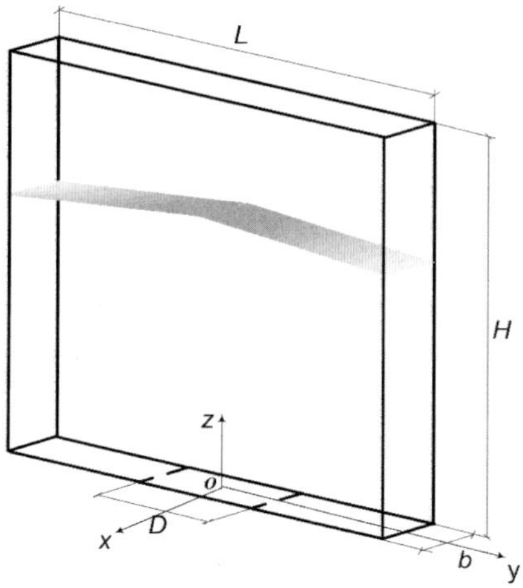

Fig. 1. The geometry of the flat-bottomed hopper

Table 1. Major data on the particles

Quantity	Value
Density (ρ)	$500\,\mathrm{kg/m^3}$
Poisson's ratio (ν)	0.3
Elasticity modulus (E)	$0.3 \cdot 10^6$ Pa
Shear modulus (G)	$0.11 \cdot 10^6$ Pa
Normal viscous damping coefficient (γ_n)	$60\,\mathrm{s}^{-1}$
Tangential viscous damping coefficient (γ_t)	$10\,\mathrm{s}^{-1}$

Granular flow in hoppers is a continuous process comprising settling of the material into the hopper and discharge. In the framework of the current investigation, settling is considered as artificial free compacting of material to the state of the rest for proper simulation or as formation of the initial conditions for discharge.

Discharge eventually means the fall of material from the hopper. It begins by opening the orifice denoted by reference time $t = 0$. Numerical results of discharge at selected time instances $t = 0.04$ s and $t = 0.2$ s are illustrated in Fig. 2. It can be observed, that the depression zone above the orifice occurs already at the beginning of the emptying process. Later, this depression deepens with progressing of the emptying. Its sides become more and more steeper and when the slope of the sides reaches the angle of repose some of the particles cascade down to the central part of the material.

The nature of granular flow in the flat-bottomed hopper corresponds to classical scheme of funnel flow exhibiting four different zones, namely, a stagnant, a plug flow zone, a converging flow zone, and a transition zone from plug flow to converging flow. The above zones may be even better contoured by

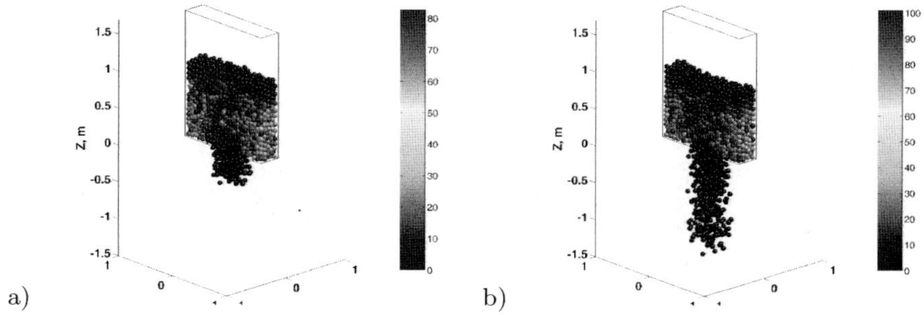

Fig. 2. Illustration of the discharge flow at various time instances: a) $t = 0.04$ s, b) $t = 0.2$ s

considering particle velocity patterns (Fig. 3). Here, clear boundaries between flowing and stagnant material are observed.

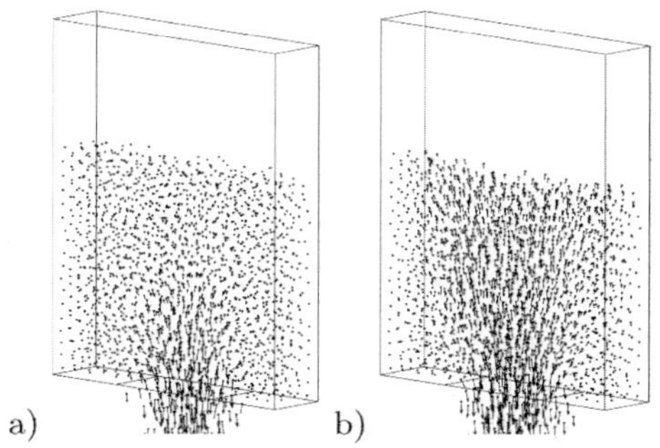

Fig. 3. The particle velocity patterns at various time instances: a) $t = 0.2$ s, b) $t = 0.6$ s

Illustration of particle motion is a direct exhibition of the result of equations of motion. However, other important data, especially those illustrating microscopic effects on macroscopic variables, may be also obtained by post-processing. Here, the bulk porosity within the hopper plays an important role. Porosity of the packed granular material is obtained by using a standard homogenization technique. The mid-plane distribution of the porosity fields during the discharge at different time instances is shown in Fig. 4. The pictures illustrate rarefaction phenomenon above the orifice and its change in time.

4 Simulation of biaxial compression test

Granular problems of a specific type are related to material testing, where compression tests are actually of major importance. Such tests could be used

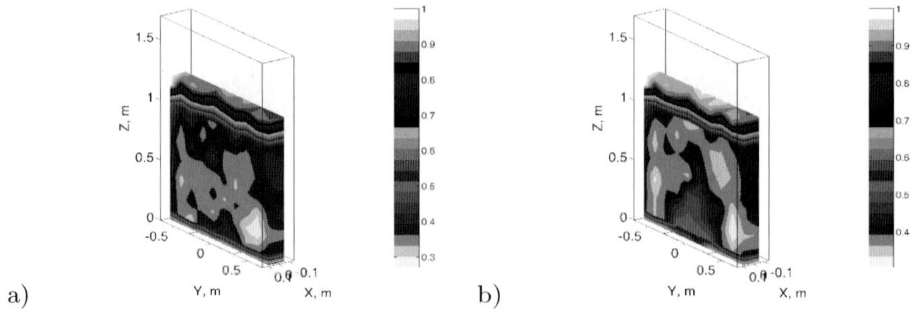

Fig. 4. The porosity fields at different time instances: a) after the filling $t = 0.0$ s, b) during the discharge $t = 0.6$ s

to determine macroscopic characteristics of media, such us Young's modulus, Poisson's ratio or angle of internal friction. Testing device presents a moving loading platform and a cylindrical pressure vessel with flexible membrane. Granular material contained in the vessel is initially subjected to the confinement pressure. Deformation behaviour of material is considered by the axial compressing load, while keeping the constant confinement pressure.

Simulation of compression is numerically more difficult compared to the granular flow because particles are subjected to a force of higher magnitudes, while reliable time integration should be performed by applying smaller time increments. The problem is three-dimensional, but the majority of numerical simulations are dealing with the 2D problem or biaxial compression test ([14]–[15]).

Two major issues have to be regarded by conducting a compression test in terms of the DEM. The first one is related to implementation of flexible boundaries and static boundary conditions, while the second is related to setting up the initial conditions. In contrast to the continuum, where the initial conditions could be set by prescribing the appropriate material parameters and initial stresses, DEM requires an independent simulation step comprising artificial generation of the physically adjustable initial state of the particles.

The present analysis addresses simulation of biaxial test using flexible rubber like membranes and generation of physically adjustable initial conditions by particle compaction. The required membrane particles and compacting scenarios are implemented into the existing DEM code.

According to a given size distribution, 2658 particles were generated and randomly embedded inside the area 5 cm by 10 cm by ensuring no overlap between particles. The particle sizes range from 0.5 mm to 3.5 mm with diameter d_{50} equal to 1.6 mm. This implies that particles of smaller diameter contain 50% of mass. The parameters used throughout the simulation period are presented in Table 2.

Compacting is performed for the case of the frictionless material by assuming zero particle-particle, particle-rigid wall and particle-membrane friction coef-

Table 2. The parameters selected for the present study

Quantity	Value
Number of particles	2658
Radii of particles	0.5–3.5Â mm
Time step (Δt)	10^{-7}s
Particle density (ρ)	$2600\,\mathrm{kg/m^3}$
Normal stiffness constant (k_n)	$1.5 \cdot 10^6$ N/m
Tangential stiffness constant (k_t)	$1.0 \cdot 10^6$ N/m
Normal viscous damping coefficient (γ_n)	$500\,\mathrm{s}^{-1}$
Tangential viscous damping coefficient (γ_t)	$500\,\mathrm{s}^{-1}$

ficients in order to ensure homogeneity of the specimen as far as possible. Three different compacting scenarios with differently manipulated loading histories on the boundaries were investigated. According to the first scenario, the generated specimen is compacted by moving all four side walls inward at the small constant wall speed 0.01 m/s to ensure quasistatic compaction conditions. According to the second scenario, flexible membranes on the sides of the specimen are introduced immediately after the generation of the particles. Using the third scenario, at the very beginning of the compaction, the rigid walls on all four sides are introduced and are moved at a prescribed speed (v = 0.01 m/s). At the displacement value 1.8 mm the two side walls are replaced by flexible membranes. Then, the pressure on the top and bottom walls and on the membranes is increased gradually up to 5 kPa. It has been found that the third scenario provides the best quality specimen, which is insensitive to unloading and has small distortions of the side boundaries.

Three compression tests using confined side pressure 5 kPa, 3 kPa and 1 kPa were simulated after compacting. Compression loading was controlled by vertical displacements of the cylinder, while keeping the constant velocity of 0.04 m/s. The particle-particle and particle-rigid wall friction coefficients were switched to $\mu = 0.5$ during compression.

The deformed specimen at ε_1 equal to 2%, 10% and 20% for the case of the confined pressure 1 kPa is shown in Fig. 5. The figures illustrate the formation of the barrel shape specimen as is usually observed in testing.

The deformation curves expressed in terms of axial stress σ_1 and strain ε_1 are presented in Fig. 6. They are obtained by applying a homogenization technique. The above graph illustrates the irregular unsmooth deformation typical for DEM analysis which could be accounted for by friction and a relatively small amount of particles. Finally, the exhibited curves are explored to determine the internal friction angle ϕ. The maximal stress σ_1 is applied for evaluation purposes and angle value $\phi = 19\,^\circ$ is obtained. It should be noted that all three confined pressures actually form a straight approximation line, what is indirect reflection of the quality of the model.

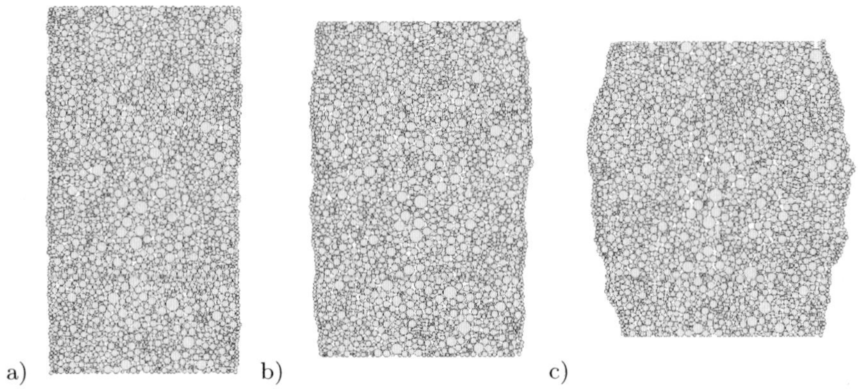

Fig. 5. Deformed specimen (confined pressure $\sigma_2 = 1kPa$): a) $\varepsilon_1 = 2\%$, b) $\varepsilon_1 = 10\%$, c) $\varepsilon_1 = 20\%$

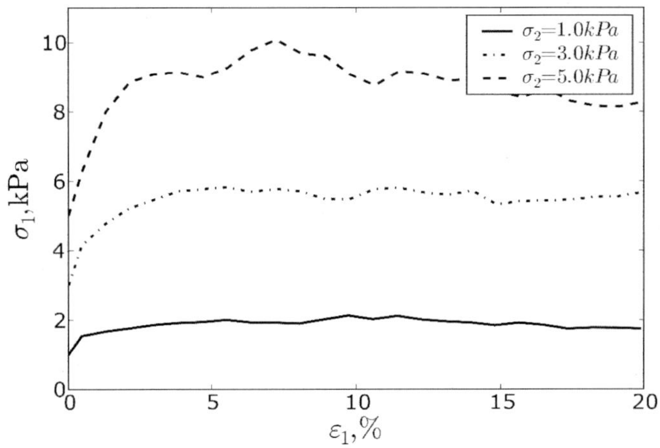

Fig. 6. Deformation curves

5 Computational aspects

The main disadvantages of the DEM, in comparison with the well-known numerical methods based on continuum approach, are related to computational capabilities limited by a huge number of particles and a relatively short time of inter-particle collision. Naturally, much effort has been to improve the efficiency of the method, and the successful software implementation may play a decisive role in it. Computational aspects of the developed DEMMAT code [9] are basically related to computation of inter-particle forces including contact searching and to time integration of the equations of motion. Detailed examination has showed, that the most time consuming step was a procedure of contact searching which can require up to 60% of the CPU time.

Computational capacity of the DEM software may be dramatically improved using the parallel implementation, while a parallel version of the DEMMAT code is recently under development. Several tests solving benchmark problems were performed on the PC cluster VILKAS at Vilnius Gediminas Technical University. The 3D particle specimen in the form of a rectangular parallelepiped under isotropic compression by rigid walls was considered. A concept of parallel computations was based on the static domain decomposition assuming the partitioning of the computational domain into subdomains, each being assigned to an individual processor.

The parallel performance of the developed code was evaluated by measuring of the speed-up. In perfect parallelisation of data, the speed-up is expected to be proportional to the number of processors. The parallel efficiency is largely predetermined by contact searching and was actually defined by the ratio of local computations over inter-processor communications. It has been detected that for the four processors the speed-up values are 3.98 and 4.01 for the samples having 20000 and 100000 particles, respectively. The increase of the number of processors up to eight yielded the speed-up value 7.49 for 20000 particles, while in the case of ten processors for 100000 particles, the speed-up was equal to 9.22. The presented results show that the implemented parallel code is well designed for the distributed memory PC clusters.

6 Concluding remarks

Numerous investigations demonstrate that the discrete element method is a powerful numerical tool to simulate the dynamic behaviuor of granular materials. It bridges the gap between continuum models and experimental investigation as well as between microscopic and macroscopic approaches. In spite of significant achievements, two major problems still need to be solved in further research. They concern the development of realistic microscopic models with realistic inter-particle data and increasing computational efficiency to deal with industrial problems.

References

1. Cundall PA, Strack ODL (1979) A discrete numerical model for granular assemblies. Géotechnique 29(1):47-65
2. Džiugys A, Peters BJ (2001) An approach to simulate the motion of spherical and non-spherical fuel particles in combustion chambers. Granular Matter 3(4):231-266
3. Luding S (2002) From microscopic simulations to macroscopic material behavior. Computer Physics Communications 147:134-140
4. Zhang N, Rosato AD (2004) Analysis of instantaneous dynamic states of vibrated granular materials. Mechanics Research Communications 31(5):524-544

5. Balevičius R, Kačianauskas R, Mroz Z, Sielamovicz I (2006) Discrete Element Method applied to multiobjective optimization of discharge flow parameters in hoppers. Structural and Multidisciplinary Optimization 31(3): 163-175
6. Tomas J (2004) Fundamentals of powder consolidation and flow. Granular Matter 6(2-3):75-86
7. Asmar BN, Langston PA, Matchett AJ, Walters JK (2002) Validation tests on a distinct element model of vibrating cohesive particle systems. Computers and Chemical Engineering 26:785-802
8. Allen MP, Tildesley DJ (1991) Computer simulation of liquids. Clarendon Press, Oxford
9. Balevičius R, Kačianauskas R, Džiugys A, Maknickas A, Vislavičius K (2005) DEMMAT code for numerical simulation of multi-particle dynamics. Information Technology and Control 34(1):71-78
10. Balevičius R, Džiugys A, Kačianauskas R (2004) Discrete element method and its application to the analysis of penetration into granular media. J. Civil Engineering and Management 10(1):3-14
11. Nedderman RM (1992) Statics and kinematics of granular materials. Cambridge University Press, New York
12. Clearly PW, Sawley ML (2002) DEM modelling of industrial granular flows: 3D case studies and the effect of particle shape on hopper discharge. Applied Mathematical Modelling 26:89-111
13. Zhu HP, Yu AB (2005) Steady-state granular flow in 3D cylindrical hopper with flat bottom: macroscopic analysis. Granular Matter 7:97-107
14. Emeriault F, Claquin C (2004) Statistical homogenization for assemblies of elliptical grains: effect of the aspect ratio and particle orientation. International Journal of Solids and Structures 41:5837-5849
15. Mirghasemi A, Rothenburg L, Matyas E (2002) Influence of particle shape on engineering properties of assemblies of two-dimensional polygon-shaped particles. Géotechnique 52(3):209-217

A Dynamical Model for the Plane Contact with Rigid Contact Areas

Rainer Keppler and Wolfgang Seemann

Institut für Technische Mechanik – Universität Karlsruhe (TH) – Germany
keppler@itm.uni-karlsruhe.de – seemann@itm.uni-karlsruhe.de

Summary. In this article a generalized elasto-plastic model for the plane frictional contact of rigid bodies is presented. The basic assumptions for this model are the rigidness of the plane contact area and locally acting friction.

1 Dynamical friction models

The idea of the following dynamical friction model is to cutoff the contact area from the body and reconnect it elastically (see Fig.1-left). This allows the body to perform pre-sliding displacements.

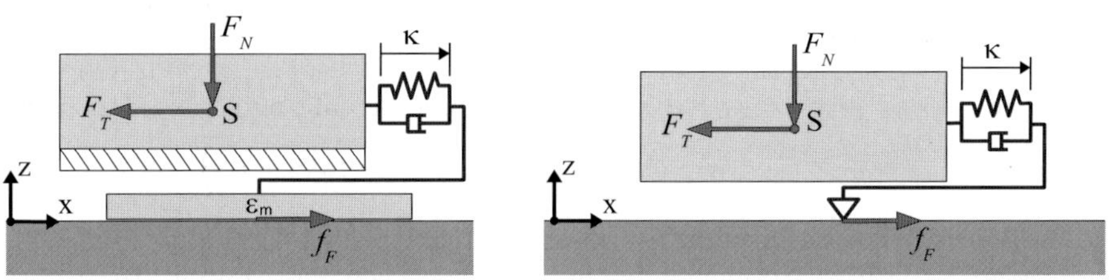

Fig. 1. Elastic interconnection (left). Half degree of freedom model (right).

1.1 The half degree of freedom model

The equations of motion for the problem pictured on the left of Fig.1 can be written in the form

$$(m - \varepsilon_m)\,\ddot{r}_{Sx} = F_T + c_0\,\kappa + c_1\,\dot{\kappa} \tag{1a}$$

$$\varepsilon_m\,(\ddot{r}_{Sx} + \ddot{\kappa}) = -c_0\,\kappa - c_1\,\dot{\kappa} + f_F, \tag{1b}$$

where $\boldsymbol{r}_S = (r_{Sx}, r_{Sz})^T$ denotes coordinates of the center of mass S with respect to the fixed reference frame. The mass of the contact surface ε_m is assumed to be small compared to the mass of the body m ($m \gg \varepsilon_m$). The relative velocity between the contact-area and the rigid halfspace is given by

Peter Eberhard (ed.), Multiscale Problems in Multibody System Contacts, 75–84.

$$\Delta u = \Delta v + \dot{\kappa}, \tag{2}$$

where Δv corresponds to the relative velocity between the upper body and the plain. In case of a sliding contact area Δu does not vanish and the frictional force f_F is modeled by

$$^{SV}f_F(\Delta u) = -\left[\mu_{sl} + (\mu_{st} - \mu_{sl})\, e^{-\frac{(\Delta u)^2}{v_S^2}}\right] F_N \frac{\Delta u}{|\Delta u|} - \mu_V\, F_N\, \Delta u. \tag{3}$$

During sticking the relative velocity Δu vanishes and the frictional force f_F is part of the friction cone $|f_F| \leq \mu_H\, F_N$. Assuming that $\varepsilon_m\, (\ddot{r}_{Sx} + \ddot{\kappa})$ is small the stiction condition can be written in form of

$$\dot{\kappa} = -\Delta v \quad \text{and} \quad |c_0\, \kappa + c_1\, \dot{\kappa}| \leq \mu_H\, F_N. \tag{4}$$

By introducing a switching parameter

$$\alpha = \alpha(\Delta u, F_\kappa) = \begin{cases} 0 & \text{if } \Delta u = 0 \text{ and } |c_0\, \kappa + c_1\, \dot{\kappa}| \leq \mu_H\, F_N \quad \text{(sticking)} \\ 1 & \text{else} \quad \text{(sliding)} \end{cases} \tag{5}$$

the sticking and sliding conditions are combined to

$$\begin{aligned} (m - \varepsilon_m)\, \ddot{r}_{Sx} &= F_T + c_0\, \kappa + c_1\, \dot{\kappa} \\ \dot{\kappa} &= -(1-\alpha) \cdot \Delta v + \alpha \cdot v_\kappa \\ \varepsilon_m\, (\ddot{r}_{Sx} + \dot{v}_\kappa) &= (1-\alpha) \cdot \varepsilon_m \left[\ddot{r}_{Sx} - \Delta\dot{v}\right] - \alpha \cdot \left[c_0\, \kappa + c_1\, v_\kappa - {}^{SV}f_F(\Delta v + \dot{\kappa})\right]. \end{aligned}$$

In the limit $\varepsilon_m \to 0$ the variable v_κ can be eliminated, which leads to the so called half degree of freedom model

$$m\, \ddot{r}_{Sx} = F_T + c_0\, \kappa + c_1\, \dot{\kappa} \tag{6a}$$

$$\dot{\kappa} = -(1-\alpha) \cdot \Delta v - \frac{\alpha}{c_1}\left(c_0\, \kappa - {}^{SV}f_F(\Delta v + \dot{\kappa})\right). \tag{6b}$$

This model contains an implicit differential equation for κ. Because of that, a linearization around stationary solutions for $\Delta v = const$ is applied. In case of sliding ($\alpha = 1$) with constant relative velocity $\Delta v = const \neq 0$ the stationary solution of the half degree of freedom model holds

$$\dot{\kappa}_s = 0 \quad \text{and} \quad \kappa_s = \frac{1}{c_0}\, {}^{SV}f_F(\Delta v). \tag{7}$$

A linearization of the half degree of freedom model around this stationary solution finally results in

$$m\, \ddot{r}_{Sx} = F_T + c_0\, \kappa + c_1\, \dot{\kappa} \tag{8a}$$

$$\dot{\kappa} = -\Delta v \cdot \left(1 - \alpha \cdot \left[1 - \nu(|\Delta v|) + \nu(|\Delta v|) \frac{c_0\, \kappa}{{}^{SV}f_F(\Delta v)}\right]\right), \tag{8b}$$

where

$$\nu(|\Delta v|) = \frac{\mu_V\, |\Delta v|\, F_N + (\mu_{Gl} + [\mu_H - \mu_{Gl}]) \cdot \exp\left(-\frac{|\Delta v|^2}{v_S^2}\right) F_N}{c_1\, |\Delta v| + \mu_V\, |\Delta v|\, F_N - 2\, \frac{\mu_H - \mu_{Gl}}{v_S^2} \cdot \exp\left(-\frac{|\Delta v|^2}{v_S^2}\right) |\Delta v|^2\, F_N}. \tag{8c}$$

1.2 The elasto-plastic model

In [2, 3] the so called elasto-plastic friction model is introduced. The model equations are given by

$$f_F = c_0\,\kappa + c_1\,\dot{\kappa}\ , \quad \dot{\kappa} = -\Delta v \cdot \left(1 - \beta(\kappa, \Delta v)\,\frac{c_0\,\kappa}{{}^S f_F(\Delta v)}\right), \tag{9}$$

with a switching parameter of the form

$$\beta = \begin{cases} 1 & \text{, if } \operatorname{sign}(\kappa) = \operatorname{sign}(\Delta v),\ \kappa_2 < |\kappa| \\ \dfrac{(|\kappa| - \kappa_1)^2(3\,\kappa_2 - \kappa_1 - 2\,|\kappa|)}{(\kappa_2 - \kappa_1)^3} & \text{, if } \operatorname{sign}(\kappa) = \operatorname{sign}(\Delta v),\ \kappa_1 \le |\kappa| \le \kappa_2 \\ 0 & \text{, else.} \end{cases}$$

One can also see that this model corresponds to the linearized half degree of freedom model (8), where $\nu(|\Delta v|) = 1$ and β is some regularization for the switching parameter α.

2 Friction models for plane contact

In the following the plane contact of two rigid bodies is investigated (see Fig. 2). The rigidity of the contact area and locally acting friction are basic assumptions of the subsequently presented model.

2.1 Kinematics

The plane contact area is denoted by Ω. Two coordinate systems, the reference frame on the plane CS_0 and the body fixed system CS_b with origins O and B are introduced. The corresponding bases are orientated in a way that, the third axis are normal to the contact area.

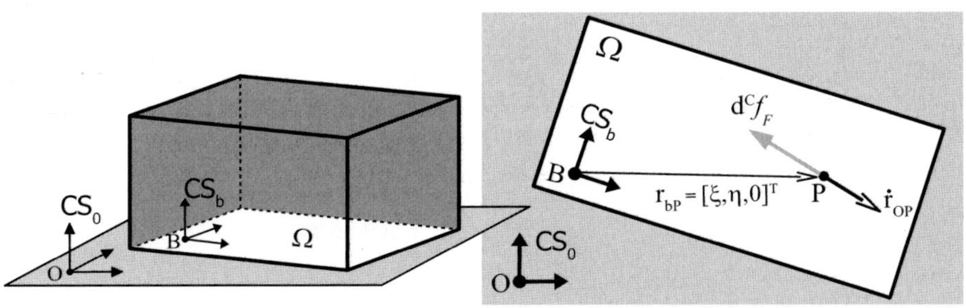

Fig. 2. Plane friction model

Because the contact area Ω is plane, only plane motions are possible and the relevant sliding velocity is of the form

$$\boldsymbol{\Delta v} = [v_1, v_2, \omega_3]^T\ , \tag{10}$$

where $\boldsymbol{v}^b$ and $\boldsymbol{\omega}^b$ denote the linear and the angular velocity of the body in coordinates of CS_b.

2.2 Sliding

For a given pressure distribution $p(\boldsymbol{r})$ the local friction force from the Coulomb model in coordinates of CS_b holds

$$d^C\boldsymbol{f}_F^b = -\mu_{sl}\,\frac{\dot{\boldsymbol{r}}_{OP}^b}{\|\dot{\boldsymbol{r}}_{OP}^b\|}\,dF_N = -\mu_{sl}\,\frac{\dot{\boldsymbol{r}}_{OP}^b}{\|\dot{\boldsymbol{r}}_{OP}^b\|}\,p(\boldsymbol{r}_{BP}^b)\,d\Omega. \tag{11}$$

An integration over Ω yields the frictional force ${}^C\boldsymbol{f}_F^b = \iint_\Omega d^C\boldsymbol{f}_F^b$ and the frictional torque ${}^C\boldsymbol{f}_T^b = \iint_\Omega \boldsymbol{r}_{BP}^b \times d^C\boldsymbol{f}_F^b$. Because of the special orientation of CS_b the coordinates $({}^C\boldsymbol{f}_F^b)_3$, $({}^C\boldsymbol{f}_T^b)_1$ and $({}^C\boldsymbol{f}_T^b)_2$ are vanishing. The remaining components are combined to the so called generalized torque for plane Coulomb friction ${}^C\boldsymbol{t}^b = [({}^C\boldsymbol{f}_F^b)_1\ ,\ ({}^C\boldsymbol{f}_F^b)_2\ ,\ ({}^C\boldsymbol{f}_T^b)_3]^T$. Using the definition (10) for the sliding velocity $\Delta\boldsymbol{v}$ one can find

$$ {}^C\boldsymbol{t}^b(\Delta\boldsymbol{v}) = -\mu_{sl}\,F_N \iint_\Omega \frac{\begin{bmatrix} v_1 - \eta\,\omega_3 \\ v_2 + \xi\,\omega_3 \\ \xi\,v_2 - \eta\,v_1 + (\xi^2+\eta^2)\,\omega_3 \end{bmatrix} \dfrac{p(\xi,\eta)}{F_N}\,d\xi\,d\eta}{\sqrt{(v_2+\xi\,\omega_3)^2 + (v_1-\eta\,\omega_3)^2}}. \tag{12}$$

By introducing the center of rotation

$$\boldsymbol{r}_{B\,cor}^b = [\xi_{cor}, \eta_{cor}, 0]^T = \frac{\boldsymbol{\omega}^b \times \boldsymbol{v}^b}{\boldsymbol{\omega}^b \cdot \boldsymbol{\omega}^b} = \left[-\frac{v_2}{\omega_3}\ ,\ \frac{v_1}{\omega_3}\ ,\ 0\right]^T \tag{13}$$

the expression (12) can be rewritten in form of

$${}^C\boldsymbol{t}^b(\Delta\boldsymbol{v}) = -\frac{\omega_3}{|\omega_3|}\,\mu_{sl}\,F_N \begin{bmatrix} I_1(\xi_{cor},\eta_{cor}) \\ I_2(\xi_{cor},\eta_{cor}) \\ a_K \cdot I_3(\xi_{cor},\eta_{cor}) \end{bmatrix}, \tag{14a}$$

where $a_K = \sqrt{\frac{1}{\pi}\iint_\Omega d\xi\,d\eta}$ is the equivalent radius of the contact area Ω and

$$I_1(\xi_{cor},\eta_{cor}) = \iint_\Omega \frac{-(\eta-\eta_{cor})}{\sqrt{(\xi-\xi_{cor})^2 + (\eta-\eta_{cor})^2}}\,\frac{p(\xi,\eta)}{F_N}\,d\xi\,d\eta \tag{14b}$$

$$I_2(\xi_{cor},\eta_{cor}) = \iint_\Omega \frac{(\xi-\xi_{cor})}{\sqrt{(\xi-\xi_{cor})^2 + (\eta-\eta_{cor})^2}}\,\frac{p(\xi,\eta)}{F_N}\,d\xi\,d\eta \tag{14c}$$

$$I_3(\xi_{cor},\eta_{cor}) = \iint_\Omega \frac{1}{a_K}\,\frac{\xi\,(\xi-\xi_{cor}) + \eta\,(\eta-\eta_M)}{\sqrt{(\xi-\xi_{cor})^2 + (\eta-\eta_{cor})^2}}\,\frac{p(\xi,\eta)}{F_N}\,d\xi\,d\eta. \tag{14d}$$

2.3 Sticking

The transition from sticking to sliding is characterized by the instantaneous appearance of the center of rotation, while the relative velocity is still zero. The set of all possible center of rotations

$$\mathcal{G} = \left\{ \pm\,\mu_{st}\,F_N \begin{bmatrix} I_1(\xi_{cor},\eta_{cor}) \\ I_2(\xi_{cor},\eta_{cor}) \\ a_K \cdot I_3(\xi_{cor},\eta_{cor}) \end{bmatrix} \middle|\ (\xi_{cor},\eta_{cor}) \in \mathbb{R}^2 \right\} \tag{15}$$

forms the so called friction limit, which represents the boundary of the sticking region $\mathcal{H}$ ($\mathbf{0} \in \mathcal{H}$ and $\partial\mathcal{H} = \mathcal{G}$) in the three-dimensional vector space of generalized torques. Because of that the friction limit $\mathcal{G}$ is a closed surface which can be uniquely parameterized

$$\mathcal{G} = \left\{ \mu_{st} F_N \begin{bmatrix} \rho_{st}(\phi,\theta)\,\sin(\theta)\,\cos(\phi) \\ \rho_{st}(\phi,\theta)\,\sin(\theta)\,\sin(\phi) \\ a_K \cdot \rho_{st}(\phi,\theta)\,\cos(\theta) \end{bmatrix} \middle| (\phi,\theta) \in (-\pi,\pi] \times [0,\pi] \right\}, \quad (16)$$

where

$$\rho_{st}(\phi,\theta) = \sqrt{I_1^2(\xi_{cor},\eta_{cor}) + I_2^2(\xi_{cor},\eta_{cor}) + I_3^2(\xi_{cor},\eta_{cor})} \quad (17)$$

$$\phi = 2\arctan\left(\frac{I_2(\xi_{cor},\eta_{cor})}{I_1(\xi_{cor},\eta_{cor}) + \sqrt{I_1^2(\xi_{cor},\eta_{cor}) + I_2^2(\xi_{cor},\eta_{cor})}}\right) \quad (18)$$

$$\theta = \arccos\left(\frac{I_3(\xi_{cor},\eta_{cor})}{\sqrt{I_1^2(\xi_{cor},\eta_{cor}) + I_2^2(\xi_{cor},\eta_{cor}) + I_3^2(\xi_{cor},\eta_{cor})}}\right). \quad (19)$$

As a measure for the distance between the actual generalized torque $\boldsymbol{t}^b$ and the friction limit $\mathcal{G}$ the quantity

$$g(\boldsymbol{t}) = \frac{\sqrt{t_1^2 + t_2^2 + \frac{1}{a_K^2} \cdot t_3^2}}{\rho_{st}(\phi_t,\theta_t)} \quad (20)$$

is defined, with $\phi_t = 2\arctan\left(\frac{t_2}{t_1+\sqrt{t_1^2+t_2^2}}\right)$, $\theta_t = \arccos\left(\frac{t_3}{\sqrt{a_K^2\,(t_1^2+t_2^2)+t_3^2}}\right)$. Using this definition sticking is uniquely classified by

$$\Delta\boldsymbol{v} = \mathbf{0} \text{ and } g\left(\boldsymbol{t}^b\right) \le \mu_{st} F_N. \quad (21)$$

These sticking conditions directly correspond to the one-dimensional case $\Delta v = 0$ and $|f_F| \le \mu_{st} F_N$.

2.4 Lubricated friction and Stribeck effect

In the same manner as in the case of the Coulomb friction, one can find that the generalized torque for lubricated friction is of the form

$${}^V\boldsymbol{t}^b(\Delta\boldsymbol{v}) = -\omega_3\,\mu_V F_N \begin{bmatrix} \iint_\Omega -(\eta-\eta_{cor})\,\frac{p(\xi,\eta)}{F_N}\,d\xi\,d\eta \\ \iint_\Omega (\xi-\xi_{cor})\,\frac{p(\xi,\eta)}{F_N}\,d\xi\,d\eta \\ \iint_\Omega [\xi\,(\xi-\xi_{cor}) + \eta\,(\eta-\eta_{cor})]\,\frac{p(\xi,\eta)}{F_N}\,d\xi\,d\eta \end{bmatrix}. \quad (22)$$

Also a suitable generalized Stribeck friction torque can be defined as

$${}^S\boldsymbol{t}^b(\Delta\boldsymbol{v}) = \left[1 + \frac{\mu_H-\mu_{Gl}}{\mu_{Gl}}\,\exp\left(-\frac{v_m^2}{v_S^2}\right)\right]\,{}^C\boldsymbol{t}^b(\Delta\boldsymbol{v}), \quad (23)$$

where v_S is some kind of Stribeck velocity and mean velocity v_m is given by

$$v_m = \sqrt{\iint_\Omega \|\dot{\boldsymbol{r}}_{k+P}^k\|^2\,\frac{p(\boldsymbol{r}_{kP}^k)}{F_N}\,d\Omega} = |\omega_3|\,\sqrt{(\xi_{cor}^2 + \eta_{cor}^2) + \boldsymbol{r}_{B\,cor}^b \cdot \boldsymbol{s}_0 + s_1},$$

with $\boldsymbol{s}_0 = \iint_\Omega [\xi,\eta,0]^T\,\frac{p(\xi,\eta)}{F_N}\,d\xi\,d\eta$ and $s_1 = \iint_\Omega (\xi^2+\eta^2)\,\frac{p(\xi,\eta)}{F_N}\,d\xi\,d\eta$.

3 Plane contact under Hertzian pressure distribution

In section 2 an arbitrary pressure distribution was used. Now the problem of plane contact under axially symmetric Hertzian pressure distribution is investigated.

3.1 Hertzian pressure distribution

For the rotationally symmetric shapefunction $f(r) = \frac{r^2}{2R}$ of the contact region, one can find from the Hertzian theory [1] that the corresponding pressure distribution takes the form

$$p(r) = \frac{3\,F_N}{2\,\pi\,a_K^2}\sqrt{1-\frac{r^2}{a_K^2}}\;,\; r \in \Omega = \{r\,|\,0 \le r \le a_K\}.$$

Furthermore, the penetration depth δ is given by $F_N = \frac{4\,E^\star}{3}\,R^{1/2}\,\delta^{3/2}$.

3.2 Sliding, Stribeck and lubricated friction

Because the problem is axially symmetric, the generalized torque for plane Coulomb friction can be written in the form

$$^{C}\boldsymbol{t}^b(\boldsymbol{\Delta v}) = -\mu_{st}\,F_N \begin{bmatrix} \frac{v_1}{\sqrt{v_1^2+v_2^2}}\,\sqrt{I_1^2(\xi_{cor},\eta_{cor})+I_2^2(\xi_{cor},\eta_{cor})} \\ \frac{v_2}{\sqrt{v_1^2+v_2^2}}\,\sqrt{I_1^2(\xi_{cor},\eta_{cor})+I_2^2(\xi_{cor},\eta_{cor})} \\ \frac{\omega_3}{|\omega_3|}\,a_K\,|I_3(\xi_{cor},\eta_{cor})| \end{bmatrix}.$$

The occuring values $\sqrt{I_1^2+I_2^2}$ and $|I_3|$ are piecewise approximated by

$$\sqrt{I_1^2+I_2^2} \approx \begin{cases} -0.294\left(\frac{r_{cor}}{a_K}\right)^3 + 1.178\left(\frac{r_{cor}}{a_K}\right) & , r_{cor} \le a_K \\ -0.020\left(\frac{a_K}{r_{cor}}\right)^4 - 0.096\left(\frac{a_K}{r_{cor}}\right)^2 + 1 & , a_K \le r_{cor} \end{cases}$$

$$|I_3| \approx \frac{3\pi}{16}\begin{cases} 0.375\left(\frac{r_{cor}}{a_K}\right)^4 - \left(\frac{r_{cor}}{a_K}\right)^2 + 1 & , r_{cor} \le a_K \\ 0.040\left(\frac{a_K}{r_{cor}}\right)^3 + 0.335\left(\frac{a_K}{r_{cor}}\right) & , a_K \le r_{cor} \end{cases},$$

where $r_{cor} = \sqrt{\xi_{cor}^2+\eta_{cor}^2} = \frac{\sqrt{v_1^2+v_2^2}}{|\omega_3|}$. In case of Stribeck friction one finds

$$^{S}\boldsymbol{t}^b(\boldsymbol{\Delta v}) = \left(1+\frac{\mu_{st}-\mu_{sl}}{\mu_{sl}}\,\exp\left(-\frac{v_1^2+v_2^2+\frac{2}{5}\,a_K^2\,\omega_3^2}{v_S^2}\right)\right)\,^{C}\boldsymbol{t}^b(\boldsymbol{\Delta v}),$$

and for the lubricated friction

$$^{V}\boldsymbol{t}^b(\boldsymbol{\Delta v}) = -\mu_V\,F_N\left[v_1, v_2, \frac{2}{5}\,a_K^2\omega_3\right]^T.$$

3.3 Sticking

Because of the rotational symmetry, the friction limit $\mathcal{G}$ does not depend on the azimuth angular ϕ and is also symmetric to the z-axis. The radius of sticking (17) can be approximated by $\rho_{st} \approx 1.20 \cdot |\cos(\theta)|^5 - 3.83 \cdot |\cos(\theta)|^4 + 4.64 \cdot |\cos(\theta)|^3 - 2.42 \cdot |\cos(\theta)|^2 + 1$, where $|\cos(\theta)| = \frac{|\boldsymbol{f}_T^b|}{R_{\boldsymbol{f}}}$ and $R_{\boldsymbol{f}} = \sqrt{a_K^2 \cdot \left|\boldsymbol{f}_F^b\right|^2 + \left|\boldsymbol{f}_T^b\right|^2}$. The sticking condition $g\left(\boldsymbol{t}^b\right) \leq \mu_{st} F_N$ can be transformed to $\frac{R_{\boldsymbol{f}}}{\mu_{st}\, a_K\, F_N} \leq \rho_{st}$ and finally approximated by

$$\frac{R_{\boldsymbol{f}}}{\mu_{st} a_K F_N} \lesssim 1.20 \left[\frac{\left|\boldsymbol{f}_T^b\right|}{R_{\boldsymbol{f}}}\right]^5 - 3.83 \left[\frac{\left|\boldsymbol{f}_T^b\right|}{R_{\boldsymbol{f}}}\right]^4 + 4.64 \left[\frac{\left|\boldsymbol{f}_T^b\right|}{R_{\boldsymbol{f}}}\right]^3 - 2.42 \left[\frac{\left|\boldsymbol{f}_T^b\right|}{R_{\boldsymbol{f}}}\right]^2 + 1. \tag{24}$$

4 Elasto-plastic model for plane friction

In the following we enhance the conventional elasto-plastic model (9) to the case of plane friction. The basic idea is to cut the contact area of the body and to reconnect it elastically (see Fig. 3).

4.1 Kinematics

First we describe the kinematics of the problem. In addition to the previously defined reference system CS_0 and body fixed system CS_b, a new coordinate system CS_Ω located in the contact area is introduced.

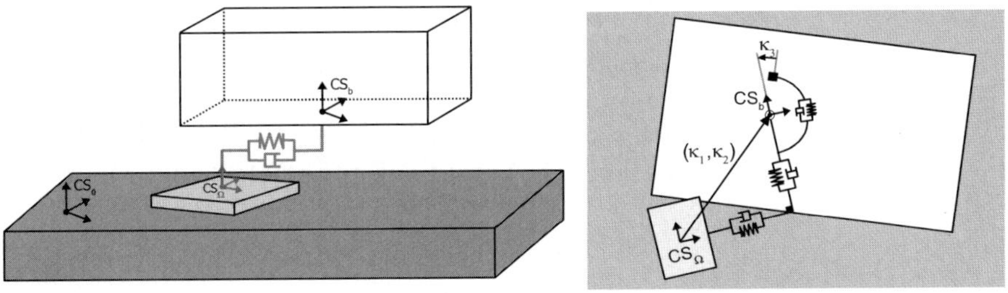

Fig. 3. Contact kinematics and the dynamic variables κ_1, κ_2 and κ_3.

The coordinate transformation from CS_b to CS_Ω depends on the three parameters $\boldsymbol{\kappa} = [\kappa_1, \kappa_2, \kappa_3]^T$ only. The transformation can be expressed in terms of the translationvector of the origins $\boldsymbol{r}_{\Omega b}^{\Omega}$ and the corresponding direction cosine transformation matrix $\mathbf{A}^{\Omega b}$

$$\boldsymbol{r}^{\Omega}_{\Omega b} = \begin{bmatrix} \kappa_1 \\ \kappa_2 \\ 0 \end{bmatrix} \quad \text{and} \quad \mathbf{A}^{\Omega b} = \begin{bmatrix} \cos(\kappa_3) & -\sin(\kappa_3) & 0 \\ \sin(\kappa_3) & \cos(\kappa_3) & 0 \\ 0 & 0 & 1 \end{bmatrix}. \tag{25}$$

The relevant friction velocity of the contact area can be written as

$$\Delta \boldsymbol{u} = \begin{bmatrix} 1 & 0 & \kappa_2 \\ 0 & 1 & -\kappa_1 \\ 0 & 0 & 1 \end{bmatrix} \left(\mathbf{A}^{\Omega b}\, \Delta \boldsymbol{v} - \dot{\boldsymbol{\kappa}}\right) = \mathbf{U} \left(\mathbf{A}^{\Omega b}\, \Delta \boldsymbol{v} - \dot{\boldsymbol{\kappa}}\right), \tag{26}$$

where $\Delta \boldsymbol{v}$ (see eq. 10) consists of the non vanishing two translational and one angular relative velocities of the body corresponding to the half space.

4.2 Half degree of freedom model for plane friction

In case of a vanishing mass of the contact area the equations of motion for the contact area are given by

$$\mathbf{D}\, \dot{\boldsymbol{\kappa}} + \mathbf{K}\, \boldsymbol{\kappa} - \boldsymbol{t}^{\Omega}_{R}(\Delta \boldsymbol{u}) = \mathbf{0},$$

where $\mathbf{D}$ and $\mathbf{K}$ need to be positive definite matrices. The conditions for sticking and sliding can be written in form of

$$\dot{\boldsymbol{\kappa}} = \mathbf{A}^{\Omega b}\, \Delta \boldsymbol{v} \quad \text{and} \quad g(\mathbf{D}\, \dot{\boldsymbol{\kappa}} + \mathbf{K}\, \boldsymbol{\kappa}) \leq \mu_H\, F_N$$

and

$$\dot{\boldsymbol{\kappa}} \neq \mathbf{A}^{\Omega b}\, \Delta \boldsymbol{v} \quad \text{and} \quad \mathbf{D}\, \dot{\boldsymbol{\kappa}} + \mathbf{K}\, \boldsymbol{\kappa} = {}^{S}\boldsymbol{t}^{\Omega}(\Delta \boldsymbol{u}).$$

By introducing a switching parameter $\alpha = \begin{cases} 0 & \text{for sticking} \\ 1 & \text{else} \end{cases}$ the two conditions can be combined

$$\dot{\boldsymbol{\kappa}} = (1 - \alpha) \cdot \mathbf{A}^{\Omega b}\, \Delta \boldsymbol{v} + \alpha \cdot \boldsymbol{v}_{\kappa} \tag{27a}$$

$$\mathbf{0} = \alpha \cdot \left(\mathbf{D}\, \boldsymbol{v}_{\kappa} + \mathbf{K}\, \boldsymbol{\kappa} - {}^{S}\boldsymbol{t}^{\Omega}(\Delta \boldsymbol{u})\right). \tag{27b}$$

Because the damping matrix $\mathbf{D}$ is positive definite, $\alpha \cdot \boldsymbol{v}_{\kappa} = -\alpha \cdot \mathbf{D}^{-1}\big(\mathbf{K}\, \boldsymbol{\kappa} - {}^{S}\boldsymbol{t}^{\Omega}(\Delta \boldsymbol{u})\big)$ follows from equation (27b) and finally the half degree of freedom model for plane friction

$$\boldsymbol{t}^{\Omega}_{R} = \mathbf{D}\, \dot{\boldsymbol{\kappa}} + \mathbf{K}\, \boldsymbol{\kappa} \tag{28a}$$

$$\dot{\boldsymbol{\kappa}} = (1 - \alpha) \cdot \mathbf{A}^{\Omega b}\, \Delta \boldsymbol{v} - \alpha \cdot \mathbf{D}^{-1} \left[\mathbf{K}\, \boldsymbol{\kappa} - {}^{S}\boldsymbol{t}^{\Omega}\left(\mathbf{U} \left[\mathbf{A}^{\Omega b}\, \Delta \boldsymbol{v} - \dot{\boldsymbol{\kappa}}\right]\right)\right]. \tag{28b}$$

4.3 The elasto-plastic model for plane friction

As in the one-dimensional case the half degree of freedom model contains an implicit differential equation of order one for the variables $\boldsymbol{\kappa}$. One could try to linearize the equations, which is quite complicated. Hence, as in Dupont

[2, 3] we just present the model and its properties. The so called elasto-plastic model for plane friction is defined by

$$\boldsymbol{t}_R^{\Omega} = \mathbf{D}\,\dot{\boldsymbol{\kappa}} + \mathbf{K}\,\boldsymbol{\kappa} \tag{29a}$$

$$\dot{\boldsymbol{\kappa}} = \mathbf{A}^{\Omega b}\,\Delta\boldsymbol{v} - \beta(d,s)\,\frac{\|\mathbf{U}\,\mathbf{A}^{\Omega b}\Delta\boldsymbol{v}\|}{\|{}^{S}\boldsymbol{t}^{\Omega}(\mathbf{U}\,\mathbf{A}^{\Omega b}\Delta\boldsymbol{v})\|}\mathbf{U}^{-1}\,\mathbf{K}\,\boldsymbol{\kappa}, \tag{29b}$$

with the switching parameter

$$\beta(d_H,s) = \begin{cases} 1 & \text{, if } s \geq 0 \text{ and } 1 < d \\ \dfrac{(d-0.99)^2(2.01-2\,d)}{0.01^3} & \text{, if } s \geq 0 \text{ and } 0.99 \leq d \leq 1 \\ 0 & \text{, else} \end{cases}$$

and $s = \boldsymbol{\kappa}\cdot\Delta\boldsymbol{v}$, $d = \frac{g(\mathbf{K}\,\boldsymbol{\kappa})}{(\mu_{Gl}+(\mu_H-\mu_{Gl})\exp(-v_m^2/v_S^2))F_N}$. The equations are posted in a way, that for constant velocities $\Delta\boldsymbol{v}$ the solution of the generalized elasto-plastic model (29) converges to the stationary solution of the half degree of freedom model (28).

5 Application to softfinger contact

The contact between a soft fingertip and a rigid object is called softfinger contact. For simulating a two finger grip of a rigid object witch such kind of contacts, the elasto-plastic model for plane friction was combined with the Hertzian model for normal contact (see [4]).

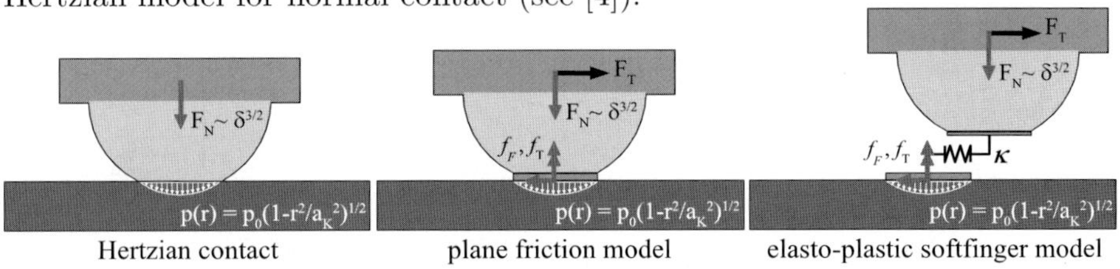

Fig. 4. The elasto-plastic softfinger model

In figure 5 some results for a typical simulation are illustrated, where a cube is grasped between two fingers. One clearly can see the effect of the friction torque caused by plane friction and also the visco-elasticity in the sticking phase.

Concluding Remarks & Acknowledgements

The presented results are part of the thesis [4], where more details and generalizations can be found. The work has been supported by the SFB 588 'Humanoid Robots – Learning and Cooperating Multimodal Robots' funded by the Deutsche Forschungs-Gemeinschaft (DFG).

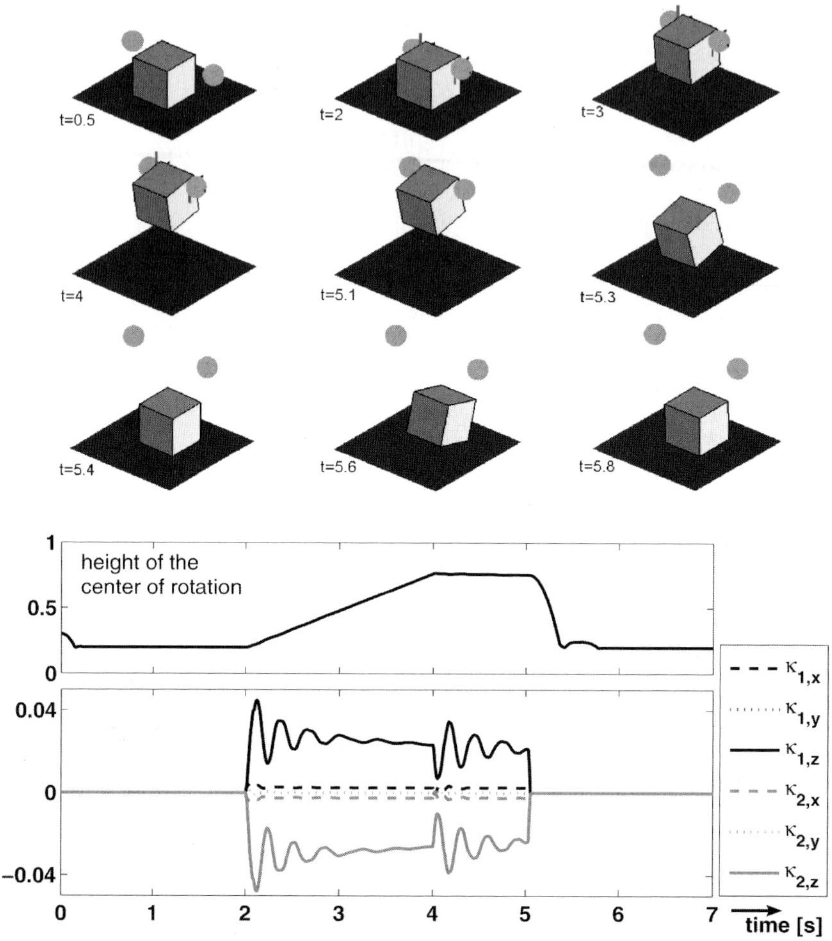

Fig. 5. Simulation of two finger grasping within MATLAB.

References

1. K.L. Johnson (1985). *Contact mechanics.* Cambridge University Press.
2. P. Dupont, B. Armstrong und V. Hayward (2000). *Elasto-Plastic Friction Model: Contact Compliance and Stiction.* Proceedings of the American Control Conference - Chicago, Illinois:1072–1077
3. P. Dupont, V. Hayward, B. Armstrong und F. Alpeter (2002). *Single State Elastoplastic Friction Models.* IEEE Transactions on Automatic Control, 47 (5):787–792
4. R. Keppler (2006) *Zur Modellierung und Simulation von Mehrkörpersystemen unter Berücksichtigung von Greifkontakt.* PhD. Thesis

Multibody Analysis of Axially Elastic Rod Chains

Erno Keskinen[1], Veli-Tapani Kuokkala[2], Taina Vuoristo[2], and Matti Martikainen[1]

[1] Tampere University of Technology, Laboratory of Machine Dynamics, P.O.Box 589, 33101, Tampere, Finland, erno.keskinen@tut.fi
[2] Tampere University of Technology, Laboratory of Material Characterization, P.O.Box 589, 33101, Tampere, Finland, veli-tapani.kuokkala@tut.fi

1 Introduction

Axially elastic rods are basic machine elements in hydraulic hammers, pilers and percussive drills [1]. The problem to analyze the motion history of such mechanisms is a very complex one, because the rods are simultaneously in large amplitude axial motion superimposed with a small amplitude elastic wave motion. The wave motion experiences division to reflected and transmitted components at each rod-rod interface depending on the current boundary stiffness [2]. The wave motion in each rod can be computed by finite elements or alternatively in space of semidefinite eigenfunctions. The feasibility of these methods in solving wave propagation problems in multi-rod systems with nonlinearly behaving rod-rod interfaces has been investigated and evaluated. The object of the experimental case study is a classical Hopkinson split bar apparatus [3] used in experimental analysis of material response to shock pulses. Another example representing a pile hammering system [4] evaluates the computational performance of the proposed approaches in long-term simulation of a complete work process.

2 Multibody modeling of a rod chain

The system to be analyzed consists of a set of rods, which are interconnected by nonlinearly behaving interface elements. The material and the cross-section are constant through the whole rod set in this case. In a general case, however, they may vary rod by rod or over each rod span. The mechanical model of a complete rod chain is illustrated in Fig. 1.

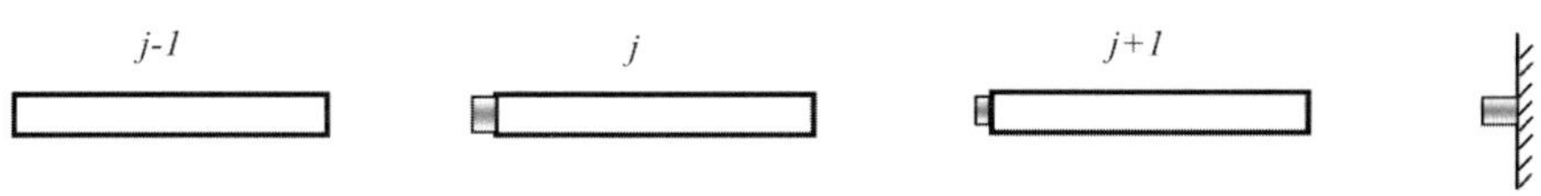

Fig. 1. A chain of rods with nonlinear interfaces.

Suppose a rod body belonging to the rod chain is dynamically loaded at left and right ends by compressive loads $L(t)$ and $R(t)$. If the rod has a constant cross-

Peter Eberhard (ed.), Multiscale Problems in Multibody System Contacts, 85–94.

section, the axial displacement motion *U(x,t)* of the rod cross-sections is governed by wave equation

$$\rho_j A_j \ddot{U}_j - E_j A_j U_j'' = 0 \tag{1}$$

subject to boundary conditions

$$E_j A_j U_j'(-\ell_j / 2) = -L(t) \tag{2a}$$
$$E_j A_j U_j'(\ell_j / 2) = -R(t) \tag{2b}$$

Due to the fact that the rod is not axially supported, it moves axially large distances as compared with the small displacements produced by the elastic strains. In order to split the motion to large and small amplitude components, a decomposition

$$U_j(x,t) = r_j + u_j(x,t) \tag{3}$$

is introduced, in which the large amplitude motion component

$$r_j = \frac{1}{\ell_j} \int_{-\ell_j/2}^{\ell_j/2} U_j(x)dx \tag{4}$$

represents the position of the instantaneous center of mass of the axially deformed rod while the small amplitude component u_j describes the vibratory motion. This is actually one form of the well-known floating frame representation, in which the frame is fixed to the dynamic center of mass of the elastic body. Based on the splitting (3) the wave equation (1) can now be transformed to separate equations for the motion of the mass center

$$\rho_j A_j \ell_j \ddot{r}_j = L_j(t) - R_j(t) \tag{5}$$

and to the nonhomogeneous wave equation

$$\rho_j A_j \ddot{u}_j - E_j A_j u_j'' = L\delta(x + \ell_j / 2) - R\delta(x - \ell_j / 2) \tag{6}$$

subject to free-free boundary conditions

$$u_j'(-\ell_j / 2) = 0 \tag{7a}$$
$$u_j'(\ell_j / 2) = 0 \tag{7b}$$

The wave propagation problem at each rod can be solved numerically using finite element discretization or in terms of eigenfunctions [5]. Both approaches can be used in total motion formulation (1)-(2) or in the floating frame formulation (5)-(7). The choice in this paper is to use finite elements in total formulation and eigenfunctions in floating frame formulation but it could be also different. What is essential, is the fact that both methods are able to describe the following phenomena:

- the motion of elastic stress waves along the bodies
- the motion of small-amplitude elastic displacement waves along the rod
- the large-amplitude rigid body motion of the rods.

3 Rod modelled by finite elements

The division of a prismatic rod into finite elements is a very straightforward procedure. Suppose one wants to use n_j elements. Then one gets n_j+1 unknown nodal displacements $u_i^j, i = 1,2,\ldots,n_j+1,$ to be solved from the dynamic equations of motion. If lumped mass description is used instead of consistent one the finite element rod model will be replaced by an equivalent spring-mass assemblage illustrated in Fig. 2.

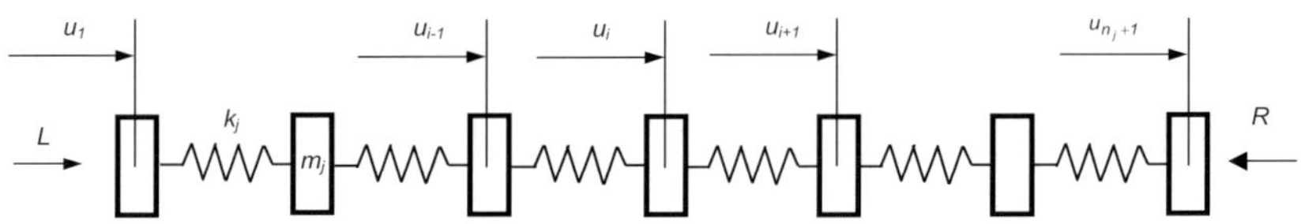

Fig. 2. Rod divided into finite elements.

If ℓ_j is the length of the rod number j in the chain, A_j the cross-sectional area and ρ_j the density, the mass of each finite element is

$$m_j = \frac{M_j}{n_j} \tag{8}$$

where the total rod mass is $M_j = \rho_j A_j \ell_j$. The axial stiffness constant is

$$k_j = n_j K_j \tag{9}$$

with $K_j = A_j E_j / \ell_j$ and E_j representing the effective elastic modulus of rod material. A free-body diagram of three neighboring elements in Fig. 3 shows how elements are interacting with each other.

The internal elastic force over ith spring element is

$$N_i^j = k_j(u_{i+1}^j - u_i^j) \qquad ; \quad i = 1,2,...,n_j \tag{10}$$

The lumped mass portions of the rod elements belonging to the nodal degrees of freedom may be divided by rules

$$\begin{aligned} m_1^j &= \tfrac{1}{2} m_j \\ m_i^j &= m_j \qquad\qquad ; \quad i = 2,3,...,n_j \\ m_{n_j+1}^j &= \tfrac{1}{2} m_j \end{aligned} \tag{11}$$

The nodal loads acting on the nodal degrees of freedom consist of internal elastic forces and external end loadings. The elastic forces are divided to the nodes by rules

$$\begin{aligned} U_1^j &= N_1^j \\ U_i^j &= N_i^j - N_{i-1}^j \qquad ; \quad i = 2,3,...,n_j \\ U_{n_j+1}^j &= -N_{n_j}^j \end{aligned} \tag{12}$$

The contribution of concentrated top and end loads to the loading vector is

$$\begin{aligned} F_1^j &= L(t) \\ F_{n_j+1}^j &= -R(t) \end{aligned} \tag{13}$$

Finally one can write the dynamic finite element equations for the rod motion

$$m_i^j \ddot{u}_i^j = F_i^j + U_i^j \qquad ; \quad i = 1,2,...,n_j + 1 \tag{14}$$

Numerical solution of system (14) for each rod j determines fully the motion history of the rod system. Once the equation system has been integrated with time, the axial stresses in the rod may be computed from the internal forces acting over the spring elements

$$\sigma_e^j = \frac{N_e^j}{A_j} = \frac{k_j}{A_j}(u_{e+1}^j - u_e^j) \tag{15}$$

in which e denotes the element number.

4 Rod modelled by eigenfunctions

As very well known vibratory motion of elastic bodies can be solved in space of eigenfunctions corresponding to the acting boundary conditions [5]. Because the motion of the rod is a combination of motion of the floating frame and axial oscillations, the eigenfunctions can be taken from the free-free boundary condition case, Eq. (5)-(7)

$$f_i^j(x) = \cos i\pi\left(x/\ell_j + ½\right) \tag{16}$$

where axial coordinate x varies in range $\left[-½\ell_j, ½\ell_j\right]$ from the rod top to the rod end. If the floating frame motion has been fixed to the dynamic center of gravity of the rod (point $x=0$), the floating frame position r and the unknown modal coordinates q_i^j, $i=1,\dots,n_j$, of the eigenmodes may be decoupled to separate differential equations, a property characteristic for semidefinite systems. The axial position of each cross-section in the rod may be computed from expansion

$$X(x,t) = x + r_j + \sum_i q_i^j(t) f_i^j(x) \tag{17}$$

where the rigid body position r_j is measuring the distance from a fixed reference point in the absolute frame. The floating frame equation of the rod is

$$M_j \ddot{r}_j = L - R \tag{18}$$

where, as in the FE formulation, L and R are the point loads acting on the rod ends. The vibratory motion is governed by uncoupled equation system

$$m_i^j \ddot{q}_i^j + k_i^j q_i^j = L - R(-1)^i \tag{19}$$

where the modal masses and stiffnesses have expressions

$$m_i^j = \frac{M_j}{2} \tag{20}$$

$$k_i^j = \frac{i^2\pi^2}{2} K_j \tag{21}$$

The dynamics of the rod motion is determined by solving differential equations (18) and (19). The stress distribution will be evaluated then by

$$\sigma(x,t) = E_j \sum_i q_i^j(t) f_i'^j(x) \tag{22}$$

For better accuracy, the boundary (end) stresses should be computed from the compressive forces over the interface elements.

5 Interface elements

In practical applications such as piling and percussive drilling [1],[4], overloading of the rods beyond yield strength can be avoided by protecting the rod ends by materials, whose yield strength is significantly lower than the one of the rod materials. These interface elements actually work also as pulse-shapers, as used often also in the Hopkinson split bar technique [7], because they modify the duration and magnitude of the stress pulses travelling through the rod system. A pulse-shaper manufactured from steel can be modelled by simple piecewise linear stress-strain rule

$$\begin{aligned} \sigma &= E\varepsilon \quad &; \quad \sigma < \sigma_y \\ \sigma &= \sigma_y + E'(\varepsilon - \sigma_y / E) \quad &; \quad \sigma > \sigma_y \end{aligned} \tag{23}$$

with elastic modulus E, yield strength σ_y and slope factor E' for the plastic zone. If polymers are used in the interfaces, the simplest way is to use the two- parameter Kelvin-Voigt model

$$\sigma = E\varepsilon + \eta\dot{\varepsilon} \tag{24}$$

For higher strains, however, many polymers behave nonlinearly. Because the stress-strain curve gets a degressive nature, a simple model with still a small number of unknown parameters to be fixed is based on power rule

$$\sigma = \sigma_u(1 - |1 - \varepsilon / \varepsilon_u|^m) \tag{25}$$

for stresses $\sigma < \sigma_u$. In contrast to polymers, many fibre or particulate reinforced composites having a soft matrix material exhibit progressive stress-strain dependence at higher strains. For composites used for instance as cover materials in paper machine rolls, a well-working model is based on a simple power rule equation

$$\sigma = \sigma_m(\varepsilon / \varepsilon_m)^\beta \tag{26}$$

with three unknown parameters σ_m, ε_m and β to be fixed. The material models are linked to rod motions by end force equations

$$R_{j-1} = L_j = \sigma(\varepsilon, \dot{\varepsilon})A \tag{27}$$

in which the strain input is computed from the rod end motions by

$$\varepsilon = \frac{u_{j-1}(\ell_{j-1}/2) - u_j(-\ell_j/2)}{h} \tag{28}$$

where h is the thickness of the interface element.

6 Stress wave analysis of the Hopkinson split bar

A Hopkinson split bar [6] has been developed to produce shock waves, which are dynamically loading the specimen, whose response behaviour is under investigation. The apparatus consists of a striker rod (or 'bar'), incident rod, transmitted rod, and optionally of a damper rod. The lay-out of the rod system is given in Fig. 3.

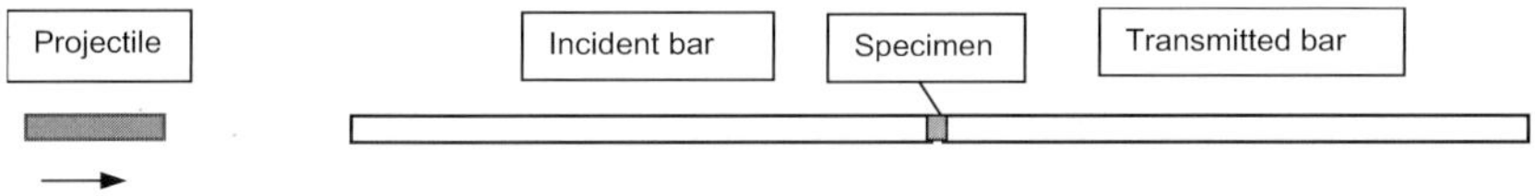

Fig. 3. Hopkinson split spar apparatus.

Two strain gages are used to measure the axial strains in the middle span of the incident and transmitted rods. The impulsive loading is produced by a projectile, which is shot by pressurized air to hit the free end of the first rod. As a consequence of the impact, the three bodies are executing a coupled motion as a multi-body system acting on each other by impulsive contact forces. Due to the axial elasticity of the rods, the rods are vibrating during the rigid body motion. Vibratory motion is related to the propagation of axial stress waves inside the rods. At the end interfaces the waves are split into reflected and transmitted components depending on the boundary condition (free end, hard contact or soft contact). The evaluation of the proposed analysis methods was carried out as a case study on an existing Hopkinson split bar apparatus in the laboratory of Materials Science at Tampere University of Technology. The Hopkinson split bar technique is widely used for material tests at strain rates in the range of 10^2 to 10^4 s^{-1}. The analyzed system consists of two *1.2 m* long AISI 4340 steel rods with a diameter of *21.77 mm* [8]. The unknown parameters of a particulate reinforced epoxy composite have been identified from the recorded response. The most important system parameter is the end velocity v of the projectile just before the hit. The characteristic properties of the specimen material and the corresponding system settings are listed in Table 1.

Behaviour	Parameters	Specimen	Projectile
Progressively elastic, nonlinearly plastic	$\sigma_m = 60$ MPa $\varepsilon_m = 0.023$ $\beta = 1.5$	$\phi = 18.97$ mm $h = 8.67$ mm	$\ell_1 = 0.2$ m $v = 3.9$ m/s $E_I = 4.54$ J

Tab. 1. Properties of specimen and pulse shaper and parameters of system.

The three different responses, the experimental one, the one computed with finite elements, and the one evaluated using modal expansions are then compared with respect to the strain outputs in the locations of the strain gages of the experimental unit.

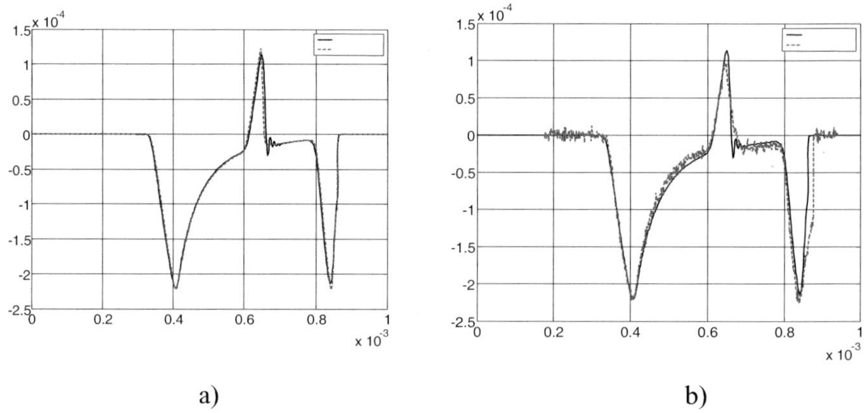

Fig. 4. Histories of middle span strains in transmitted bar: a) eigenfunction solution = dashed line, finite element solution = uniform line, b) experimental solution = dashed line, finite element solution = uniform line.

In the first comparison, the response curves of strains in the transmitted bar corresponding to eigenfunction (dashed line) and FE (uniform line) responses have been plotted in Fig. 4a. The numerically obtained response curves show an excellent match.

In the second comparison, the response curves of strains in the transmitted bar corresponding to experimental (dashed line) and FE (uniform line) responses have been plotted in Fig. 4b. Again, the experimental and computed responses are very close to each other.

7 Analysis of pile hammering process

A complete piling system consists of submodels for the pile, ram, ram lifting actuator, and pile guiding system. Pile cushion body is also needed for softer force transmission from ram to pile top. In shock wave analysis the pile has to be modelled as an axially elastic beam. The division of a prismatic pile into finite elements is a well-known method in foundation engineering [4], [9]. As the hitting force from ram to pile is transmitted through a set of anvil, cap and cushion bodies, the whole spring-mass assemblage representing the system is rather complicated as illustrated in Fig. 5.

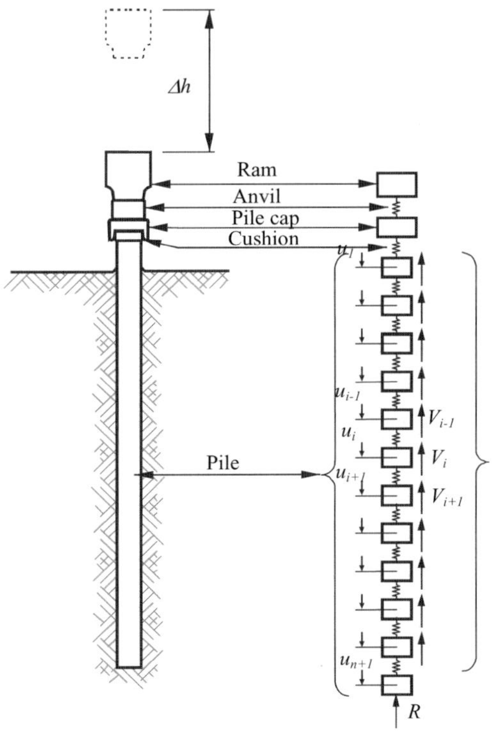

Fig. 5. Pile divided into finite elements

The test system simulated consists of a *10 m* long steel pile of mass *1000 kg* driven by ram body of *1000 kg*. The soil type is hard cohesive clay. Ram is lifted *1 m* in *0.5 s* time period and left to fall freely for hitting the pile top, after which the sequence is repeated. Record length is *100 s* and the time step used is fixed to $2 \cdot 10^{-5} s$. The pile has been divided into *20* elements in finite element solution and modelled using *6* free-free eigenfunctions. The pile penetration plotted in Fig. 6 shows excellent matching of responses obtained by finite elements (a) and eigenfunctions (b) during time evolution of the piling process.

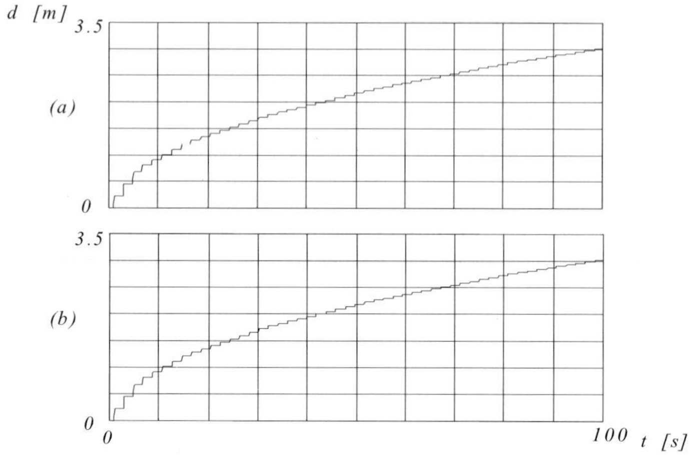

Fig. 8. Comparison of pile penetration histories for a) *6* eigenfunctions and b) *20* finite elements.

References

1. Keskinen, E.K., Keskiniva, M., Raitaniemi M., and Raunisto Y. (1996) "Design and Analysis of a Multipurpose Double-Acting Hammering Unit for Hydraulic Excavators." Proc. of the 3rd biennal ASME ESDA '96 conf., Montpellier .

2. Doyle, J.F. (1997) Wave Propagation in Structures, Springer, New York.

3. Kolsky, H. (1949), "An investigation of the mechanical properties of materials at very high rates of loading," Proceedings of the Physical Society Section B, 62, pp 676-700.

4. Koerner, R.M. (1984) Construction and Geotechnical Methods in Foundation Engineering. McGraw-Hill, N.Y.

5. Meirovitch, L. (1986) Elements of Vibration Analysis, McGraw-Hill, Singapore.

6. Nemat-Nasser S. (editor) (2000) "High strain rate testing", in ASM handbook Vol. 8, Mechanical testing and evaluation, ASM International, Materials Park, Ohio, pp. 425-496.

7. Frew D. J., Forrestal M. J., Chen W. (2002) "Pulse shaping techniques for testing brittle materials with a Split Hopkinson pressure bar" Experimental Mechanics, 42, pp. 93-106.

8. Vuoristo, T. and Kuokkala, V. -T. (2002) "Creep, recovery and high strain rate response of soft roll cover materials" Mechanics of Materials, 34, pp. 493-504.

9. Bowles, J.E. (1988) Foundation Analysis and Design, McGraw-Hill, Singapore.

On a Continuous Transfer of History Variables for Frictional Contact Problems Based on Interpretations of Covariant Derivatives as a Parallel Translation

Alexander Konyukhov and Karl Schweizerhof

Institut für Mechanik, Universität Karlsruhe
Kaiserstrasse 12, 76131, Karlsruhe, Germany
Konyukhov@ifm.uni-karlsruhe.de Schweizerhof@ifm.uni-karlsruhe.de

Summary. Regularization methods based on the penalization of the tangent displacements are among the most exploited techniques in combination with finite element methods to model frictional interactions. Usually the global tangent displacements are described via convective coordinates which are e.g. used in a finite element discretization of the contact surface. These displacements serve to compute the tangent tractions in the case of sticking via a regularization procedure as well as in the case of sliding via a return-mapping scheme. The convective coordinates of the contact point as well as the corresponding tangent tractions are considered as history variables and have to be stored for each load step. In this contribution, we discuss the particular issue of continuous transfer for history variables in the case of large deformation problems adapted for the covariant contact description developed in Konyukhov and Schweizerhof [4]. Some specific examples are chosen to illustrate the effect of incorrect transfer for both non-frictional and frictional problems and, therefore, the necessity of the proposed techniques.

1 Introduction

The formulation of the frictional contact conditions in rate form is an essential part in the development of efficient algorithms. Several approaches are known in literature [6], [10]: these are a) the Lagrangian multiplier method, b) various regularization techniques such as the penalty method and the augmented Lagrangian method, and c) combined methods such as Nitsches method. In the general case of large deformation contact problems, contacting bodies are meshed with dissimilar meshes. The necessary information about contact interactions such as contact tractions for both sticking and sliding cases, and sliding displacements are assigned to contact pairs, e.g., to nodal contact pairs within the node-to-node approach, the latter of course only in the case of small deformation. Separate assignments either to a master segment, or to a slave

Peter Eberhard (ed.), Multiscale Problems in Multibody System Contacts, 95–101.

segment are normally chosen for large displacement problems, mainly due to robustness reasons. Thus, a problem of information passing, or, so-called, the problem of transfer of history variables between master and slave segments, neighboring masters segments or neighboring slave segments occurs. Of course, this transfer has to be done in accordance with both the equilibrium conditions on the contact surface, and the corresponding interface conditions. The first conditions are normally directly fulfilled in contact mechanics, while in fluid-structure interaction problems, they are treated separately known as conservative load transfer conditions, see [2]. The second, interface conditions include the treatment of a constitutive law defined between surfaces, e.g., the Coulomb friction law, together with a proper numerical scheme, such as the return-mapping scheme.

2 Transfer algorithm for a trial force

The description in convective variables has been introduced in contact mechanics for careful treatment of frictional contact problems in particular to avoid artefacts due to improper discretization [7]. The tangent surface displacements are described via convective variables and, in due course, they are used to compute the tangent tractions in the case of sticking via a regularization procedure as well as in the case of sliding via a return-mapping scheme. Within the master-slave approach, the convective coordinates are defined, of course, on the master side, while the information about the contact forces should be supplied for both, the master and the slave sides. Problems arise, when a slave segment is crossing several master segments during deformation, see Fig. 1.

Let us assume, that at step (n) a slave point $S^{(n)}$ is projected onto the master segment $\mathbf{A}^{(n)}$, giving a contact point $C^{(n)}$, defined via convective coordinates ξ_A^1, ξ_A^2. At load step $(n+1)$, the same slave point is moved into position $S^{(n+1)}$ and projected onto the neighboring master segment $\mathbf{B}^{(n+1)}$ giving a contact point $\mathbf{C}^{(n+1)}$, defined via convective coordinates ξ_B^1, ξ_B^2. Since the convective coordinates ξ_A^1, ξ_A^2, resp. ξ_B^1, ξ_B^2 are defined only locally on each contact element, a problem concerning the correct transfer of history variables while crossing the element boundaries arises for large displacement problems. A typical example is the computation of components for a trial tangent force

$$T_i^{(n+1)} = T_i^{(n)} - \epsilon_T a_{ij} \Delta\xi^j \tag{1}$$

leading to a maximum jump in tangent forces as $2\epsilon_T$, see details in [5].

Various approaches can be exploited to solve this problem. Wriggers et. al. [9] proposed an algorithm for the 2D case, based on the usage of the path length of the projection point C. In this case the full geometrical information about the path passed by the slave point has to be stored. Puso and Laursen [8] proposed to determine increments of convective coordinates in the geometric form for the 3D case. In this case 3D coordinates of slave nodes are

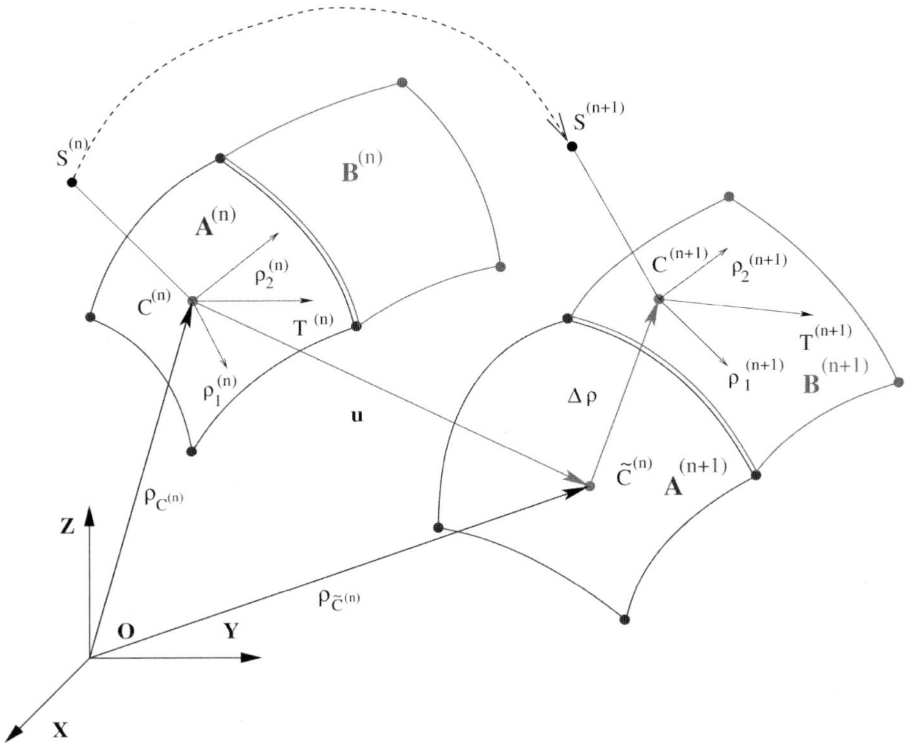

Fig. 1. Contact point moving across element boundaries. Covariant derivatives. Sketch of integration scheme.

stored, but the geometry of the contact segments is no longer captured. As an alternative we consider a continuous transfer algorithm of first order with respect to convective coordinates based upon the interpretation of covariant derivatives as a parallel translation. A pair of points $S^{(n)}$ and $C^{(n)}$ defines then a spatial coordinate system corresponding to the closest point projection. For the terminology and detailed information we refer to the covariant description given in [4]. The following steps lead to a transfer algorithm for trial forces:

a) *The tangent components* of a traction vector $T^{(n)}$ on the master segment $\mathbf{A}^{(n)}$ have to be transformed in tensor form into *the tangent components* on the master segment $\mathbf{B}^{(n)}$.
b) The increment vector $\boldsymbol{\Delta\rho}$ has to be expressed in the surface metrics of the current contact segment $\mathbf{B}^{(n+1)}$.

The increment vector $\boldsymbol{\Delta\rho}$ is obtained in the global reference Cartesian system as

$$\boldsymbol{\Delta\rho} = \boldsymbol{\rho}_{C^{(n+1)}}|_{\xi_A^1,\, \xi_A^2} - (\boldsymbol{\rho}_{C^{(n)}} + \mathbf{u}_{C^{(n)}})\,|_{\xi_B^1,\, \xi_B^2}, \tag{2}$$

where the first term is representing the current position of the contact point $C^{(n+1)}$, while the terms in brackets can be recomputed knowing only two previous convective coordinates ξ^1_A, ξ^2_A as well as the previous segment **A**. According to the aforementioned steps, the continuous analog of a trial step in eqn. (1) is build as follows

$$T^j_{B,\,(n+1)} = T^i_{A,\,(n)}(\boldsymbol{\rho}^A_i \cdot \boldsymbol{\rho}^j_B) - \epsilon_T(\Delta\boldsymbol{\rho} \cdot \boldsymbol{\rho}^j_B), \tag{3}$$

where $\boldsymbol{\rho}^j_B$ are current contravariant basis vectors computed for segment **B** at a point ξ^1_B, ξ^2_B; $\boldsymbol{\rho}^A_i$ are covariant basis vectors computed for the previous segment **A** at a point ξ^1_A, ξ^2_A; the tangent traction vector components T^j are taken in contravariant form for both current and previous segments. The standard return mapping algorithm for the Coulomb friction is applied then to compute the real frictional force.

The update algorithm for sliding displacements, see details in [5], has to be adjusted according to the continuous transfer algorithm, namely, the new coordinates of the elastic region have to be recomputed with new sliding displacements as

$$\Delta\xi^i_{sl} = (\Delta\boldsymbol{\rho}_{sl} \cdot \boldsymbol{\rho}^j_B), \tag{4}$$

where $\Delta\boldsymbol{\rho}_{sl}$ is computed in analogy to eqn. (2), but in the center of the elastic region ξ^1_{sl}, ξ^2_{sl}.

The algorithm is straightforward if history variables are stored at the master segment, as for the transfer of information between neighboring master segments within a node-to-segment approach. Within other approaches, the current transfer technique has to be modified including an algorithmic coupling with searching procedures.

We focus here on details of the transfer algorithm only. The full algorithm for contact problems based on an iterative Newton scheme, the structure of the corresponding tangent matrices and residuals and details of the finite element implementation are given in [3], [4], [1].

3 Numerical example.

In this section, we consider a special contact case, for which the application of the described scheme is absolutely necessary. An elastic strip **AD**, see Fig. 2 ($E = 2.1 \cdot 10^4$, $\nu = 0.3$, thickness $h = 0.5$, length $L = 24$) is positioned at the beginning of a channel with width $a = 13$. The channel itself is modeled by two rigid blocks $\mathbf{B_1}$ and $\mathbf{B_2}$. For contact the side part of the channel is modeled as a rigid surface described by an analytical function. The strip is loaded incrementally by a prescribed displacement v at the center until it is inserted into the channel, see Fig. 3. The strip here is modeled with 24 linear solid-shell elements, see Harnau et al [1], and due to symmetry only one half of the system is modeled. The crucial point during the analysis is the sliding of a sharp corner **C** over the element boundaries **1**, **2**, **3**, see Fig. 3. A load-displacement curve computed for the loading point is chosen as the representative parameter to study the effect of proposed approach. The following parameters are taken: load increment $v = 0.0025$, penalty parameters $\varepsilon_N = 2.1 \cdot 10^5$, $\varepsilon_T = 2.1 \cdot 10^5$ and friction coefficient $\mu = 0.2$. A straightforward analysis without the continuous transfer scheme leads to a disconvergence for frictional contact problems, see Fig. 4. For non frictional problem, though it is a converging case, the load-displacement curve contains jumps caused by the slave node passing from one master segment to another. The application of the continuous transfer scheme, however, allows to obtain the full force-displacement curve, even in the part when the strip is fully inserted into the channel, see the straight part of the curves in Fig. 4. For comparison, the analysis is carried out also with different friction coefficients $\mu = 0.1$ and $\mu = 0.3$.

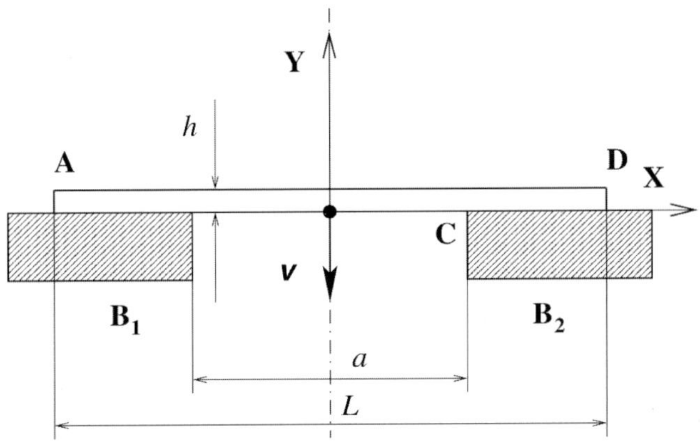

Fig. 2. Drawing of an elastic strip into a channel with sharp corners. Definition of geometrical parameters.

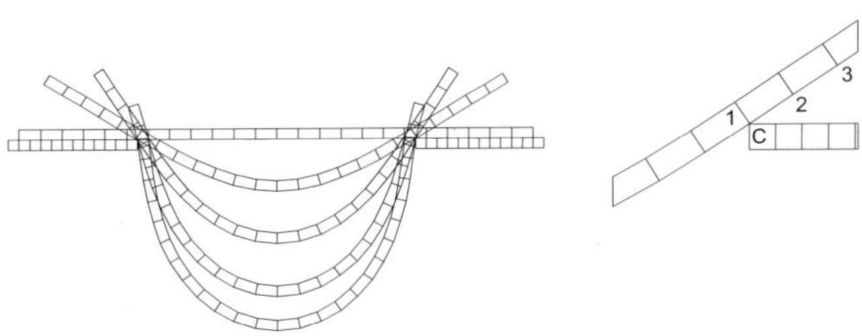

Fig. 3. Sequence of deformations for the elastic strip. Nodes are sliding over the sharp corner **C**.

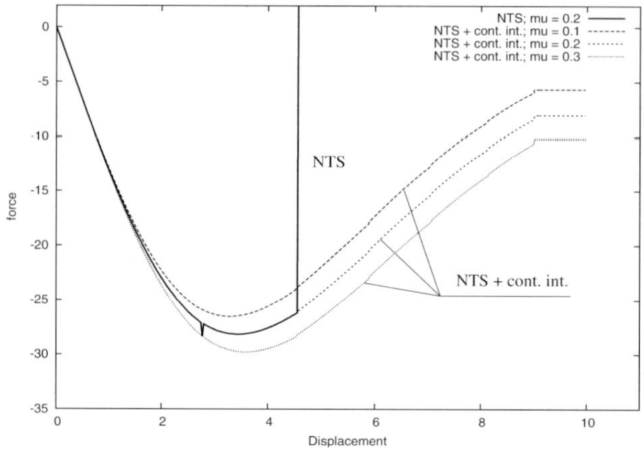

Fig. 4. Drawing of an elastic strip over a sharp corner. Load-displacement curve. Frictional case – various friction coefficients. Straightforward NTS approach and continuous transfer scheme.

4 Conclusions

A continuous transfer algorithm for history variables based on the interpretation of covariant derivatives as a parallel translation is developed for frictional contact problems. The necessity of such a procedure is shown by a particular numerical example. The possibility to transfer other history variables in a similar fashion is also discussed for the example of updated sliding displacements; for further examples we refer to [5].

References

1. Harnau M, Konyukhov A, Schweizerhof K (2005) Algorithmic aspects in large deformation contact analysis using "Solid-Shell" elements. Comput and Struct 83(21-22): 1804–1823.
2. Jaiman RK, Jiao X, Geubelle PH, Loth E (2005) Assessment of conservative load transfer for fluid-solid interfaces with non-matching meshes. Int J Numer Meth Engng, 64:2014–2038.
3. Konyukhov A, Schweizerhof K (2004) Contact formulation via a velocity description allowing efficiency improvements in frictionless contact analysis. Comput Mech 33:165–173.
4. Konyukhov A, Schweizerhof K (2005) Covariant description for frictional contact problems. Comput Mech 35:190–213.
5. Konyukhov A, Schweizerhof K (2005) A special focus on 2D formulations for contact problems using a covariant description. Int J Numer Meth Engng 66:1432–1465.

6. Laursen TA (2002) Computational Contact and Impact Mechanics. Fundamentals of Modeling Interfacial Phenomena in Nonlinear Finite Element Analysis. Springer: New-York, Heidelberg, Paris.
7. Laursen TA, Simo JC (1993) A continuum-based finite element formulation for the implicit solution of multibody large deformation frictional contact problems. Int J Numer Meth in Engng 35:3451–3485.
8. Puso MA, Laursen TA (2002) A 3D contact smoothing method using Gregory patches. Int J Numer Meth Engng, 54:1161-1194.
9. Wriggers P, Krstulovic-Opara L, Korelc J (2001) Smooth C^1-interpolations for two-dimensional frictional contact problems. Int J Numer Meth Engng, 51:1469–1495.
10. Wriggers P (2002) Computational Contact Mechanics, Wiley: Chichester.

First Integrals of Equations of Motion of a Heavy Rotational Symmetric Body on a Perfectly Rough Plane

Alexander Kuleshov

Department of Mechanics and Mathematics, Moscow State University, Main building of MSU, Leninskie gory, Moscow 119992, Russia
kuleshov@mech.math.msu.su

Summary. We consider the problem of the motion of a heavy dynamically symmetric rigid body bounded by a surface of rotation on a fixed perfectly rough horizontal plane. The integrability of this problem was proved by S.A. Chaplygin [1]. Chaplygin has found that the equations of motion of given mechanical system have, besides the energy integral, two first integrals, linear in generalized velocities. However, the explicit form of these integrals is known only in the case, when the moving body is a nonhomogeneous dynamically symmetric ball. In the case, when the moving body is a round disk or a hoop, the integrals, linear in the velocities, are expressed using hypergeometric series [1],[2],[3] . In the paper of Kh.M. Mushtari [4] the investigation of this problem was continued. For additional restrictions, imposed on the surface of moving body and its mass distribution, Mushtari has found two particular cases, when the motion of the body can be investigated completely. In the first case the moving rigid body is bounded by a surface formed by rotation of an arc of a parabola about the axis, passing through its focus, and in the second case, the moving body is a paraboloid of rotation. For other bodies, bounded by a surface of rotation and moving without sliding on a horizontal plane, the explicit form of additional first integrals is unknown.

In this work we find some new cases when all the integrals of the problem can be expressed explicitly, when the surface of moving body satisfies to a Mushtari condition. The set of surfaces of moving bodies satisfying to this condition is described.

1 Formulation of the problem and the equations of motions

Suppose a rigid body of the mass m, symmetrical in shape and mass distribution about the $G\zeta$ axis, while its centre of mass is situated at G, leans on the fixed horizontal plane Oxy at the point M. We will use the following notation: θ is the angle between the axis of symmetry of the body and the vertical, β is the angle between the meridian $M\zeta$ of the body and a certain fixed meridian plane, and α is the angle between the horizontal tangent MQ of the meridian

Peter Eberhard (ed.), Multiscale Problems in Multibody System Contacts, 103–110.

$M\zeta$ and the Ox axis. The position of the body will be completely defined by the angles α, β and θ and the x and y coordinates of the point M.

In addition, we will introduce the system of coordinates $G\xi\eta\zeta$ moving both in space and in the body so that the $G\xi$ axis always lies in the plane of vertical meridian, while the axis $G\eta$ is perpendicular to this plane (Fig. 1). Let the velocity $\mathbf{v}$ of the centre of mass G, the angular velocity vector $\boldsymbol{\omega}$ of the body, the angular velocity vector $\boldsymbol{\Omega}$ of the trihedron $G\xi\eta\zeta$, and the reaction of the plane $\mathbf{R}$ are specified in the system of coordinates $G\xi\eta\zeta$ by the components v_ξ, v_η, v_ζ; p, q, r; Ω_ξ, Ω_η, Ω_ζ and R_ξ, R_η, R_ζ, respectively. The moment of inertia of the body with respect to the $G\xi$ and $G\eta$ axes is denoted by A_1 and its moment of inertia with respect to the axis of symmetry is denoted by A_3.

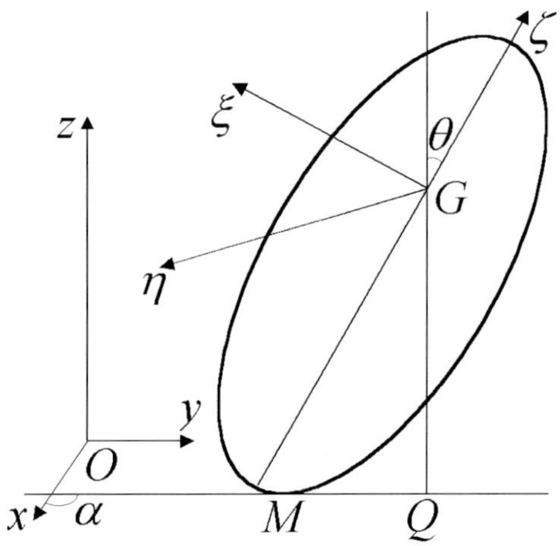

Fig. 1.

Note [1]-[2], that the distance GQ from the centre of mass and to the Oxy plane will be a function of the angle θ i.e. $GQ = f(\theta)$. The coordinates ξ, η and ζ of the point M where the body touches the plane in the system of coordinates $G\xi\eta\zeta$ will also be functions of only the angle θ, where $\eta = 0$, and (for more details see [1])

$$\xi = -f(\theta)\sin\theta - f'(\theta)\cos\theta, \quad \zeta = -f(\theta)\cos\theta + f'(\theta)\sin\theta \tag{2.1}$$

Since the $G\zeta$ axis is fixed in the body, we have $\Omega_\xi = p$ and $\Omega_\eta = q$. The plane $G\xi\zeta$ will all the time be vertical; therefore, $\Omega_\zeta - \Omega_\xi \cot\theta = 0$. The velocity of the point of contact M is equal to zero, and consequently

$$v_\xi + q\zeta = 0, \quad v_\eta + r\xi - p\zeta = 0, \quad v_\zeta - q\xi = 0$$

The theorem of the motion of the centre of mass as a projection on the axis $G\eta$ and the theorem of the change in the angular momentum for the axes $G\xi$ and $G\zeta$ after simple transformations, give

$$\frac{d\,(p\zeta - r\xi)}{dt} - pq\,(\zeta\cot\theta + \xi) = \frac{R_\eta}{m},$$
$$A_1\frac{dp}{dt} + (A_3 r - A_1 p\cot\theta)\,q = -\zeta R_\eta, \quad A_3\frac{dr}{dt} = \xi R_\eta \tag{2.2}$$

Omitting henceforth the special case when θ = const and taking into account that $q = -d\theta/dt$, we can eliminate R_η from system (2.2) and obtain

$$A_1\frac{dp}{d\theta} + A_3\frac{\zeta}{\xi}\frac{dr}{d\theta} = -A_1 p\cot\theta + A_3 r,$$
$$\zeta\frac{dp}{d\theta} - \frac{A_3 + m\xi^2}{m\xi}\frac{dr}{d\theta} = -\left(\zeta\cot\theta + \xi + \zeta'\right)p + \xi' r. \tag{2.3}$$

Hence, from system (2.3) we can determine two first integrals of the equations of motion of the body that are linear in p and r. Presently an explicit form of these integrals can only be obtained in the case, when a moving body is a nonhomogeneous dynamically symmetric ball. Indeed, let a dynamically symmetric ball of the radius R moves on a horizontal plane. We suppose, that the centre of mass of a ball does not coincide with the geometrical centre. Then the centre of mass will be located on the axis of symmetry of a ball. Let d is a distance from the centre of mass of a ball to its geometrical centre. Then we have

$$f(\theta) = R - d\cos\theta, \quad \xi = -R\sin\theta, \quad \zeta = d - R\cos\theta$$

and from the system (2.3) we can find two first integrals (for more details see [1])

$$A_1 p\sin\theta + A_3 r\left(\cos\theta - \frac{d}{R}\right) = c_1 = \text{const}, \tag{2.4}$$

$$r\sqrt{A_1 A_3 + mR^2\left(A_1\sin^2\theta + A_3\left(\cos\theta - \frac{d}{R}\right)^2\right)} = c_2 = \text{const}. \tag{2.5}$$

Integral (2.4) is known as the Jellett integral [5]. It expresses the law of preserving of scalar product of the vector of angular momentum of a ball and the vector $\boldsymbol{GM}$. Integral (2.5) is usually called as the Chaplygin integral. Its physical meaning is unknown.

In this paper we point out some new cases when, in the problem of the motion of a dynamically symmetric rigid body, one can obtain first integrals that are linear in p and r in explicit form.

2 Derivation of the first integrals

The system of equations (2.3) can be represented in the form

$$\frac{d\tau}{d\theta} = h(\theta)r, \qquad \frac{dr}{d\theta} = z(\theta)\tau. \tag{3.1}$$

Here we have used the following notation:

$$\tau = m\sqrt{\Delta}\left[A_1 p \sin\theta + \frac{(A_3\zeta - A_1\xi')\sin\theta}{\xi + \zeta'} r\right],$$

$$h(\theta) = m\sqrt{\Delta}\left(A_3 \sin\theta + \frac{d}{d\theta}\left(\frac{(A_3\zeta - A_1\xi')\sin\theta}{\xi + \zeta'}\right)\right), \qquad z(\theta) = \frac{\xi(\xi + \zeta')}{\Delta^{3/2}\sin\theta},$$

$$\Delta = A_1 A_3 + A_1 m\xi^2 + A_3 m\zeta^2.$$

We will assume that the surface of the moving body and the mass distribution in it are such that the following condition is satisfied

$$\frac{(A_3\zeta - A_1\xi')\sin\theta}{\xi + \zeta'} = A_3(\cos\theta + \ell), \tag{3.2}$$

where ℓ is an arbitrary constant.

Condition (3.2) in slightly different form was written for the first time in the paper by Kh.M. Mushtari [4]. It was shown there that when this condition is satisfied the moving body could make a permanent rotations with constant angular velocity about the axis of symmetry: $r = r_0 = \text{const}$. We will call this condition as Mushtari condition. In our case, when this condition is satisfied we have $h(\theta) \equiv 0$, and from the first equation of system (3.1) we obtain

$$\tau = \tau_0 = \text{const}.$$

After some simplification, this integral can be presented in the form:

$$\sqrt{A_1 A_3 + A_1 m\xi^2 + A_3 m\zeta^2}\,[A_1 p \sin\theta + A_3 r(\cos\theta + \ell)] = c_1. \tag{3.3}$$

Note, that the structure of the integral (3.3) reminds the Chaplygin integral (2.5).

After finding integral (3.3) it is easy to obtain the other first integral in explicit form. Since $\tau = \tau_0 = \text{const}$, then from the second equation of system (3.1), we have

$$\frac{dr}{d\theta} = z(\theta)\tau_0.$$

Consequently, Eqs (3.1), in addition to integral (3.3), also allow of the following integral

$$r - mc_1 \int z(\theta)d\theta = c_2. \tag{3.4}$$

Hence, when condition (3.2) is satisfied the equations of motion of a dynamically symmetric body on a perfectly rough plane allow of first integrals (3.3) and (3.4). We will search now the shape of the moving body which satisfies condition (3.2).

3 Determination of the shape of the body surface

Consider condition (3.2) and assume initially, that $\ell = 0$. Substituting expressions (2.1) for ξ, ζ and their derivatives into condition (3.2) and introducing the dimensionless parameter $k = A_3/A_1$, we find that the function $f(\theta)$, which defines the shape of the body surface, must satisfy the equation:

$$(k-1)f'' \sin\theta\cos\theta - kf' + (k-1)f\sin\theta\cos\theta = 0. \tag{4.1}$$

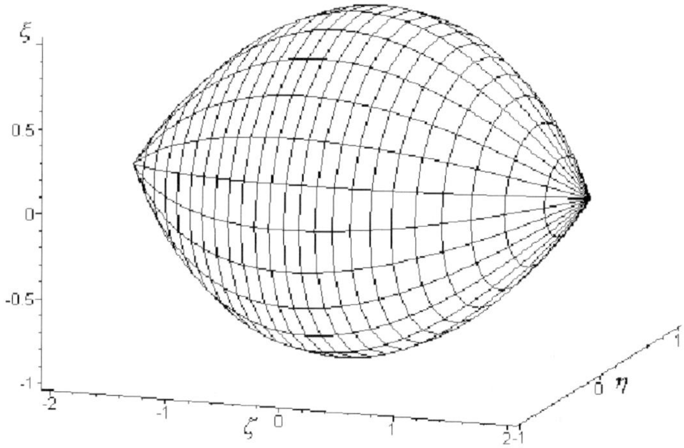

Fig. 2.

It is easy to show, that Eq.(4.1) has two particular solutions

$$\begin{aligned} &1)\quad A_3 = \frac{2}{3}A_1, \quad f(\theta) = \frac{\lambda}{\sin\theta}, \quad \xi = \frac{\lambda\cos^2\theta}{\sin^2\theta} - \lambda, \quad \zeta = -\frac{2\lambda\cos\theta}{\sin\theta}, \\ &2)\quad A_3 = 2A_1, \quad f(\theta) = \frac{\lambda}{\cos\theta}, \quad \xi = -\frac{2\lambda\sin\theta}{\cos\theta}, \quad \zeta = \frac{\lambda\sin^2\theta}{\cos^2\theta} - \lambda, \end{aligned} \tag{4.2}$$

where λ is an arbitrary constant. These particular solutions were pointed out for the first time by Mushtari in [4]. In the first case the moving body is bounded by a surface formed by rotation of an arc of a parabola about the axis, passing through its focus (Fig. 2), and in the second case, the moving body is a paraboloid of rotation (Fig. 3).

We will investigate what other solution this equation has. We will put

$$f(\theta) = \frac{g(\theta)}{\cos\theta}.$$

Then, we can write Eq.(4.1) for the function $g(\theta)$ as follows:

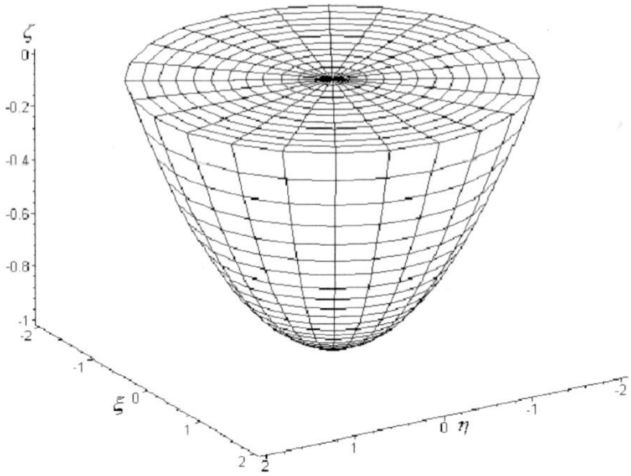

Fig. 3.

$$g'' + 2\left[\frac{\kappa}{\sin\theta\cos\theta} - \frac{\cos\theta}{\sin\theta}\right]g' + \frac{2\kappa}{\cos^2\theta}g = 0; \quad \kappa = \frac{(k/2-1)}{k-1}.$$

Making the change of independent variable in Eq.(4.3) in accordance with the formula

$$w = 1/\cos^2\theta,$$

we reduce it to the Gauss hypergeometric equation [6]-[7]

$$w(1-w)\frac{d^2g}{dw^2} + \left[2 - \left(\frac{3}{2} + \kappa\right)w\right]\frac{dg}{dw} - \frac{\kappa}{2}g = 0.$$

Hence, Eq.(4.1) has the non-trivial particular solution

$$f_0(\theta) = \frac{1}{\cos\theta}F\left(\frac{1}{2}, \kappa, 2; \frac{1}{\cos^2\theta}\right)$$

and the general solution of Eq.(4.1) can be written as follows:

$$f(\theta) = f_0(\theta)\left(\lambda + \mu\int\frac{(\tan\theta)^{k/(k-1)}}{f_0^2(\theta)}d\theta\right), \tag{4.4}$$

where λ and mu are the arbitrary constants.

Here $F = F(a, b, c; z)$ is the Gauss hypergeometric function [6]-[7]. We will show how one can obtain particular solutions (4.2) from the general solution (4.4). To do this we will make two assumptions. Suppose $\mu = 0$ in formula (4.4). Then the function $f(\theta)$ is given by the formula

$$f(\theta) = \frac{\lambda}{\cos\theta}F\left(\frac{1}{2}, \kappa, 2; \frac{1}{\cos^2\theta}\right)$$

We will assume that the function F in this expression can be represented by a finite sum, i.e. one of its parameters is equal to a negative integer or zero [6]-[7]. In the case considered this means $\kappa = -N$, where N is a natural number or zero. Hence we obtain

$$k = 2(N+1)/(2N+1) \tag{4.5}$$

Using the formula (see [6]-[7])

$$F(a,b,c;z) = (1-z)^{-a} F\left(a, c-b, c; z(z-1)^{-1}\right)$$

we can also conclude that the hypergeometric series will be a finite sum, if the number $c-b$ is equal to a negative integer. In the case considered this condition gives $2-\kappa = -N$, whence we find

$$k = 2(N+1)/(2N+3) \tag{4.6}$$

When $N=0$, from Eq.(4.5) we obtain $k=2$, and it follows from Eq.(4.6) that $k=2/3$. Consequently, the particular solutions (4.2), given for the first time by Mushtari in [4], are obtained from the general solution (4.4) when $\mu = 0$ and $N = 0$. In the general case the set of functions which satisfy Eq.(4.1) is defined by formula (4.4) and has the power of a continuum.

We will now assume that the constant $\ell \neq 0$ in condition (3.2). In this case the equation for determining the function $f(\theta)$ will have the form

$$\sin\theta\left((k-1)\cos\theta + k\ell\right) f'' - k(1+\ell\cos\theta) f' + (k-1)\sin\theta\cos\theta f = 0. \tag{4.7}$$

Making the replacement of the independent variable in (4.7) in accordance with the formula

$$w = \cos^2(\theta/2)$$

we reduce it to the equation

$$w(w-1)(w-a_1)\frac{d^2 f}{dw^2} + \Big[(\alpha_1+\beta_1+1)w^2 - \\ - \Big[\alpha_1+\beta_1+1+a_1(\gamma_1+\delta_1)-\delta_1\Big]w + a_1\gamma_1\Big]\frac{df}{dw} + (\alpha_1\beta_1 w - q_1) f = 0 \tag{4.8}$$

known as Heun's equation [8] with the parameters

$$a_1 = -\frac{1}{4\gamma_1}, \quad \alpha_1 = -\beta_1 = 1, \quad q_1 = -\frac{1}{2},$$

$$\gamma_1 = \frac{1}{2(1+k(\ell-1))}, \quad \delta_1 = \frac{1}{2(1-k(\ell+1))}$$

When $|a_1| \geq 1$ and $\gamma_1 \neq 0, -1, -2, -3, \ldots$ the solution of Eq.(4.8) can be represented in the form of a series

$$F(a_1, q_1, \alpha_1, \beta_1, \gamma_1, \delta_1; w) = \sum_{n=0}^{\infty} c_n w^n \tag{4.9}$$

the coefficients of which are defined by the recurrence formulae

$$c_0 = 1, \quad a_1 \gamma_1 c_1 = q_1$$

$$a_1 (n+1)(\gamma_1 + n) c_{n+1} = \left[a_1 (\gamma_1 + \delta_1 + n - 1) + \alpha_1 + \beta_1 - \delta_1 + n + \frac{q_1}{n}\right] n c_n -$$

$$- \left[(n-1)(n-2) + (n-1)(\alpha_1 + \beta_1 + 1) + \alpha_1 \beta_1\right] c_{n-1}$$

This series necessarily converges when $|w| \leq 1$. Note, however, that the solutions of the Heun equation (4.8) are much less well understood than hypergeometric functions. No general integral representation for them is known, for instance.

Hence, when $\ell = 0$ the surface of the moving body, what satisfies condition (3.2), is determined using the hypergeometric function (see formula (4.4)), and when $\ell \neq 0$ it is determined using Heun's function (4.9).

References

1. Chaplygin S.A. (2002) Regular & Chaotic Dynamics 7:119–130
2. Appell P. (1900) Rendiconti del circolo matematico di Palermo 14:1–6
3. Vierkandt A. (1892) Monatshefte für Mathematik und Physik 3:31–54, 97–134
4. Mushtari Kh.M. (1932) Mat. Sbornik 39:105–126 (Russian)
5. Jellett J.H. (1872) A treatise on the Theory of Friction. MacMillan, London Dublin
6. Gradshtein I.S., Ryzhik I.M. (1970) Tables of Integrals, Sums, Series and Products. Academic Press, New York
7. Whittaker E.T., Watson G.N. (1927) A course of modern analysis. Cambridge University Press, Cambridge
8. Ronveaux A. (1995) Heun's Differential Equations. Oxford University Press, Oxford

Complementary Models for Log Grinding

Sirpa Launis[1], Erno Keskinen[2], Claude Bohatier[3], and Frédéric Dubois[3]

[1] John Deere Forestry
Lokomonkatu 21, Box 474, FIN-33101 Tampere, Finland
LaunisSirpaS@johndeere.com
[2] Tampere University of Technology, Institute of Machine Design
P.O. Box 589, FIN-33101 Tampere, Finland erno.keskinen@tut.fi
[3] LMGC, UMR CNRS 5508, Université Montpellier II
cc 048, 34095 Montpellier cedex 5, France bohatier@lmgc.univ-montp2.fr, dubois@lmgc.univ-montp2.fr

Summary. Pulp for making paper is produced in the groundwood (GW) process from logs. In batch-type GW-machines the control of the log loading is a demanding task because of the complicated log group dynamics where also the heterogeneous wood material has a great influence on the batch behaviour. It is essential to achieve better understanding of the regularities occurring in the process in order to improve the loading control and thus the quality and efficiency of pulp production. For this purpose a numerical model of the complete process has been built up and the dynamic system behaviour has been studied by time domain simulation. In this paper the focus is in the description of the log batch behaviour. Three separate approaches have been adopted to model the batch.

1 Introduction

Groundwood process is a common method to produce pulp for papermaking from logs. In the process, batches of coaxial logs are fed with hydraulically actuated shoe body between guiding walls against a rotating grinding stone, as illustrated in figure 1. The feeding system has a feedback loop to maintain constant conditions in the wood-stone interface but as the behaviour of the log batch is very complicated, the control of the force is a difficult problem. In order to improve the quality control of the production and to reduce the energy required by the process, the grinding machine construction should be developed so that the nonlinearities in the log feeding can be compensated by the load control. To support the development work, a coupled simulation model of the complete process has been built up including the mechanical parts of the machine, the hydraulic and electrical actuators and their feedback controllers, the processed raw material, i.e. the logs, and the various interactions between all of these components. By computer simulation it is possible to study such things, as the effect of the log size on the process, the effect of structure design and, moreover, to develop and test different actuator controllers [1].

Peter Eberhard (ed.), Multiscale Problems in Multibody System Contacts, 111–121.

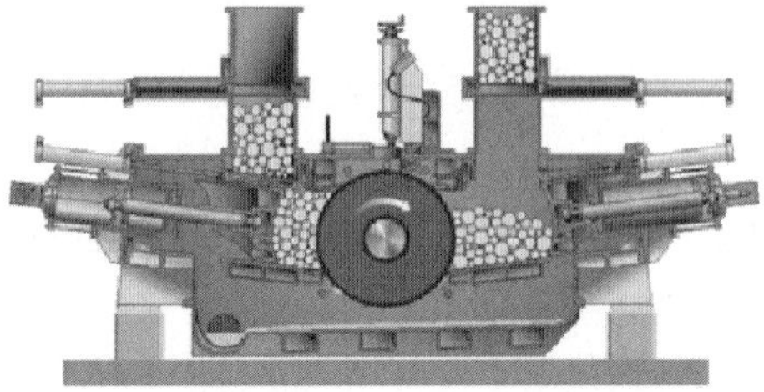

Figure 1. Typical pocket-type pressure groundwood grinder (Metso Co).

The focus of this paper is the description of the dynamical behaviour of the log batch and the material removal from the batch during grinding. We describe three approaches that have been adopted to model the log batch behaviour. Different approaches are necessary because the log processing is a multiscale problem as described in the following. Our approach is limited to the plane representation of the log behaviour. The chosen plane is the one in perpendicular to the grains of the wood.

2 Complementary Log Batch Models

The behaviour of the log batch during grinding is complicated because both small and large time scale phenomena are present in the system. In the beginning of the log loading sequence the batch is loose. As measured within short time periods the contact forces governing log movements are changing discontinuously when collisions and separations are taking place. In order to study local arching and collapses, the log group has to be considered as a multibody system (figure 2a). During the loading the logs are gradually packed tightly and under heavy compression (figure 2b). In this case the fluctuation caused by void space in the wood batch is minimized and the wood batch progresses towards the grinding stone more like a uniform body than a group of separate logs. The long period time history of the resulting normal force against the stone follows a regular trend although the fluctuation is relatively high due to the remaining multibody dynamical effects of the log group. Additional scale to the problem is added by the phenomena happening near the grinding contact, the material flow in the contact zone etc (figure 2c). In this sense, the log batch behaviour is characterized as a multiscale problem and complementary batch models are needed to find out information of the different phenomena.

We have chosen three separate approaches to study the different scales of the log batch problem :

1. The phenomenological approach with multibody representation for individual logs.
2. The phenomenological approach with one-dimensional continuum material model for wood.
3. The mesoscopic approach where the logs are discretised to smaller particles.

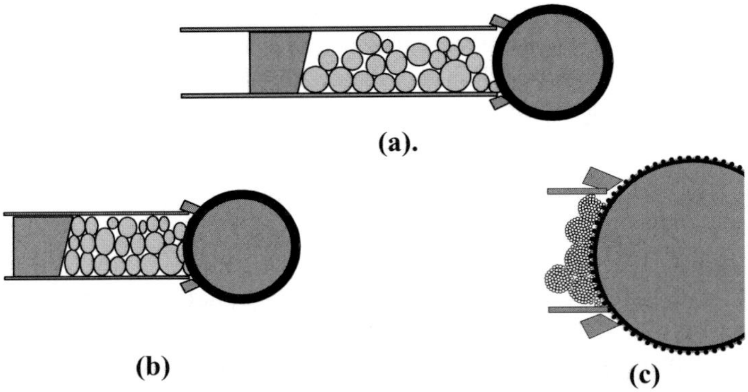

Figure 2. Different stages of log batch grinding sequence: (a) loose batch, (b) tightly packed batch and (c) log crushing in grinding contact zone.

2.1 Phenomenological Approach with Multibody Representation for Logs

The first of the approaches describes the log batch as a multi-body object with various interactions between logs and structural parts of the machine. By "phenomenological" we mean that the grinding contact is described with a phenomenological model for abrasive wear. Logs behave as rigid bodies whereas contacts between them are nonlinearly elastic. Also damping is introduced in the contact representation. The equation of motion of log *i* is given by

$$\mathbf{M}_i\ddot{\mathbf{q}}_i = \mathbf{G}_i + \sum_{j=1}^{n,\, j\neq i}\left(\mathbf{N}_{l\_ij} + \mathbf{Q}_{l\_ij} + \mathbf{T}_{l\_ij}\right) + \sum_{j=1}^{3}\left(\mathbf{N}_{s\_ij} + \mathbf{Q}_{s\_ij} + \mathbf{T}_{s\_ij}\right) + \mathbf{N}_i + \mathbf{Q}_i + \mathbf{T}_i, \tag{1}$$

where the state vector $\mathbf{q}_i = \{x_i \quad y_i \quad \theta_i\}^T$, mass matrix $\mathbf{M}_i = [m_i \quad m_i \quad J_i]_{diag}$, $\mathbf{G}_i$ is the gravitation force vector, $\mathbf{N}_{l\_ij}$, $\mathbf{Q}_{l\_ij}$ and $\mathbf{T}_{l\_ij}$ are the force and torque components in the contact between log *i* and *j*, $\mathbf{N}_{s\_i}$, $\mathbf{Q}_{s\_i}$ and $\mathbf{T}_{s\_j}$ are the components in the contact between log *i* and the structural parts, i.e. the shoe and the chamber walls whereas $\mathbf{N}_i$, $\mathbf{Q}_i$ and $\mathbf{T}_i$ are the components of the grinding contact .

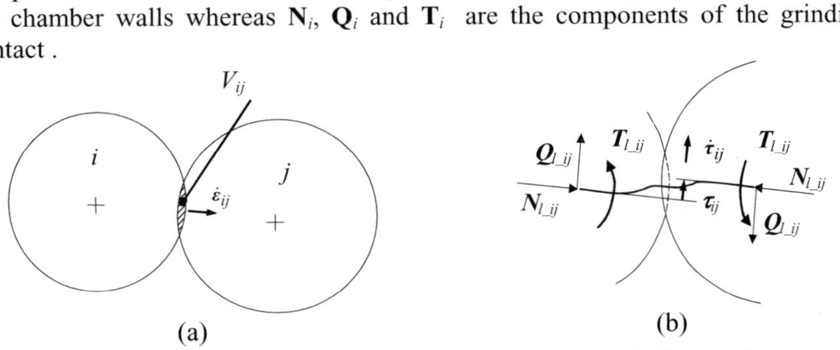

Figure 3: (a) Compression in normal direction. (b) Shear deflection in tangential direction.

The contact force calculation presented in the following is based on a soft crush model between two inhomogeneous spheres [2,3]. The theory is similar to the Winkler elastic foundation model in [4]. The resultant normal force in the log-log interface may be calculated from

$$N_{l\_ij} = A_{proj} B \frac{\Delta V_{ij}}{V_{ij}} + c\dot{\varepsilon}_{ij}, \tag{2}$$

where A_{proj} is the projected compression area in the contact, B is the bulk modulus of the wood material, $\Delta V/V$ is the fractional change in volume due to compression in the contact, c damping coefficient and $\dot{\varepsilon}$ the rate of compression. V is based on the dynamical state of the logs. A simplification has been made, that the overlapping volume of two logs is equal to the total volume change. The direction of the force N_{l_ij} is normal to the line that goes through the intersection points of the fraction lines of the intersecting logs.

The shear force in the log-log contact is approximated by adapting the equations given in [5]. The maximum shear force is given by

$$Q_{\max\_ij} = \mu_s g(|\overline{v}|) N_{l\_ij}, \tag{3}$$

where μ_s is the static friction coefficient, $g(\overline{v})$ is a dimensionless friction function which gives the friction force dependence of sliding speed between contacting logs. Equation 3 is equal to the static friction force when relative velocity between the log surfaces is zero. Linear relation given by Eq. 3 is known to be valid only at small tangential forces. Due to the effect of partial slip the tangential elasticity has a degressive nature. The actual shear force is calculated

$$Q_{l\_ij} = Q_{\max\_ij} \overline{Q}_{l\_ij}, \tag{4}$$

where $\overline{Q}_{l\_ij}$ is the dimensionless partial shear force function given by

$$\overline{Q}_{l\_ij} = \operatorname{sgn}(\overline{\tau}_{ij}) \left[1 - \left(1 - |\overline{\tau}_{ij}|\right)^m\right], \tag{5}$$

where the dimensionless tangential deflection is the ratio of the actual shear deflection τ_{ij} and maximum shear deflection $\overline{\tau}_{ij} = \tau_{ij} / \tau_{\max\_ij}$. The exponent parameter m can be used for adjusting the order of degression.

The interaction between individual logs and grinding stone is covered by wear equation for abrasive wear, where the rate of material removal is proportional to the sliding distance of the stone and the normal force between the stone and each log. It is assumed that the stone surface conditions remain constant during simulated grinding periods, which is quite realistic in modern machines where stone surface is washed and sharpened frequently with online high-pressure water showers. The amount of removed material in grinding as well as contact penetration volume between different parts, which is needed in contact force calculation, are solved from log positions by discretising (using the Delaunay triangulation) the cross-sectional shape logs and using numerical methods to solve the overlapping area between contacting elements. An implicit time-integration method is used to solve the system equations. The resultant normal force between the stone and a log is given by wear equation for abrasive wear [6]

$$N_i = \frac{V_i H}{k_{abr} \tau}, \tag{6}$$

where V is the removed volume, H is the hardness of the wood material, τ is the sliding distance and k_{abr} is a non-dimensional wear coefficient, a material and surface pair dependent factor and may be identified with measurements for example with pin-disk machine. V is based on the logs' iterated dynamical state. The normal force, N, and the total tangential force, Q, have linear connection

$$Q_i = \frac{c_e k_{abr}}{H} N_i, \tag{7}$$

where c_e is the special energy consumption per unit volume [1].

2.2 Phenomenological Approach with One-Dimensional Continuum Material Model for Wood

In the second approach, a fluid like representation is used for the compressible wood flow under material removing conditions. The material model for wood is symmetrically viscoelastic homogeneous one-dimensional continuum with average bulk modulus $\hat{E}$ and viscosity η [7]. We use one of the simplest viscoelastic material models, which is the Kelvin-Voigt spring-dashpot model [1, 8], consisting of a spring and dashpot in parallel. The normal force distribution in the batch follows an exponential law meaning that the ratio between the total grinding force N applied on the grinding stone and pushing force F of the shoe body is given by

$$\frac{N}{F} = e^{-\alpha L}, \tag{8}$$

where $\alpha = \dfrac{2\mu}{h}$, factor μ combines the effects of internal (wood-wood) and external (wood-wall) frictions and h is the height of the chamber. Relation (8) shows clearly that a remarkable loss of efficiency in force transmission is present in the pushing system. This phenomenon is known as the silo-effect in the mechanics of granular materials (Janssen's model). Interesting observation available here, from the designers' point of view, is that the ratio between the length and height of the batch affects to the compression distribution.

In the pushing system, the wood continuum ought to be pushed along the grinding pocket in such a way that the wearing process in wood-stone contact will continue in constant conditions. The movements and state of compression of the wood batch are depending on the time evolution of the balance between the rates the wood is pushed and machined. What dynamically happens has to be solved from the dynamic state equation governing the state of compression in the wood batch [1]. The rate of the normal force N is given by

$$\dot{N} = \frac{A\alpha}{e^{\alpha L} - 1}\left\{\hat{E}\left[\dot{x} - \left(1 - \frac{N}{A\hat{E}}e^{\alpha L}\right)\dot{s}\right] + \eta(\ddot{x} - \ddot{s})\right\}, \tag{9}$$

where $\dot{x}$ is the pushing speed and $\dot{s}$ is the wear rate. State equation (9) of compression force connects the dynamics of pushing and grinding processes together providing a well-established basis for process-dynamical simulations of groundwood systems.

The interaction between stone and wood is described as an abrasive wearing process. The momentum transfer from stone to wood batch is governed by energy equation and the material transfer out of the batch is governed by a similar wear equation to (6). The relation between the normal force and the tangential force in grinding becomes

$$Q = c_e k_w N\,. \tag{10}$$

2.3 Mesoscopic Approach

In the third approach, i.e. in the mesoscopic model, the chosen scale is the one of "grain bundles". In this prospective approach, the logs, in their initial configuration, are formed by small circles each of which represents a bundle of grains. The bundles are held together by cohesion forces. The numerical simulations are based on the Contact Dynamics (CD) method [9,10] implemented in the LMGC90 program (http://www.lmgc.univ-montp2.fr/~dubois/LMGC90). Even if the original CD method considers rigid bodies, and frictional contact, it is relevant to model complex deformation of a granular medium as a whole. The CD method is based on a dedicated non smooth dynamical motion equation, which, once discretized, is able to cope with multiple contact events in a time step. The integration scheme is implicit and frictional contact laws are not regularized. Classically contact condition should be written as the called Signorini condition.

Here we need to transform these conditions due to the presence of cohesion [11]. We adapt a simple model dedicated to wet granular:

$$\begin{cases} R_n - c \geq 0,\ g \geq 0,\ (R_n - c)g = 0, & if\ g \leq g_c \\ R_n = 0, & if\ g > g_c \end{cases} \tag{11}$$

which means that when the gap between two particles is small enough the particles will attract each other until the gap closes. This model was adapted so that the material removal is accomplished when the contact force between a grain bundle and the grinding stone exceed the cohesion forces, the bundle is torn off from the log. Once a cohesion link breaks it cannot evolve again but the contact law changes to the non-cohesive one. In tangential direction the Coulomb friction force is applied by one particle to another.

What ever is the cohesive model, its parameters choice is not obvious for the wood material because the cohesion force is difficult to measure or to find in the common literature. More research work is still needed in the modelling the initial state of the grain bundles i.e. the description of the residual stresses of the logs in the mesoscopic model. In addition, the model lacks the torsion but its role is probably not very important in case of soft fibres.

3 Application Examples

3.1 Design of Test Bench

A small scale laboratory test-bench has been developed to verify the models. The test bench utilizes the inverse load sensing method for elastic structures. The simulation model was used to design the strain gage positioning in the test bench [12]. In figure 5 is illustrated the simulated heavy compression state of the log batch in the test setting and corresponding loadings of the guiding walls. The thickness of the guiding walls was dimensioned to be such that the strains would be inside a well-measurable range.

Similarly, the loadings of the real machine structures may be studied by dynamic simulations and further by giving the resulting forces as a force input for example for a finite element analysis of the structures.

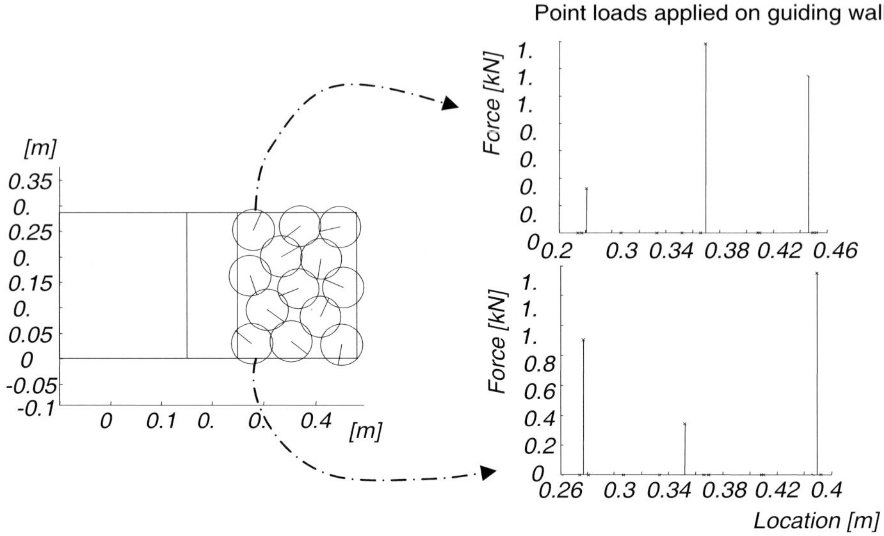

Figure 5. Test bench loadings were computed by simulation.

3.2 Alternations in Pushing Force and Model Based Control

In [1] was presented three different PD-controllers which were based on the phenomenological one-dimensional continuum model. The first of the developed controllers is aimed to maintain constant the total normal force in the batch-grinding stone contact, the second one to maintain constant production rate and the third one to maintain constant grinding stone shaft line power. In figure 6 are shown the corresponding simulated hydraulic pressures during the grinding of one batch with each control schemes. The continuum material model was used in these simulations.

The resulting curves are quite similar with each controller. Noticeable here is that the simulated curves are very smooth. This is not the case in figure 7, where is shown a measured curve of the hydraulic pressure in the Metso Co. pilot grinder where test measurements were driven with ordinary (not model based) power control. There can be distinguished five separate batches from the curve in figure 7. The fluctuation in the hydraulic pressure of the pushing cylinder is high. This is due to the multibody effects and crushing of the logs which cannot be modelled with the continuum material model. If these effects would be taken into account in the control system, the unnecessarily high pushing forces could be reduced. For example, in figure 8 there is a pushing force curve, which was simulated with the multibody representation for logs. In this simulation case, constant speed was maintained in the pushing and no material was removed from the batch. In the curve we can see a sudden drop somewhere around two and half seconds. Before the drop there is a significantly steep rise in the force. These phenomena are due to a bridge formation and its collapse.

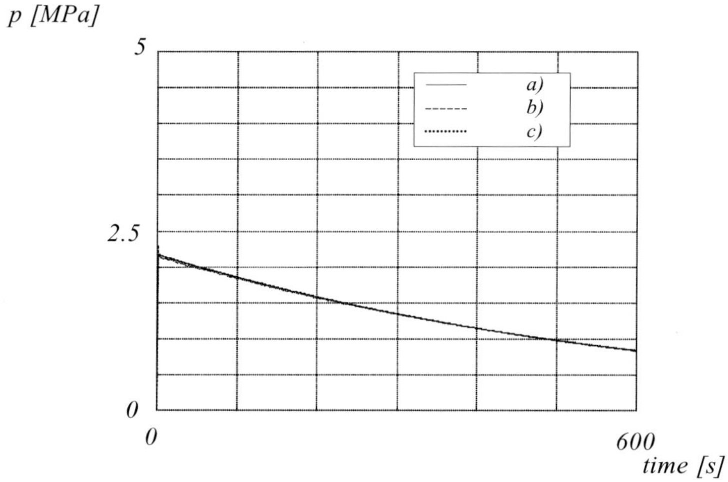

Figure 6. Simulated hydraulic pressure using model based controllers.

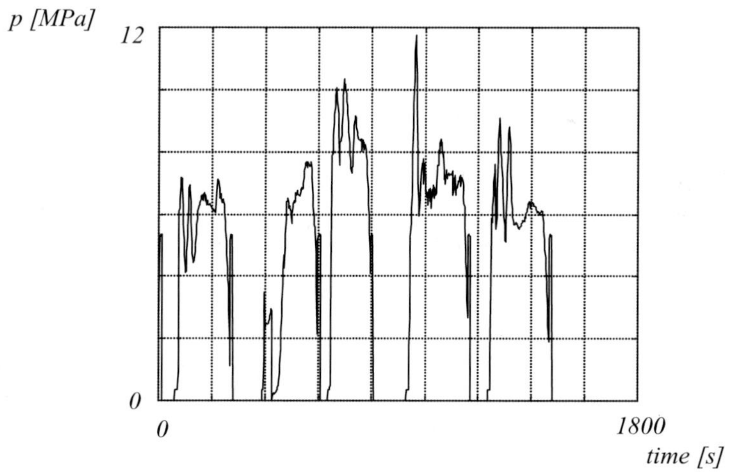

Figure 7. Measured hydraulic pressure in Metso Co. pilot grinder with power control.

If the pushing control were based on an average model and traced, does the pushing force actually follow a corresponding predicted force curve, we could forecast from sudden rises of the force, that we are building a bridge and lifting up the pushing force thus unnecessarily, because the force is not transmitted to the grinding stone. In this case it would be wiser to pull the shoe body backwards instead in order to collate the bridge formation.

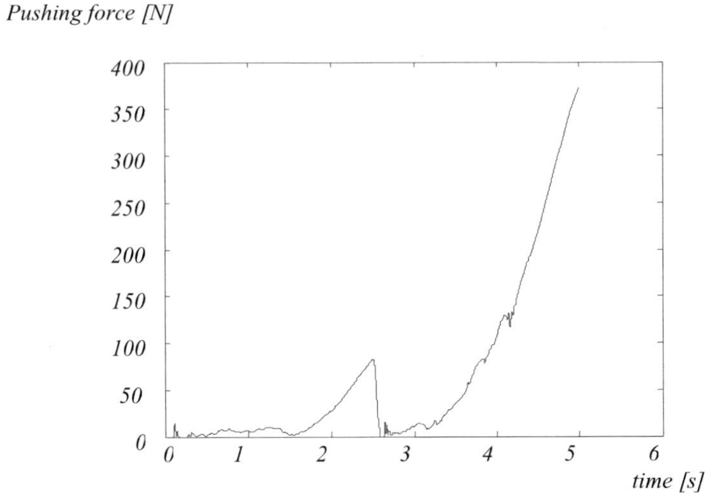

Figure 8. Simulated constant velocity pushing force of unground batch.

4 Conclusion

The dynamical behaviour of the log batch and the material removal from the batch during the groundwood process has been modelled with three different approaches in order to study the phenomena of different scales taking place during the batch-wise grinding sequence. The first of the approaches describes the batch as a multibody object in the scale of individual logs to study the group behaviour of the logs in the batch. In the second approach, a fluid like representation is used for the compressible wood flow under material removing conditions. This model explains the regular trend of the resulting normal force against the stone during the long period time history. In the third, mesoscopic, approach, the chosen scale is the one of "grain bundles". This scale has been chosen in particular to study the material flow and crushing in the contact zone. The models are implemented into the process level model on computer to use virtual simulation as a tool to develop the process control and machine structures. Some suggestions of the future development line of the log pushing controllers were made.

References

1. Launis S., Keskinen E., Cotsaftis M. and Raitaniemi M. (1999): Nonlinear Dynamics and Control of Log Pushing in Groundwood Systems. Proc. of the IASTED International Conference Intelligent Systems and Control, Santa Barbara, California, pp 102-108
2. Johnson K. L. (1985): Contact mechanics. Cambridge University Press, Cambridge

3. Duran J. (2000): Sands, Powders and Grains, An Introduction to the Physics of Granular Materials. Springer-Verlag, New York
4. Lubkin J. L. (1962): Contact Problems, in Handbook of Engineering Mechanics by Flügge,W. (Ed.). McGraw-Hill, New York
5. Keskinen E. K. (1994): A Contact Dynamics Formulation for Distributed Simulation of Flexible Hydromechanical Systems. Tampere University of Technology Publications 134, Tampere, Finland
6. Bhushan B. (1999): Principles and Applications of Tribology. John Wiley & Sons Inc., U.S.A
7. Launis, S., Keskinen, E.K. & Cotsaftis, M. (2001): Dynamics of Wearing Contact in Groundwood Manufacturing System, First M.I.T. Conference on Computational Fluid and Solid Mechanics, Cambridge, Massachusett, USA.
8. Lakes, R. S. (1999): Viscoelastic Solids. CRC Press LLC, Florida
9. Moreau, J.J. (2002): Numerical Dynamics of Granular Materials, in Contact Mechanics, edited by Martins, J.A.C., Monteiro Marques, Proceedings of the 3rd Contact Mechanics International Symposium, Peniche, Portugal, 2001. Kluwer Academic Press, Netherlands.
10. Jean M. (1999): The non-smooth contact dynamics method. Comput. Methods Appl. Mech. Engrg. 177, pp 235-257
11. Chetouane, B., Dubois, F., Vinches, M. and Bohatier, C. (2004): NSCD Discrete Element Method for Modeling Masonry Structures, International Journal for Numerical Methods in Engineering 2004; 04:1–27.
12. Launis S., Kivinen J.-M., Miettinen J. and Keskinen E. (2005): Equipment for Experimental Testing of Log Batch Compression. Proc of the 17th International Wood Machining Seminar, Rosenheim, Germany.

The Admissible Set for a Robotic System with Unilateral Constraint

Caishan Liu[1,2], Zhen Zhen[3], and Bin Chen[1]

[1] Department of Mechanics & Engineering Science, Peking University, Beijing, 100871, China liucs@pku.edu.cn
[2] State key Laboratory of Turbulence and Complex Systems
[3] Beijing Institute of Graphic Communication

Summary. This paper focuses on investigating systematically the phenomenon of the bouncing motion when a robotic manipulator slides on a rough surface. From the viewpoint of rigid body dynamics, this phenomenon is related to the dynamical properties of a multi-rigid-body system subject to unilateral constraints with friction. Applying the LCP (Linear Complementary Problem) theory, we can classify the bouncing motion into two cases: one that is due to the action of inertia of system; the other that is due to the singularities of rigid-body model induced by friction forces. As an example of a planar multibody system with single unilateral constraint, the admissible set for a two-link manipulator is studied in detail. Meanwhile, The numerical results show that the paradoxes in rigid body model can occur even if the value of coefficient of friction is very small.

1 Introduction

Unilateral constraints combined with friction can result in many complex dynamical phenomena. A typical example relating to them is the control of keeping a robotic manipulator sliding on a rough surface which can be hampered by the instability bouncing motion [1].

It is well-known that there are singularities in using the rigid body model for the dynamical simulation and control of the systems with contact and friction. Some authors attribute these problems to the Painleve paradox[2, 3, 4]. Then, one might want to know whether the bouncing motion arising in the robotic system is related to the problem of Painleve paradox.

To answer this question should need to understand the physical origin of paradoxical situations in rigid body model. In principle, there are two cases in rigid body model relating to the situations of paradox: one that is termed as indeterminacy where the dynamical equations have multiple solutions; the other that is termed as inconsistency where the dynamical equations have no solution at all. Many authors[5, 6, 7, 8, 9, 10, 11] have confirmed that there is a shock at the contact point once the case of inconsistency occurs, which is

Peter Eberhard (ed.), Multiscale Problems in Multibody System Contacts, 123–133.

also addressed as impacts without collisions or tangential impacts[3, 8].

Of course, to explain such paradoxical situations in rigid body model requires the methods of deformable elements of discretizing the contact area. Such works can be found in the references[13, 14, 15, 16, 17]. Although there are more or less discrepancies among these numerical results of adopting different physical models, yet, the similar global properties for paradoxical situations have been verified : for the indeterminate cases, no normal force exists, and contact is immediately broken. For the inconsistent cases, the contact forces have the characteristic of impulse, and the contact will finally separate.

Obviously, no matter what case is in the paradoxical situation of rigid body model, the detachment at contact point will appear all to nothing. Therefore, we can, at least to some extent, attribute the bouncing motion appearing in robotic system to the occurrence of paradoxes in rigid body model. In fact, Leine[10] is the first one that put the hopping motion of a Friction Impact Oscillator connecting with the Painleve paradoxes.

Then, if we can establish a criterion for identifying the paradoxical situation, it will be easy to distinguish whether the bouncing motion in the robotic system is induced by friction force or not. Fortunately, This can be easily implemented by linear complementary theory developed in[4, 5]. In particular, such criterion for a multibody system with single unilateral constraint can be expressed as a simple algebraic formulas.

From the viewpoint of the control of robotic system, if we want to keep the end of the manipulator smoothly sliding on a rough surface, the issue of the controllability for such system is very important for the design of the controller. However, we still have not a clear notion of the controllability for the mechanical system with unilateral constraints[17]. Usually, we have to establish the reachable sets to characterize the properties of controllability for such systems[18].

Thus, the purpose of this paper has two folds: First, we want to establish a unified criterion for identifying when the detachment will appear in a planar multibody system with single unilateral constraint. Furthermore, an admissible set where the robotic system can smoothly slide on a rough surface will be defined in terms of the linear complementary theory. Second, we want to investigate the influence of friction on the admissible sets of a certain system that a two-link manipulator touches on a moving belt. Meanwhile, we will note whether the paradoxical situations in rigid body model only occur in a large value of coefficient of friction for such system.

2 The Admissible Set in Configuration Space

Suppose that the planar multi-rigid body system moves on a rough plane. The inertial Cartesian coordinate frame$(\boldsymbol{i}_1, \boldsymbol{i}_2)$ is fixed on the contact plane at point O, where $\boldsymbol{i}_1$ and $\boldsymbol{i}_2$ represent the unit vector in tangential and normal direction, respectively. The coordinate components of the contact point in the inertial frame can be expressed by generalized coordinates $(q_1, q_2, \cdots, q_n)$

$$x_i = x_i(q_1, q_2, \cdots, q_n), \text{ where } (i = 1, 2) \tag{1}$$

In terms of Euler-Lagrange Equations, the dynamical equations can be written by

$$\ddot{\boldsymbol{q}} = \boldsymbol{M}^{-1}\boldsymbol{K}\boldsymbol{F} + \boldsymbol{M}^{-1}(-\boldsymbol{R} + \boldsymbol{W}) \tag{2}$$

where, $\boldsymbol{M}_{n\times n}$ is symmetric positive mass matrix, $\boldsymbol{W}_{n\times 1}$, $\boldsymbol{R}_{n\times 1}$ and $\boldsymbol{F}_{2\times 1}$ are the column matrix of generalized active forces, the inertial forces and the contact forces, respectively, $\dot{\boldsymbol{q}}_{n\times 1}$ is the generalized speed, $\boldsymbol{K}_{n\times 2}$ is the Jacobian matrix of the transformation equations (1) from the contact coordinates to the generalized coordinates.

In order to solve the dynamical equations (2), one should determine the values of the contact forces at contact point. Thus, some complementary equation has to be provided

Differentiating Eq.1 twice, we have the acceleration vectors in tangential and normal direction at contact point

$$\ddot{\boldsymbol{x}} = \boldsymbol{K}^{\boldsymbol{T}}\ddot{\boldsymbol{q}} + \boldsymbol{S} \tag{3}$$

where, $\boldsymbol{S}(= \begin{bmatrix} S_1 \ S_2 \end{bmatrix})$ is the affixation terms produced by the second derivatives of the transformation equations (1)

$$S_i = \sum_{j=1}^{n}\sum_{k=1}^{n} \frac{\partial^2 x_i}{\partial q_j \partial q_k}\dot{q}_j\dot{q}_k.$$

Substitution of (2) into (3) leads to

$$\ddot{\boldsymbol{x}} = \boldsymbol{K}^{\boldsymbol{T}}\boldsymbol{M}^{-1}\boldsymbol{K}\boldsymbol{F} + \boldsymbol{K}^{\boldsymbol{T}}\boldsymbol{M}^{-1}(-\boldsymbol{R} + \boldsymbol{W}) + \boldsymbol{S} \tag{4}$$

Assume that the contact point initially slides on the rough plane. Dry friction at the contact point is represented by Coulomb law of sliding friction with constant friction coefficientμ_0. Then the tangential friction force and the normal force have the following relationship

$$F_1 = \mu \cdot F_2 \tag{5}$$

where $\mu = \mu_0$ if $\dot{x}_1 < 0$ and $\mu = -\mu_0$ if $\dot{x}_1 > 0$.

Define $\boldsymbol{Q}_{2\times 2} = \boldsymbol{K}^{\boldsymbol{T}}\boldsymbol{M}^{-1}\boldsymbol{K}$ and it is a symmetric positive matrix as $\boldsymbol{M}$ is a symmetric positive matrix and $\boldsymbol{K}$ is full-rank if the number of generalized coordinates is not less than two (otherwise, the system will have no freedom).

Then, we can express the differential equations (4) in the tangential and normal direction as following

$$\ddot{x}_1 = C(\boldsymbol{q}, \mu) \cdot F_2 + D(\boldsymbol{q}, \dot{\boldsymbol{q}}) \tag{6}$$

$$\ddot{x}_2 = A(\boldsymbol{q}, \mu) \cdot F_2 + B(\boldsymbol{q}, \dot{\boldsymbol{q}}) \tag{7}$$

where

$$A(\boldsymbol{q}, \mu) = \mu Q_{21} + Q_{22}, \quad B(\boldsymbol{q}, \dot{\boldsymbol{q}}) = \boldsymbol{K}_2^T \boldsymbol{M}^{-1}(-\boldsymbol{R} + \boldsymbol{W}) + S_2$$

$$C(\boldsymbol{q}, \mu) = \mu Q_{11} + Q_{12}, \quad D(\boldsymbol{q}, \dot{\boldsymbol{q}}) = \boldsymbol{K}_1^T \boldsymbol{M}^{-1}(-\boldsymbol{R} + \boldsymbol{W}) + S_1$$

$\boldsymbol{K}_1$ and $\boldsymbol{K}_2$ represent the first and second column of matrix $\boldsymbol{K}$.

Based on the complementary equation (7), we can provide the information of contact forces for the dynamical equations (2). Meanwhile, under the assumption of impenetrability of unilateral constraint, the normal component of acceleration and normal component of contact force should satisfy with the following complementary condition

$$\ddot{x} \geq 0, \quad F_2 \geq 0 \text{ and } \ddot{x} \cdot F_2 = 0 \tag{8}$$

Equation (7) combining with (8) is the standard form of complementary problem. Since both A and B may be positive or negative due to the friction, there are four possibilities corresponding to the calculation of contact forces: if $A > 0$ but $B < 0$, there is a unique solution and the slip at contact point will keep; if $A > 0$ and $B \geq 0$, there is still a unique solution but the detachment at contact point will occur; if $A \leq 0$ and $B < 0$, no solution exists which corresponds to the case of inconsistency; if $A \leq 0$ and $B \geq 0$, there are two solutions satisfied with the complementary relationship which corresponds to the case of indeterminacy.

Therefore, the unique case of the solution of Eq.7 that can keep slip at contact point is that $A > 0$ and $B < 0$. Then we have a definition of the admissible set for a planar multi-rigid body system with unilateral constraints.

Definition: *Assuming that $\boldsymbol{s}(\cdot) = (\boldsymbol{q}(\cdot), \dot{\boldsymbol{q}}(\cdot))$ represents the state of the system, $\boldsymbol{s}(\cdot)$ is admissible at time t where $t \in (\tau_0, \tau_0 + \epsilon)$ if and only if $\boldsymbol{s}(t) \in \Phi$, where $\Phi = \{\boldsymbol{s}(t) | A > 0, B < 0\}$. Φ is termed as an admissible set in the state space of the system.*

By observing the components of A, we can find that its value is just related to the configuration of the system and the coefficient of friction. Thus, there are some regions, which is defined by $A \leq 0$, is not reachable in the space of the configuration of a multibody system with unilateral constraints. In other words, one can not keep the system touching smoothly on the rough surface only by applying external forces if the configuration of the system is not changed.

3 The Two-Link Robot Touching on a Moving Belt

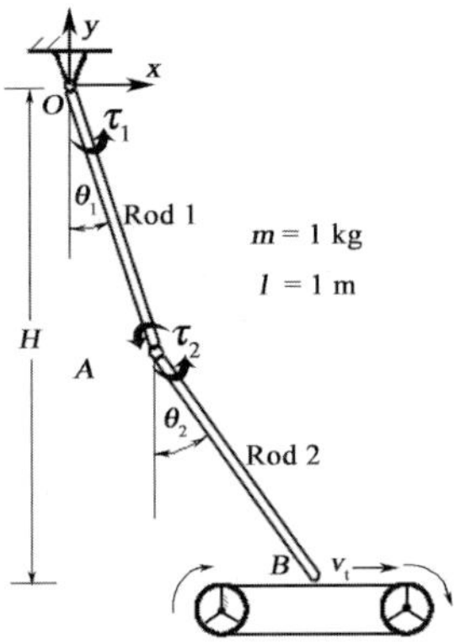

Fig. 1. A two-link manipulator touches a moving belt

As an example for above theory, in this section we will study the admissible set of a two-link manipulator touching a moving belt(Fig.1). The speed of the moving belt is v_t. The two links are equal with the same length l and mass m. The generalized coordinates are selected as θ_1 and θ_2. The inertial coordinate frame Oxy is attached at the joint O. The vertical distance from point O to the rough surface is H. The contact forces in the tangential and normal direction are F_x, F_y, respectively. τ_1 and τ_2 are the torsion acting on the joint O and the joint A. The positive values of θ_1 and θ_2 are prescribed along the inverse direction of clock-time.

The kinetic energy of the system is

$$T = \frac{2}{3}ml^2\dot{\theta}_1^2 + \frac{1}{6}ml^2\dot{\theta}_2^2 + \frac{1}{2}ml^2\dot{\theta}_1\dot{\theta}_2\cos(\theta_1 - \theta_2) \tag{9}$$

The relationship of the inertial coordinates of the contact point with the generalized coordinates of the system is

$$x = l(\sin\theta_1 + sin\theta_2), \quad y = -H = -l(\cos\theta_1 + cos\theta_2) \tag{10}$$

In terms of Euler-Lagrange Equations, the matrixes of the coefficient in the dynamical equations (2) have the following form

$$\boldsymbol{M} = \begin{bmatrix} 4ml^2/3 & ml^2\cos(\theta_1 - \theta_2)/2 \\ ml^2\cos(\theta_1 - \theta_2)/2 & ml^2/3 \end{bmatrix} \quad \boldsymbol{K} = \begin{bmatrix} l\cos\theta_1 & l\sin\theta_1 \\ l\cos\theta_2 & l\sin\theta_2 \end{bmatrix}$$

$$\boldsymbol{W} = \begin{bmatrix} \tau_1 - \tau_2 - 3mgl\sin\theta_1/2 & \tau_2 - mgl\sin\theta_2/2 \end{bmatrix}^T$$

$$\boldsymbol{S} = \begin{bmatrix} -l(\dot{\theta}_1^2\sin\theta_1 + \dot{\theta}_2^2\sin\theta_2) & l(\dot{\theta}_1^2\cos\theta_1 + \dot{\theta}_2^2\cos\theta_2) \end{bmatrix}^T$$

$$\boldsymbol{R} = \left[ml^2\dot{\theta}_2^2 \sin\theta_1 - \theta_2/2 \; ml^2\dot{\theta}_1^2 \sin\theta_1 - \theta_2/2\right]^T$$

In terms of Eq.10, the relative velocity in tangential direction at contact point can be expressed as

$$\dot{x}_r = l(\dot{\theta}_1 \cos\theta_1 + \dot{\theta}_2 \cos\theta_2) - v_t \tag{11}$$

In terms of Eq.7, the complementary equation for calculating the normal contact force can be written by the following expression

$$\ddot{y} = AF_y + B \tag{12}$$

where,

$$A = l\sin\theta_1 \frac{B_1}{A_1} + l\sin\theta_2 \frac{B_2}{A_2} \qquad B = l\sin\theta_1 \frac{C_1}{A_1} + l\sin\theta_2 \frac{C_2}{A_2} + l\dot{\theta}_1^2 \cos\theta_1 + l\dot{\theta}_2^2 \cos\theta_2$$

The coefficient A_1, A_2, B_1, B_2, C_1 and C_2 are shown in Appendix A

Obviously, if $A < 0$, the complementary equation (12) can not provide a reasonable solution for the dynamical simulation. The critical friction, which makes the coefficient A just equal zero, can be obtained

$$\mu_m = \left| \frac{4[3\cos\theta_1 - \theta_2 \sin\theta_1 \sin\theta_2 - ({\sin\theta_1}^2 + 4{\sin\theta_2}^2)]}{\sin 2\theta_1 - 5\sin 2\theta_2} \right| \tag{13}$$

Meanwhile, θ_1 and θ_2 will be satisfied with the following geometrical relationship if the contact is kept.

$$l(\cos\theta_1 + \cos\theta_2) = H \tag{14}$$

For each height of H, we can obtain a minimum of the critical coefficient of friction. Fig.2 shows the changes of such minimum with the variation of H. We can find the paradoxical situation in rigid body model for such system can occur even if the value of coefficient of friction is very small. In particular, the critical value can approach zero if the height of the system approaches $2l$.

The admissible set of the two-link manipulator will be changed with the shift of the value of coefficient of friction. In following of this section, we will study the influence of the coefficient of friction on the admissible set. In terms of the geometrical relationship expressed in Eq.14, we can express θ_2 as a function with respect to θ_1, and restrict $\theta_2 > 0$ by considering the symmetric configuration of the system. Thus, the changes of the coefficient A and B relating to the four cases of the solutions in complementary equation can be reflected on the phase plane of $(\theta_1, \dot{\theta}_1)$. We denote that: M_s^1 is related to $A > 0$ and $B > 0$; M_s^2 is related to $A > 0$ and $B < 0$; M_s^3 is related to $A < 0$ and $B < 0$; M_s^4 is related to $A < 0$ and $B < 0$.

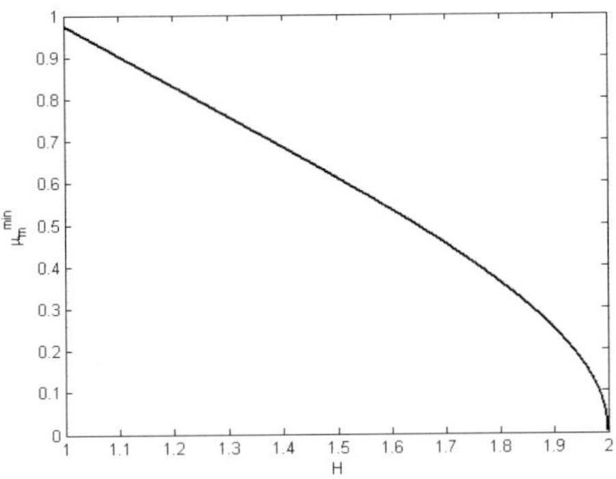

Fig. 2. the changes of the minimum of critical coefficient of friction with the height H

Let the physical parameters of the two-link manipulator as: $H = 5/3m, l = 1m, m = 1kg, g = 9.8m/s^2, \tau_1 = 0Nm, \tau_2 = 0Nm, v_t = 0m/s$. The value of θ_1 is just limited in the scope of $[-0.8411, 0.8411]$, which can keep the end of the manipulator touching on the belt in terms of geometrical configuration of the system. The minimum of critical friction coefficient is 0.4807 which is obtained from Eq.13.

In the following, we set three typical values of coefficient of friction to study its influences on the admissible set. In the first example, the value of coefficient of friction equals 0.5, which is just greater than the minimum of critical coefficient of friction. The different colors in Fig.3 shows the regions corresponding to the four cases of the solution of the complementary equation.

The line of $\dot{\theta} = 0$ represents the static equilibrium positions of the manipulator which correspond to the initial configurations when the manipulator is put on the moving belt with zero velocity. The stability of these static equilibrium positions can be identified by the joint angular acceleration whose sign is represented by different arrows in Fig.3.

$$\ddot{\theta}_1 = -\frac{B_1 B}{A_1 A} + \frac{C_1}{A_1} \tag{15}$$

Obviously, in the range of $[-0.842, 0.1468]$ and $[0.65, 0.841]$ for θ_1, the static equilibrium positions are stable as the accelerations on the both sides of the line $\dot{\theta}_1 = 0$ approach these equilibrium positions. But, if the value of θ_1 belongs to $[0.1468, 0.65]$, the static equilibrium positions are unstable since such equilibrium points are saddle points according to the direction of the acceleration. In such case, the initial configuration of the manipulator will be destroyed even if it touches the belt with zero speed.

From Fig.3, we can find the region of Painleve paradoxes is very small and only related to the value of $\dot{\theta}_1$ (the region is limited by two vertical lines) when the rod slips along left ($\dot{\theta}_1$ is negative).

The structure of the admissible set may become somewhat complex with the increment of the frictional coefficient. In the second example, we set the value of coefficient of friction to equal 0.8, which make the region of Painleve paradoxes include the special configuration where the two links of the manipulator are alignment. The angular for such configuration is

$$\theta^{\star} = \theta_1 = \theta_2 = \arccos \frac{H}{2l} \tag{16}$$

We can obtain the critical friction coefficient $\mu_0^{\star}$ for such special configuration.

$$\mu_0^{\star} = \frac{\sqrt{4l^2 - H^2}}{H} \tag{17}$$

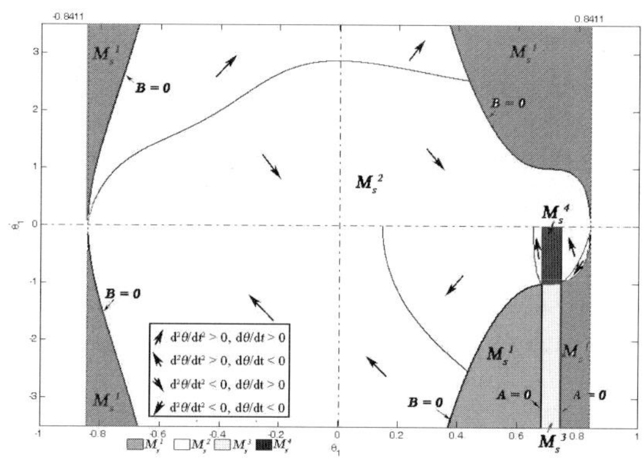

Fig. 3. Four regions of the solution of complementary equation($\mu_0 = 0.5$)

For the given physical parameters, $\mu_0^{\star} = 0.6330$, $\theta^{\star} = 0.5850$. We set the the value of coefficient of friction as $\mu = 0.8$. Fig.4 shows the four different regions corresponding to cases of the solution of the complementary equation. It is obvious that the region for the stable static equilibrium positions and the admissible set in the state space are reduced compared with the case of $\mu = 0.5$. Meanwhile, the Painleve paradox will appear in the two sides of the special configuration.

In the third example, we set the value of coefficient of friction as $\mu = 1.2$. The region that $A < 0$ even appears in the left side of the $\theta_1 - \dot{\theta}_1$ phase plane (shown in Fig.5). The admissible set in the state space for this case is less than the two situations of before. If the value of coefficient of friction is enlarged further, the region of Painleve paradox will expand and even occupy the most part of the $\theta_1 - \dot{\theta}_1$ phase plane.

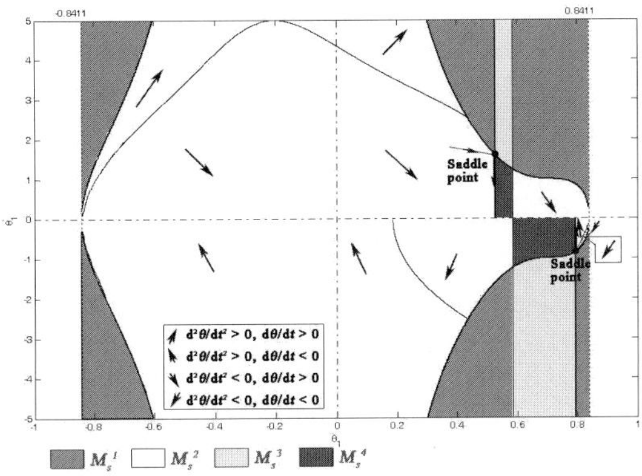

Fig. 4. Four regions of the solution of complementary equation ($\mu_0 = 0.8$)

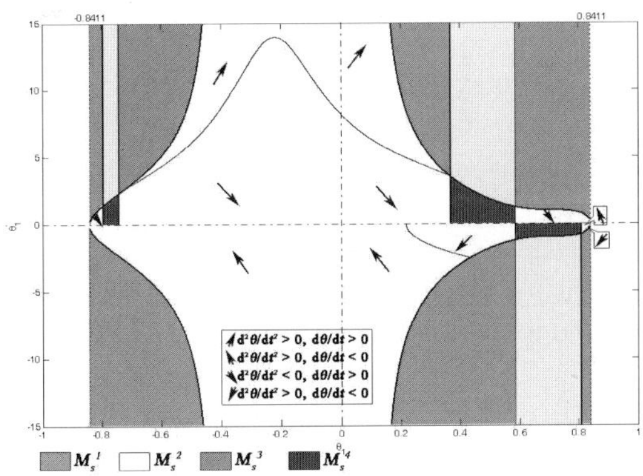

Fig. 5. Four regions of the solution of complementary equation ($\mu_0 = 1.2$)

4 Conclusion

There have two causes for the bouncing motion when a robotic manipulator contact with a rough surface: One that is due to the inertia action of the system, such as the inertia forces and external active forces; the other that is due to the Painleve paradoxes. Based on the LCP theory, we can established a criterion for identifying whether the manipulator can move smoothly on a unilateral constraint.

The two-link manipulator contact with a moving belt is studied in detail. For such example, the paradoxical situations in rigid body model can appear even in a very small value of the coefficient of friction, which depends on the configuration of the system. The admissible set for such system is decreased with the increase of the value of coefficient of friction.

Appendix A

$$A_1(\theta_1,\theta_2) = -\frac{4}{3}ml^2 + \frac{3}{4}ml^2\cos^2(\theta_1-\theta_2) < 0 \quad A_2(\theta_1,\theta_2) = \frac{1}{4}A_1$$

$$B_1(\theta_1,\theta_2,\mu) = \frac{3}{2}l\cos(\theta_1-\theta_2)(\mu\cos\theta_2+\sin\theta_2) - l(\mu\cos\theta_1+\sin\theta_1)$$

$$B_2(\theta_1,\theta_2,\mu) = \frac{3}{8}l\cos(\theta_1-\theta_2)(\mu\cos\theta_1+\sin\theta_1) - l(\mu\cos\theta_2+\sin\theta_2)$$

$$C_1(\theta_1,\theta_2,\dot\theta_1,\dot\theta_2) = \frac{3}{8}ml^2\dot\theta_1^2\sin(2(\theta_1-\theta_2)) + \frac{1}{2}ml^2\dot\theta_2^2\sin(\theta_1-\theta_2) - \frac{3}{4}mgl\sin\theta_2\cos(\theta_1-\theta_2) + \frac{3}{2}mgl\sin\theta_1 - \tau_1 + \left[\frac{3}{2}\cos(\theta_1-\theta_2)+1\right]\tau_2$$

$$C_2(\theta_1,\theta_2,\dot\theta_1,\dot\theta_2) = -\frac{3}{32}ml^2\dot\theta_2^2\sin(2(\theta_1-\theta_2)) - \frac{1}{2}ml^2\dot\theta_1^2\sin(\theta_1-\theta_2) - \frac{9}{16}mgl\sin\theta_1\cos(\theta_1-\theta_2) + \frac{1}{2}mgl\sin\theta_2 + \frac{3}{8}\cos\theta_2\tau_1 - \left[\frac{3}{8}\cos(\theta_1-\theta_2)+1\right]\tau_2$$

Acknowledgment

The support of the National Science Foundation of China (0502009,60334030) is gratefully acknowledged.

References

1. Brogliato B (1999) Nonsmooth Mechanics, 2nd edition. Springer, London, 1999.
2. Ivanov A P (2003) Singularities in the dynamics of systems with non-ideal constraints, J. Appl. Math. Mech 67(2): 185-192
3. Brach R M (1997) Impact Coefficients and Tangential Impacts Trans. ASME. J of App Mech 64: 1014-1016.1997
4. Pfeiffer F, Glocker CH (1998) Unilateral multibody contact. Netherlands: Kluwer Academic Publishers
5. Payr M, Glocker C (2005) Oblique Frictional Impact of a Bar: Analysis and Comparison of Different Impact Laws, Nonlinear Dynamics 41: 361-383
6. David E. Stewart (2000) Rigid-Body Dynamics with Friction and Impact, SIAM 42(1): 3-39
7. Génot F, Brogliato B. (1999) New results on Painlevé paradoxes. European Journal of Mechanics A/Solids 18:653-677
8. Moreau J J, Jean M (1996) Numerical treatment of contact and friction: The contact dynamicsmethods, Engineering and System Design Analysis 4: 201-208
9. Lotstedt P (1981) Coulomb friction in two-dimensional rigid-body systems. ZAMM, 61: 605-615
10. R.I. Leine, B. Brogliato, H. Nijmeijer (2002) Periodic motion and bifurcations induced by the Painlevé paradox, European Journal of Mechanics A/Solids 21: 869-896

11. Yu W, Mason M T (1992) Two-dimensional rigid-body collisions with friction,J Appl Mech 59: 635-642
12. Lecornu L (1905) Sur la loi de Coulomb. Comptes Rendu des Séances de l'Academie des Sciences 140 (6):847-848
13. Baraff D (1991) Coping with Friction for Non-penetrating Rigid Body Simulation, Computer Graphics 25(4): 31-40
14. Zhao Zhen, Chen Bin, Caishan Liu and Jin Hai (2004) Impact Model Resolution On Painlevé's Paradox. ACTA Mechanica Sinica 20(6):659-660
15. Peng S, Kraus P, Kumar V, Dupont P (2001) Analysis of rigid-body dynamic models for simulation of systems with frictional contacts, Journal of Applied Mechanics 68: 118-128
16. Ivanov A P (1997) The problem of constrainted Impact, J Appl Maths Mechs 61(3):341-253
17. B. Brogliato (2003) Some Perspectives on the Analysis and Control of Complementarity Systems, IEEE Transaction on automatic control 48(6): 918-935
18. Nijmeijer H, van der Shaft A J (1990) Nonlinear Dynamical Control Systems, New York: Springer-Verlag

Contact Models for Very Loose Granular Materials

Stefan Luding

Particle Technology, Nanostructured Materials, DelftChemTech,
TU Delft, Julianalaan 136, 2628 BL Delft, The Netherlands
s.luding@tudelft.nl

Summary. One challenge of todays research on particle systems is the realistic simulation of granular materials consisting of many thousands of particles with peculiar contact interactions. In this study, molecular dynamics (MD, also called discrete element method, DEM) is introduced for the simulation of many-particle systems. A wide class of realistic contact models is presented, involving dissipation, adhesion, plastic deformation, friction, rolling- and torsion resistance.
The effect of the contact properties on a simple compaction test is discussed with the goal to achieve as small as possible packing densities. With contact forces only, packing volume fractions down to 0.42 can be achieved, while somewhat longer ranged adhesion forces allow for volume fractions as low as 0.34.

1 Introduction

Molecular Dynamics (MD) or Discrete Element Models (DEM) are solving the equations of motion for all particles in a system, where the contact forces are the only physical laws that have to be defined beforehand. A straightforward approach towards the understanding of the macroscopic material behavior of fine granular materials like powders, by just modeling and simulating all particles in a big system, is not possible due to the huge number of particles typically involved. Therefore, one has to reduce the size of the system under consideration, so that a microscopic simulation of all particles becomes feasible. The goal is to understand the macroscopic flow behavior from such small scale models – both from simulations and from experiment – and to provide (macroscopic) constitutive relations for standard tools like the finite-element method (FEM) suited to deal with large-scale systems. Tools to perform a so-called micro-macro transition are developed [1], and the goal is to relate the macroscopic flow behavior to the microscopic contact properties.

For powders, as an example, the particle properties and interaction laws are inserted into a discrete particle molecular dynamics and lead to the collective behavior of the dissipative, frictional, adhesive many-particle system.

Peter Eberhard (ed.), Multiscale Problems in Multibody System Contacts, 135–150.

From the particle simulation, one can extract, e.g., the coordination number or the pressure of the system as a function of density. In the following, normal and tangential interactions, like adhesion, plastic contact deformation, friction, rolling- and torsion resistance are discussed for spherical model particles. Examples of a compression test are presented for which the previously defined contact model parameters are varied.

2 The Soft Particle Molecular Dynamics Method

Particle simulations like MD or DEM [1, 2, 3, 4, 5, 6, 7] can complement experiments on small "representative volume elements" (REV). Alternative methods like contact dynamics (CD) or cell- and lattice gas-methods are not discussed here.

2.1 Discrete Particle Model

The elementary units of granular materials are mesoscopic grains, which deform under stress. Since the realistic modeling of the deformations of the particles is much too complicated, we relate the interaction force to the overlap δ of two particles, see Fig. 1. In tangential direction, some forces also depend on the tangential displacement since the beginning of the contact, like it is the case for torques, e.g., due to friction or rolling resistance. Note that the evaluation of the inter-particle forces based on the overlap may not be sufficient to account for the inhomogeneous stress distribution inside the particles and possible multi-contact effects. Consequently, the results presented here are of the same quality as the simplifying assumptions about the force-overlap relations made.

2.2 Equations of Motion

If all forces $\boldsymbol{nf}_i$ acting on the particle i, either from other particles, from boundaries or from external forces, are known, the problem is reduced to the integration of Newton's equations of motion for the translational and rotational degrees of freedom:

$$m_i \frac{\mathrm{d}^2}{\mathrm{d}t^2} \boldsymbol{nr}_i = \boldsymbol{nf}_i + m_i \boldsymbol{ng} \ , \quad \text{and} \quad I_i \frac{\mathrm{d}^2}{\mathrm{d}t^2} \boldsymbol{n\varphi}_i = \boldsymbol{nq}_i \tag{1}$$

with the mass m_i of particle i, its position $\boldsymbol{nr}_i$ the total force $\boldsymbol{nf}_i = \sum_c \boldsymbol{nf}_i^c$ acting on it due to contacts with other particles or with the walls, the acceleration due to volume forces like gravity $\boldsymbol{ng}$, the spherical particles moment of inertia I_i, its angular velocity $\boldsymbol{n\omega}_i = \mathrm{d}\boldsymbol{n\varphi}_i/\mathrm{d}t$ and the total torque $\boldsymbol{nq}_i$.

The equations of motion are thus a system of $\mathcal{D} + \mathcal{D}(\mathcal{D}-1)/2$ coupled ordinary differential equations to be solved in $\mathcal{D}$ dimensions, with $\mathcal{D} = 2$ or

3. With tools from numerical integration, as nicely described in textbooks as [8, 9], this is a straightforward exercise. The typically short-ranged interactions in granular media, allow for further optimization by using linked-cell or alternative methods [8, 9] in order to make the neighborhood search more efficient. In the case of long-range interactions, (e.g., charged particles or van der Waals type forces) this is not possible anymore, so that either a cut-off or more advanced methods for optimization have to be applied – for the sake of brevity, we restrict ourselves to the cut-off method below.

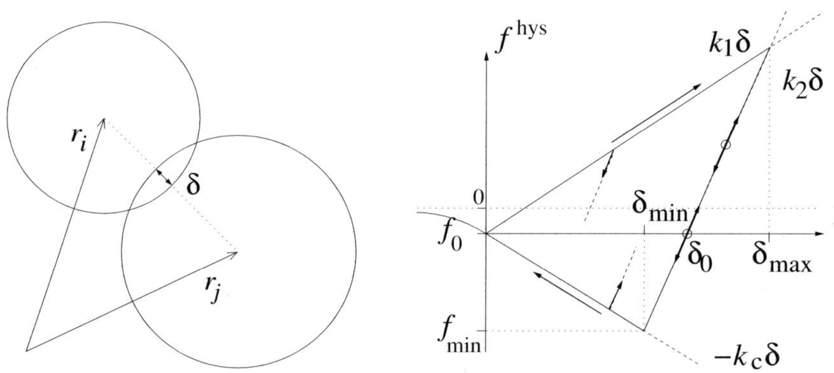

Fig. 1. (Left) Two particle contact with overlap δ in normal direction. (Right) Schematic graph of the piece-wise linear, hysteretic, adhesive force-displacement model in normal direction.

2.3 Normal Contact Force Laws

Two spherical particles i and j, with radii a_i and a_j, respectively, interact only if they are in contact so that their overlap

$$\delta = (a_i + a_j) - (\boldsymbol{nr}_i - \boldsymbol{nr}_j) \cdot \boldsymbol{nn} \tag{2}$$

is positive, $\delta > 0$, with the unit vector $\boldsymbol{nn} = \boldsymbol{nn}_{ij} = (\boldsymbol{nr}_i - \boldsymbol{nr}_j)/|\boldsymbol{nr}_i - \boldsymbol{nr}_j|$ pointing from j to i. The force on particle i, from particle j, at contact c, can be decomposed into a normal and a tangential part as $\boldsymbol{nf}^c := \boldsymbol{nf}_i^c = f^n \boldsymbol{nn} + f^t \boldsymbol{nt}$, where $\boldsymbol{nn} \cdot \boldsymbol{nt} = 0$.

Linear Normal Contact Model

The simplest normal contact force model, which takes into account excluded volume and dissipation, involves a linear repulsive and a linear dissipative force

$$f^n = k\delta + \gamma_0 v_n \ , \tag{3}$$

with a spring stiffness k, a viscous damping γ_0, and the relative velocity in normal direction $v_n = -\boldsymbol{n}\boldsymbol{v}_{ij} \cdot \boldsymbol{n}\boldsymbol{n} = -(\boldsymbol{n}\boldsymbol{v}_i - \boldsymbol{n}\boldsymbol{v}_j) \cdot \boldsymbol{n}\boldsymbol{n} = \dot{\delta}$.

This so-called linear spring dashpot (LSD) model allows to view the particle contact as a damped harmonic oscillator, for which the half-period of a vibration around an equilibrium position with a certain contact force, can be computed [10]. The typical response time on the contact level is

$$t_c = \frac{\pi}{\omega} \ , \quad \text{with} \quad \omega = \sqrt{(k/m_{12}) - \eta_0^2} \ , \tag{4}$$

the eigenfrequency of the contact, the rescaled damping coefficient $\eta_0 = \gamma_0/(2m_{ij})$, and the reduced mass $m_{ij} = m_i m_j/(m_i + m_j)$. From the solution of the equation of a half period of the oscillation, one also obtains the coefficient of restitution

$$r = v_n'/v_n = \exp\left(-\pi\eta_0/\omega\right) = \exp\left(-\eta_0 t_c\right) \ , \tag{5}$$

which quantifies the ratio of normal relative velocities after (primed) and before (unprimed) the collision. For a more detailed discussion of this and other, more realistic, non-linear contact models, see Ref. [10].

The contact duration in Eq. (4) is also of practical technical importance, since the integration of the equations of motion is stable only if the integration time-step Δt_{MD} is much smaller than t_c. Note that t_c depends on the magnitude of dissipation: In the extreme case of an overdamped spring, t_c can become very large (which would render the contact behavior artificial [19]). Therefore, the use of neither too weak nor too strong dissipation is recommended.

Adhesive, Plastic, Hysteretic Normal Contact Model

Here we apply a variant of the linear hysteretic spring model [10, 11, 12], as an alternative to the frequently applied spring-dashpot models. This model is the simplest version of some more complicated nonlinear-hysteretic force laws [11, 13, 14], which reflect the fact that at the contact point, plastic deformations may take place and attractive (adhesive) forces exist. The adhesive, plastic (hysteretic) force-law can be written as

$$f^{\mathrm{hys}} = \begin{cases} k_1\delta & \text{for loading,} & \text{if } k_2(\delta-\delta_0) \geq k_1\delta \\ k_2(\delta-\delta_0) & \text{for un/reloading,} & \text{if } k_1\delta > k_2(\delta-\delta_0) > -k_c\delta \\ -k_c\delta & \text{for unloading,} & \text{if } -k_c\delta \geq k_2(\delta-\delta_0) \end{cases} \tag{6}$$

with $k_1 \leq k_2$, see Fig. 1. During the initial loading the force increases linearly with the overlap δ, until the maximum overlap δ_{max} is reached (which has to be kept in memory as a history parameter). The line with slope k_1 thus defines the maximum force possible for a given δ. During unloading the force drops from its value at δ_{max} down to zero at overlap $\delta_0 = (1 - k_1/k_2)\delta_{\mathrm{max}}$, on the line with slope k_2, so that δ_0 resembles the *plastic contact deformation.*

Reloading at any instant leads to an increase of the force along the line with slope k_2, until the maximum force is reached; for still increasing δ, the force follows again the line with slope k_1 and $\delta_{\max}$ has to be adjusted accordingly.

Unloading below δ_0 leads to negative, *attractive forces* until the minimum force $-k_c\delta_{\min}$ is reached at the overlap $\delta_{\min} = (k_2 - k_1)\delta_{\max}/(k_2 + k_c)$. This minimum force, i.e., the maximum attractive force, is obtained as a function of the model parameters k_1, k_2, k_c, and the history parameter $\delta_{\max}$. Further unloading leads to attractive forces $f^{\text{hys}} = -k_c\delta$ on the adhesive branch with slope $-k_c$. The highest possible attractive force, for given k_1 and k_2, is reached for $k_c \to \infty$, so that $f^{\text{hys}}_{\max} = -(k_2 - k_1)\delta_{\max}$. Since this would lead to a discontinuity at $\delta = 0$, it is avoided by using finite k_c.

The lines with slope k_1 and $-k_c$ define the range of possible force values and departure from these lines takes place in the case of loading and unloading, respectively. Between these two extremes, unloading and reloading follow the line with slope k_2. Possible equilibrium states are indicated as circles in Fig. 1, where the upper and lower circle correspond to a pre-stressed and stress-free state, respectively. Small perturbations lead, in general, to small deviations along the line with slope k_2 as indicated by the arrows in Fig. 1.

A non-linear un/reloading behavior would be more realistic, however, due to a lack of detailed experimental informations, the piece-wise linear model is used as a compromise. One reasonable refinement, which accounts for an *increasing stiffness* with deformation, is a k_2 value dependent on the maximum overlap. This also implies relatively small and large plastic deformations for weak and strong contact forces, respectively. The model, as proposed recently [15], requires an additional model parameter,

$$\delta^*_{\max} = \frac{k_2}{k_2 - k_1}\phi_f \frac{a_1 + a_2}{2} , \tag{7}$$

with the dimensionless plasticity depth, ϕ_f, defined relative to the average radius. If the penetration is larger than a fraction ϕ_f of the (average) particle radius, the constant stiffness k_2 is used [1]. For smaller penetration, $k_2(\delta_{\max})$ interpolates between k_1 to k_2:

$$k_2(\delta_{\max}) = \begin{cases} k_2 & \text{if } \delta_{\max} \geq \delta^*_{\max} \\ k_1 + (k_2 - k_1)\delta_{\max}/\delta^*_{\max} & \text{if } \delta_{\max} < \delta^*_{\max} \end{cases} , \tag{8}$$

and k_2 in Eq. (6) is replaced by $k_2(\delta_{\max})$ from Eq. (8).

While in the case of collisions of particles with large relative velocities and thus deformations, dissipation takes place due to the hysteretic nature of the force-law, reasonably strong dissipation of small amplitude deformations is achieved by adding the viscous, velocity dependent dissipative force from Eq. (3) to the hysteretic force, such that $f^n = f^{\text{hys}} + \gamma_0 v_n$.

[1] Note that a limit to the slope k_2 is needed for practical reasons. If k_2 would not be limited, the contact duration could become very small so that the time step would have to be reduced below reasonable values.

In summary, the adhesive, plastic, hysteretic normal contact model contains the five parameters k_1, k_2, k_c, ϕ_f, and γ_0 that respectively account for loading-reloading-stiffness and plastic deformation, adhesion, plastic overlap-range of the model, and viscous dissipation .

Long Range Normal Forces

Medium range van der Waals forces can be taken into account in addition to the hysteretic force such that $f^n = f^{\mathrm{hys}} + \gamma_0 v_n + f^{\mathrm{vdW}}$ with, for example, a Lennard-Jones Potential, leading to the force as function of distance:

$$f^{\mathrm{vdW}}(r) = -(4\varepsilon/r_0)\left[6(r_0/r)^7 - 12(r_0/r)^{13}\right] . \tag{9}$$

In order to have a continuous force-displacement relation and to limit the range of the force, usually, a cut-off is introduced, so that

$$f^{\mathrm{vdW}} = f^{\mathrm{vdW}}(r) - f^{\mathrm{vdW}}(r_c) \ , \ \text{for } r < r_c \ , \tag{10}$$

and $f^{\mathrm{vdW}} = 0$ elsewhere. The new parameters necessary for this force are an energy scale ε, a typical length scale r_0 and the cut-off length r_c. As long as r_c is not too large as compared to the particle diameter, the methods for short range interactions still can be applied to such a medium range interaction model – only the linked cells have to be larger than twice the cut-off radius. When r_0 is smaller than the particle diameter, the repulsive part of the force becomes irrelevant due to the repulsive contact model.

2.4 Tangential Contact Force Laws

For the tangential degrees of freedom, there are three different force- and torque-laws to be implemented: (i) friction, (ii) rolling resistance, and (iii) torsion resistance. For *friction*, the relative tangential velocity of the contact points,

$$\boldsymbol{n}v_t = \boldsymbol{n}v_{ij} - \boldsymbol{n}n(\boldsymbol{n}n \cdot \boldsymbol{n}v_{ij}) \ , \tag{11}$$

is to be considered for the force and torque computations in the following subsections, with the total relative velocity of the particle surfaces at the contact

$$\boldsymbol{n}v_{ij} = \boldsymbol{n}v_i - \boldsymbol{n}v_j + a_i \boldsymbol{n}n \times \boldsymbol{n}\omega_i + a_j \boldsymbol{n}n \times \boldsymbol{n}\omega_j \ . \tag{12}$$

Thus, the frictional force and torque are active when the two particles are rotating in parallel. The forces on the contacting particles are equally strong, but opposite, i.e., $\boldsymbol{n}f_j^t = -\boldsymbol{n}f_i^t$, while the corresponding torques are parallel but not necessarily equal in magnitude, i.e., $\boldsymbol{n}q_i^{\mathrm{friction}} = -a_i \boldsymbol{n}n \times \boldsymbol{n}f_i$, and $\boldsymbol{n}q_j^{\mathrm{friction}} = (a_j/a_i)\boldsymbol{n}q_i^{\mathrm{friction}}$. Note that forces and torques together conserve the total angular momentum, see Ref. [10].

For *rolling resistance*, the rolling velocity

$$\boldsymbol{n}v_r = -a_i \boldsymbol{n}n \times \boldsymbol{n}\omega_i + a_j \boldsymbol{n}n \times \boldsymbol{n}\omega_j \ , \tag{13}$$

is to be considered, which activates torques when two particles are rotating anti-parallel with spins in the tangential plane. These torques act against rolling and are equal in magnitude and opposite in direction, i.e., $\boldsymbol{n}q_i^{\text{rolling}} = -\boldsymbol{n}q_j^{\text{rolling}} = a_{ij}\,\boldsymbol{n}n \times \boldsymbol{n}f_r$, with the reduced radius $a_{ij} = 2a_ia_j/(a_i + a_j)$, and the quasi-force $\boldsymbol{n}f_r$. This quasi-force is equal for both particles and does not act on the center of mass so that the total momentum (translational and angular) is conserved.

For *torsion resistance*, the relative spin along the normal direction

$$\boldsymbol{n}v_o = (a_i\,\boldsymbol{n}n \cdot \boldsymbol{n}\omega_i - a_j\,\boldsymbol{n}n \cdot \boldsymbol{n}\omega_j)\,\boldsymbol{n}n \ , \tag{14}$$

is to be considered, which activates torques when two particles are rotating anti-parallel with spins parallel to the normal direction. These torsion torques are also equal in magnitude and directed opposite in direction, i.e., $\boldsymbol{n}q_i^{\text{torsion}} = -\boldsymbol{n}q_j^{\text{torsion}} = a_{ij}\,\boldsymbol{n}f_o$, with the quasi-force $\boldsymbol{n}f_o$ that also does not change the translational momentum, but results in torques that conserves the total angular momentum.

The implementation of the tangential force computations for $\boldsymbol{n}f_t$, $\boldsymbol{n}f_r$, and $\boldsymbol{n}f_o$ as based on $\boldsymbol{n}v_t$, $\boldsymbol{n}v_r$, and $\boldsymbol{n}v_o$, respectively, is assumed to be *identical*, i.e., even the same subroutine is used, however, with different parameters as specified below. The difference is that friction leads to a force in the tangential plane (changing both translational and angular momentum), while rolling- and torsion-resistance lead to a quasi-forces in the tangential plane and the normal direction, respectively, changing angular momentum only. For more details on tangential contact models, friction, rolling and torsion, see Refs. [17, 18].

Frictional Tangential Contact Model

The tangential force is coupled to the normal force via Coulombs law, i.e. $f^t \le \mu^s f^n$, where for the limit case one has dynamic friction with $f^t = \mu^d f^n$. The dynamic and the static friction coefficients follow, in general, the relation $\mu^d \le \mu^s$. The static situation requires an elastic spring in order to allow for a restoring force, i.e. a non-zero remaining tangential force in static equilibrium due to activated Coulomb friction.

If a repulsive contact is established, and thus one has $f^n > 0$, the tangential force is active. In the presence of adhesion, Coulombs law has to be slightly modified in so far that f^n is replaced by $f^n + k_c\delta$. With other words, the reference criterion for a contact is no longer the zero force level, but it is the adhesive, attractive force level along $-k_c\delta$. Coulombs law in the presence of adhesion thus reads $f^t \le \mu^s(f^n + k_c\delta)$ for the static case and $f^t = \mu^d(f^n + k_c\delta)$ for the dynamic, sliding case.
If a contact is active, one has to project (or better rotate) the tangential spring into the actual tangential plane, since the frame of reference of the contact

may have rotated since the last time-step. The new tangential spring is:

$$\boldsymbol{n}\xi = \boldsymbol{n}\xi' - \boldsymbol{n}n(\boldsymbol{n}n \cdot \boldsymbol{n}\xi') \ , \tag{15}$$

where $\boldsymbol{n}\xi'$ is the old spring from the last iteration. This action is relevant only for an already existing spring; if the spring is new, the tangential spring-length is zero anyway, however, its change is well defined even for the first, initiation step. In order to compute the changes of the tangential spring, a tangential test-force is first computed as the sum of the tangential spring force and a tangential viscous force (in analogy to the normal viscous force)

$$\boldsymbol{n}f_0^t = -k_t\, \boldsymbol{n}\xi - \gamma_t \boldsymbol{n}v_t \ , \tag{16}$$

with the tangential spring stiffness k_t, the tangential dissipation parameter γ_t, and $\boldsymbol{n}v_t$ from Eq. (11). As long as $|\boldsymbol{n}f_0^t| \leq f_C^s$, with $f_C^s = \mu^s(f^n + k_c\delta)$, one has static friction and, on the other hand, if the limit $|\boldsymbol{n}f_0^t| > f_C^s$ is reached, sliding friction is active with magnitude $f_C^d = \mu^d(f^n + k_c\delta)$. (As soon as $|\boldsymbol{n}f_0^t|$ becomes smaller than f_C^d, static friction is active again.) In the former, static case, the tangential spring is incremented

$$\boldsymbol{n}\xi' = \boldsymbol{n}\xi + \boldsymbol{n}v_t\, \Delta t_{\mathrm{MD}} \ , \tag{17}$$

to be used in the next iteration in Eq. (15), and the force $\boldsymbol{n}f^t = \boldsymbol{n}f_0^t$ from Eq. (16) is used. In the latter, sliding case, the tangential spring is adjusted to a length which is consistent with Coulombs condition

$$\boldsymbol{n}\xi' = -\frac{1}{k_t} f_C^d\, \boldsymbol{n}t \ , \tag{18}$$

with the tangential unit vector, $\boldsymbol{n}t = \boldsymbol{n}f_0^t/|\boldsymbol{n}f_0^t|$, defined by Eq. (16), and thus the magnitude of the Coulomb force is used. Inserting $\boldsymbol{n}\xi'$ from Eq. (18) into Eq. (16) leads to $\boldsymbol{n}f_0^t \approx f_C^d \boldsymbol{n}t - \gamma_t \boldsymbol{n}v_t$. Note that $\boldsymbol{n}f_0^t$ and $\boldsymbol{n}v_t$ are not necessarily parallel in three dimensions. However, the mapping in Eq. (18) works always, rotating the new spring such that the direction of the frictional force is unchanged and, at the same time, limiting the spring in length according to Coulombs law. In short notation the tangential contact law reads

$$\boldsymbol{n}f^t = f^t \boldsymbol{n}t = +\min\left(f_C, |\boldsymbol{n}f_0^t|\right) \boldsymbol{n}t \ , \tag{19}$$

where f_C follows the static/dynamic selection rules described above. The torque on a particle due to frictional forces at this contact is $\boldsymbol{n}q^{\mathrm{friction}} = \boldsymbol{n}l_i^c \times \boldsymbol{n}f_i^c$, where $\boldsymbol{n}l_i^c$ is the branch vector, connecting the center of the particle with the contact point.

The four parameters for the friction law are k_t, μ_s, $\phi_d = \mu_d/\mu_s$, and γ_t, accounting for tangential stiffness, the static friction coefficient, the dynamic friction ratio, and tangential viscosity, respectively. Note that the tangential force described above is identical to the classical Cundall-Strack spring only in the limits $\mu = \mu^s = \mu^d$, i.e., $\phi_d = 1$, and $\gamma_t = 0$. The sequence of computations and the definitions and mappings into the tangential direction can be used in three dimensions as well as in two.

Rolling Resistance Contact Model

The three new parameters for rolling resistance are k_r, μ_r, and γ_r, while ϕ_d is used from the friction law. The new parameters account for rolling stiffness, the static rolling "friction" coefficient, and rolling viscosity, respectively. In the subroutine called, the rolling velocity $\boldsymbol{nv}_r$ is used instead of $\boldsymbol{nv}_t$ and the computed quasi-force $\boldsymbol{nf}_r$ is used to compute the torques, q^{rolling}, on the particles, see above.

Torsion Resistance Contact Model

The three new parameters for rolling resistance are k_o, μ_o, and γ_o, while ϕ_d is used from the friction law. The new parameters account for torsion stiffness, the static torsion "friction" coefficient, and torsion viscosity, respectively. In the subroutine, the torsion velocity $\boldsymbol{nv}_o$ is used instead of $\boldsymbol{nv}_t$ and the projection is a projection along the normal unit-vector. The computed quasi-force $\boldsymbol{nf}_o$ is then used to compute the torques, q^{torsion}, on the particles.

2.5 Background Friction

Note that the viscous dissipation takes place in a two-particle contact. In the bulk material, where many particles are in contact with each other, this dissipation mode is very inefficient for long-wavelength cooperative modes of motion [16, 19]. Therefore, an additional damping with the background can be introduced, so that the total force on particle i is

$$\boldsymbol{n}f_i = \sum_j \left(f^n \boldsymbol{nn} + f^t \boldsymbol{nt}\right) - \gamma_b \boldsymbol{n}v_i \ , \tag{20}$$

and the total torque

$$\boldsymbol{n}q_i = \sum_j \left(\boldsymbol{n}q^{\text{friction}} + \boldsymbol{n}q^{\text{rolling}} + \boldsymbol{n}q^{\text{torsion}}\right) - \gamma_{br} \boldsymbol{n}\omega_i \ , \tag{21}$$

with the damping artificially enhanced in the spirit of a rapid relaxation and equilibration. The sum in Eqs. (20) and (21) takes into account all contact partners j of particle i, but the background dissipation can be attributed to the medium between the particles. Note that the effect of γ_b and γ_{br} should be checked for each simulation in order to exclude artificial over-damping.

3 Compaction Simulation Results

In this section, a compression test is presented, where the particles are positioned on a square-lattice in a cubic system with periodic boundary conditions, in order to avoid wall effects. The system is first allowed to evolve to a disordered state, by attributing random velocities to all particles. The density is then increased by slowly increasing the particle size while the system volume $V = L^3$, with $L = 0.025\,$m, is kept constant, and reporting density and energies.

3.1 Model Parameters

The systems examined in the following contain $N = 1728$ particles with equal radii a. In the simulation, the radii change according to the relation

$$\frac{da}{dt} = g_a \ , \tag{22}$$

with the growth rate $g_a = 2.10^{-7}\,\mathrm{ms}^{-1}$, if not explicitly specified. The growth is stopped when a target volume fraction $\nu = NV(a)/V$ is reached, with the particle volume $V(a) = (4/3)\pi a^3$. The particle mass $m(a) = \rho V(a)$, with the material density ρ, changes with the radius during the growth period. A typical set of material parameters is given in table 1. The choice of numbers and units is such that the particles correspond to micro-meter sized, (overly) soft aluminum spheres. The stiffness magnitude (this is not the material bulk modulus, but a contact property) used thus appears much too small for this material – however, dependent on the volume fraction (or the external) pressure, the material deformation (overlap) can be realistic if the simulations are performed so slow that rate effects are small and overlaps are not becoming too large. A simple rescaling of time brings the material parameters into the reasonable range – see rightmost column in table 1.

Using the parameter $k = k_2$ in Eq. (4) leads to a typical contact duration (half-period) $t_c \approx 2.27\,10^{-4}\,$s, for a normal collision with $\gamma = 0$. Accordingly, an integration time-step of $t_{\mathrm{MD}} = 2\,10^{-6}\,$s is used, in order to allow for a 'safe' integration of contacts. Note that not only the normal "eigenfrequency" but also the eigenfrequencies in tangential and rotation direction have to be considered as well as the viscous response times $t_\gamma \approx m/\gamma$. All of the eigenfrequencies should be considerably larger than t_{MD}, whereas the viscous response times should be even larger, so that $t_\gamma > t_c > t_{\mathrm{MD}}$. The discussion of all the effects due to the interplay between the model parameters is far from the scope of this paper, however.

Property	Symbol	Values	t-rescaled
Time Unit	t_u	1 s	1 μs
Initial particle radius	a_0	0.5 μm	
Growth rate	g_a	0.2 μm/s	0.2 m/s
Particle radius	$a(t) = a_0 + g_a t$		
Material density	ρ	2000 kg/m^3	
Elastic stiffness	$k = k_2$	10^{-7} kg/s^2	10^5 kg/s^2
Plastic stiffness	k_1/k	0.2	
Adhesion "stiffness"	k_c/k	1.0	
Friction stiffness	k_t/k	0.2	
Rolling stiffness	k_r/k	0.2	
Torsion stiffness	k_o/k	0.2	
Plasticity depth	ϕ_f	0.05	
Coulomb friction coefficient	$\mu = \mu_d = \mu_s$	1	
Dynamic to static Friction ratio	$\phi_d = \mu_d/\mu_s$	1	
Rolling "friction" coefficient	μ_r	0.1	
Torsion "friction" coefficient	μ_o	0.1	
Normal viscosity	$\gamma = \gamma_n$	$2\,10^{-13}$ kg/s	$2\,10^{-7}$ kg/s
Friction viscosity	γ_t/γ	0.25	
Rolling viscosity	γ_r/γ	0.25	
Torsion viscosity	γ_o/γ	0.25	
Background viscosity	γ_b/γ	0.10	
Background viscous torque	γ_{br}/γ	0.05	
Lennard Jones energy	ε	$0.\,10^{-15}$ J	$0.\,10^{-3}$ J
Lennard Jones distance	$r_0/(2a)$	0.5	
Lennard Jones cut-off	$r_c/(2a)$	1.5	

Table 1. The microscopic material parameter values used if not explicitly specified. The third column contains those values that are different due to rescaling of the unit of time, i.e., when seconds are read as μs.

3.2 Compression simulations

When compressing the system (by growing the particles) the first quantity of interest is the density (volume fraction) ν. For a set of frictionless hard spheres, the maximum volume fraction is $\nu_{\max} \approx 0.74$, when all spheres are optimally arranged on a crystal lattice. Random packings can reach volume fractions between 0.63 and 0.69, dependent on the degree of local crystallization. When friction is switched on and also the other force laws are used, much smaller volume fractions between 0.55 and 0.58 are expected as indicated by the difference between open and solid symbols in Fig. 2 (Left).

Before the results of the compression simulations are discussed, one has first to decide on a criterion whether a packing is stable and quasi-static or not. In the following, the ratio of kinetic to potential energy is used, $e = E_{kin}/E_{pot}$, and the densities are reported when $e = 1$, 10^{-1}, 10^{-2}, and 10^{-3}, as given in the inset of Fig. 2. Since the particles are continuously growing, the system has no chance to become static at the lowest density (relaxation times are extremly large at this point). However, the decay of e close to this point is very rapid and indicates at least the possibility of a static, stable packing configuration. A more detailed study of alternative boundary conditions is in progress [2].

For fixed friction coefficient, $\mu = 1$, increasing rolling- and torsion-coefficients lead to lower densities. For the higher values of μ_r and μ_o, re-organization can appear more violently during ongoing compaction, leaving the system with somewhat higher density. For fixed finite rolling- and torsion coefficients, $\mu_r = \mu_o = 0.1$, the density is close to the reference without tangential forces and torques. With increasing friction coefficient μ the density drops. But the highest values of $\mu \geq 0.5$ do not necessarily lead to lower densities, as one could have expected. Again, the more violent reorganization events could be responsible.

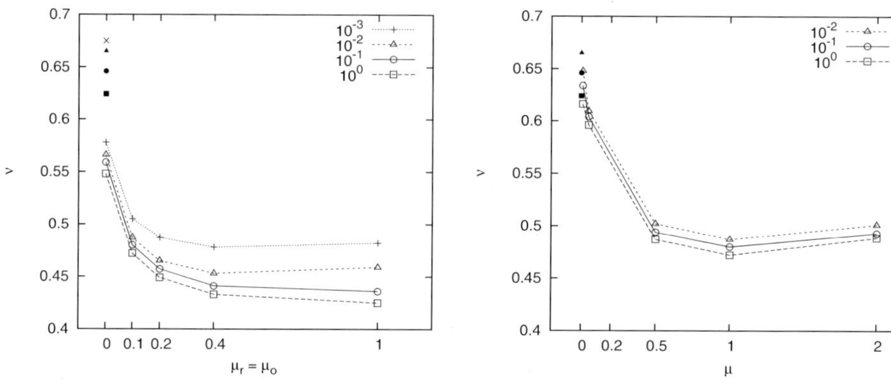

Fig. 2. (Left) Densities (volume fractions) at which the energy ratio reaches the values e as given in the inset. The parameters are given in table 1, only the values of rolling- and torsion-coefficients are varied while $mu = 1$ is kept constant. The lines are a guide to the eye and the solid points are the reference data for $\mu = \mu_r = \mu_o = 0$. (Right) Rolling and torsion-coefficients are $\mu_r = \mu_o = 0.1$ and the friction coefficient μ is varied. Lines and solid points are the same as in the left panel.

From various simulations with different material parameters (data not shown), one can conclude that higher viscosity parameters lead to faster relaxation – too high viscosity on the other hand leads to spurious effects. As long as the growth rate was small enough, the achieved low densities were comparable in magnitude, with no evident dependency on the viscosity parameters. Note, however, that all viscosities were active – in normal, tangential, rolling,

---

[2] The simple approach of a system that is compressed by pressure controlled walls does not work, since the packing becomes then inhomogeneous: very low density in the center, higher density at the walls and very high densities in the corners.

and torsion direction as well as for the background damping. The relative magnitude of the viscosities, as given in table 1, is found reasonable; however, a more systematic study might reveal a better, more realistic, combination of viscosity values.

Based on the variation of the friction-, rolling-, and torsion-coefficients, the lowest volume fraction to be expected for a stable packing can be extrapolated from Fig. 2 (Left) to be about $\nu_{\min} \approx 0.42$. Too small friction coefficients are always related to rather high densities. On the other hand, extremely high friction-coefficients do not necessarily lead to lower densities due to a different reorganization dynamics. A test simulation with $\mu = \mu_r = \mu_o = 10$ did not lead to lower densities as one could have hoped.

3.3 Compaction with long range force

In order to achieve even lower densities as in the previous subsection, all parameters from table 1 are used, only the Lennard-Jones energy parameter is varied using the values $\varepsilon = 10^{-21}$, 10^{-20}, 10^{-19}, $10^{-18}\,\mathrm{m^2\,kg\,s^{-2}}$. The potential energy now also involves the longrange potential and is thus not an adequate criterion for comparison to the kinetic energy. From the kinetic energy, the simulation with $\varepsilon = 10^{-20}\,\mathrm{m^2\,kg\,s^{-2}}$ seemed to relax most rapidly. The minimal density found was $\nu_{\min} \approx 0.34$ – further detailed studies about the value where the packing remains stable are necessary.

A qualitative comparison of the packing structures in Fig. 3 shows somewhat larger holes in the packing with long range forces (visible due to the dark particles in the back). A more quantitative analysis using the pair-correlation function reveals that a considerable fraction of particles that are close to each other without longrange forces stick together when long range forces are active. More detailed parameter studies are in progress.

Thus with some attractive long range force, the minimal packing density can be considerably decreased relative to the lowest densities achieved with contact forces only. However, very strong attractive forces do not necessarily lead to better relaxation behavior. Therefore, more detailed studies under better controlled conditions will be necessary to achieve stable configurations with extremely low density and to better understand the compaction and reorganization dynamics.

4 Conclusion

The present study is a summary of the soft particle force models involving elasticity, plastic contact deformation, adhesion, friction, and rolling- as well as torsion resistance. A set of parameters is given and several criteria and rules for parameter selection are discussed. Using friction and rolling-/torsion-resistance, stable static packings could be reached with densities (volume fractions) somewhat above $\nu_{\min} \approx 0.4$. When also an attractive, longer ranging

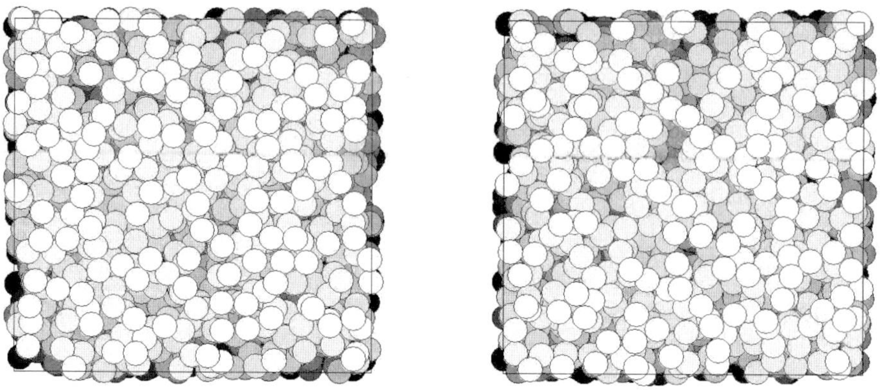

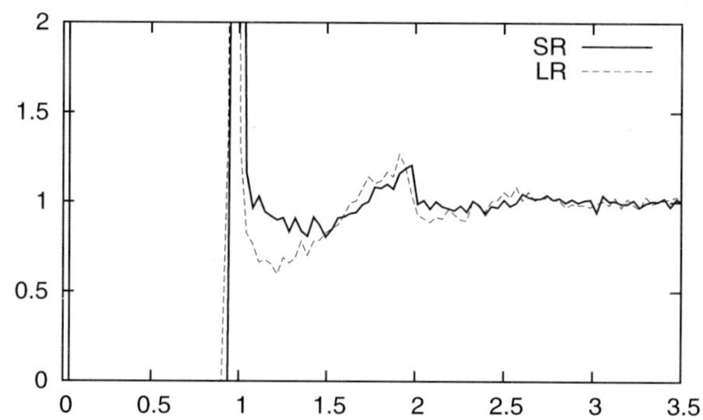

Fig. 3. Snapshots from simulations without (Left) and with (Right) longer ranged attractive forces at volume fraction $\nu = 0.34$. The left packing is not yet stable at this density whereas the right packing is. The greyscale indicates the distance from the viewer – more distant particles are darker. The lines indicate the (periodic) boundary of the system; note that dark particles at the system boundaries are visible due to the periodic boundary conditions – not due to holes in the system. (Bottom) Pair correlation functions for the simulations with (LR) and without (SR) longrange forces.

force was added, the minimal possible density was below $\nu_{\min} \approx 0.34$, however, in this case the criterion for a static, stable configuration has to be improved.

Even though molecular dynamics particle methods are a helpful tool for the understanding of granular systems, the quality of the results depends strongly on the contact models used. The set of contact models presented here, besides many model assumptions, still involves a large number of parameters. Some of them are less important for physical properties and behavior of the system than others – the latter, most relevant parameters have to be identified and their interplay has to be better understood.

The qualitative particle-modeling approach of the early years has now developed into the attempt of a quantitative predictive modeling of the diverse modes of complex behavior in granular media. The quantitative comparison with experiments and validation of the models is the task for the near future. The measurement of low packing fractions in adhesive, frictional fine powders is one of the possible experiments to be examined in more detail – a challenge for particle contact modeling.

Acknowledgements

Besides valuable discussions with many colleagues, we acknowledge the financial support of the Deutsche Forschungsgemeinschaft (DFG) and the Stichting voor Fundamenteel Onderzoek der Materie (FOM), financially supported by the Nederlandse Organisatie voor Wetenschappelijk Onderzoek (NWO).

References

1. P. A. Vermeer, S. Diebels, W. Ehlers, H. J. Herrmann, S. Luding, and E. Ramm, editors. *Continuous and Discontinuous Modelling of Cohesive Frictional Materials*, Berlin, 2001. Springer. Lecture Notes in Physics 568.
2. P. A. Cundall and O. D. L. Strack. A discrete numerical model for granular assemblies. *Géotechnique*, 29(1):47–65, 1979.
3. Y. M. Bashir and J. D. Goddard. A novel simulation method for the quasi-static mechanics of granular assemblages. *J. Rheol.*, 35(5):849–885, 1991.
4. H. J. Herrmann, J.-P. Hovi, and S. Luding, editors. *Physics of dry granular media - NATO ASI Series E 350*, Dordrecht, 1998. Kluwer Academic Publishers.
5. C. Thornton. Numerical simulations of deviatoric shear deformation of granular media. *Géotechnique*, 50(1):43–53, 2000.
6. C. Thornton and L. Zhang. A dem comparison of different shear testing devices. In Y. Kishino, editor, *Powders & Grains 2001*, pages 183–190, Rotterdam, 2001. Balkema.
7. M. Lätzel, S. Luding, H. J. Herrmann, D. W. Howell, and R. P. Behringer. Comparing simulation and experiment of a 2d granular couette shear device. *Eur. Phys. J. E*, 11(4):325–333, 2003.
8. M. P. Allen and D. J. Tildesley. *Computer Simulation of Liquids.* Oxford University Press, Oxford, 1987.
9. D. C. Rapaport. *The Art of Molecular Dynamics Simulation.* Cambridge University Press, Cambridge, 1995.
10. S. Luding. Collisions & contacts between two particles. In H. J. Herrmann, J.-P. Hovi, and S. Luding, editors, *Physics of dry granular media - NATO ASI Series E350*, page 285, Dordrecht, 1998. Kluwer Academic Publishers.
11. O. R. Walton and R. L. Braun. Viscosity, granular-temperature, and stress calculations for shearing assemblies of inelastic, frictional disks. *J. Rheol.*, 30(5):949–980, 1986.
12. Jürgen Tomas. Particle adhesion fundamentals and bulk powder consolidation. *KONA*, 18:157–169, 2000.

13. C. Y. Zhu, A. Shukla, and M. H. Sadd. Prediction of dynamic contact loads in granular assemblies. *J. of Applied Mechanics*, 58:341, 1991.
14. M. H. Sadd, Q. M. Tai, and A. Shukla. Contact law effects on wave propagation in particulate materials using distinct element modeling. *Int. J. Non-Linear Mechanics*, 28(2):251, 1993.
15. S. Luding, K. Manetsberger, and J. Müllers. A disrete model for long time sintering. *Journal of the Mechanics and Physics of Solids*, 53(2):455, 2005.
16. S. Luding, E. Clément, A. Blumen, J. Rajchenbach, and J. Duran. The onset of convection in molecular dynamics simulations of grains. *Phys. Rev. E*, 50:R1762, 1994.
17. G. Bartels, T. Unger, D. Kadau, D. E. Wolf, and J. Kertesz. The effect of contact torques on porosity of cohesive powders. *Granular Matter*, 7:139, 2005.
18. E. Dintwa, M. van Zeebroeck, E. Tijskens, and H. Ramon. Torsion of viscoelastic spheres in contact. *Granular Matter*, 7:169, 2005.
19. S. Luding, E. Clément, A. Blumen, J. Rajchenbach, and J. Duran. Anomalous energy dissipation in molecular dynamics simulations of grains: The "detachment effect". *Phys. Rev. E*, 50:4113, 1994.

A Multigrid Approach in the Numerical Problem of Tangential Contact

Alexander Lünenschloß

Dynamics Group
Dept. of Mechanical Engineering
TU Darmstadt
Hochschulstr. 1, 64289 Darmstadt, Germany
luenenschloss@dyn.tu-darmstadt.de

Summary. This work presents the numerical simulation of rolling bodies in three dimensions with multigrid-methods and boundary discretization. The use of different scales during the calculation overcomes known problems, regarding convergence velocity and required time for summations. The final objective is the study of roughness in dry, tangential contact. Discretization up to 300×300 points are shown.

Geometrical assumptions and formulation of boundary conditions follow the pioneering work in numerics of contact by *Kalker* [1]. He developed different algorithms for various simplifications, and most of them are based on maximizing complementary energy. In this case there are proofs for existence and uniqueness of solutions. A serious disadvantage is a relationship of cubic order between number of unknowns and calculation time. This problem makes it impossible to handle fine resolutions, even with fast computers. A set of multigrid methods, introduced by *Venner et. al.* for contact with separating film of liquid [2] is used to solve this problem.

Contact areas between rough surfaces can be structured finely, therefore, the method of grid-transformation must be chosen carefully. The common geometry, used in literature, leads to an incorrect mapping of areas, thus existing methods are reformulated with suitable geometry. The problems appeared also in normal contact, were the adapted methods have been applied successfully [3].

In contrast to finite elements, the discretization of the boundary method can be restricted to the potential contact area. That way discretizations up to 300×300 points are accessible, which is sufficient for study of disturbed surfaces. The influence coefficients, that appear in the boundary approach, can be saved in reduced form under assumption of smooth macro-geometries.

1 Physical Model

The problem of normal contact is treated separately from tangential contact, which is possible for quasi-identical materials, e. g. the bodies fulfill a specific condition [1]. In the following we suppose an already known pressure field p and contact area C in the common tangential plane. In this plane material

Peter Eberhard (ed.), Multiscale Problems in Multibody System Contacts, 151–161.

particles are identified by their body j, position ${}^{j}\mathbf{x}$ regarding rigid bodies and a small displacement ${}^{j}\mathbf{u}$ due to elasticity. Hence, the actual position depending on time t becomes

$$ {}^{j}\mathbf{y}({}^{j}\mathbf{x}, t) = {}^{j}\mathbf{x}(t) + {}^{j}\mathbf{u}({}^{j}\mathbf{x}, t). \tag{1} $$

The relative velocity is called

$$ \text{slip } \mathbf{s} := {}^{1}\dot{\mathbf{y}} - {}^{2}\dot{\mathbf{y}}. \tag{2} $$

If the velocity-field does not depend on time,

$$ \mathbf{s} = {}^{1}\dot{\mathbf{x}} - {}^{2}\dot{\mathbf{x}} + \frac{\partial\, {}^{1}\mathbf{u}}{\partial\, {}^{1}\mathbf{x}}\, {}^{1}\dot{\mathbf{x}} - \frac{\partial\, {}^{2}\mathbf{u}}{\partial\, {}^{2}\mathbf{x}}\, {}^{2}\dot{\mathbf{x}}. \tag{3} $$

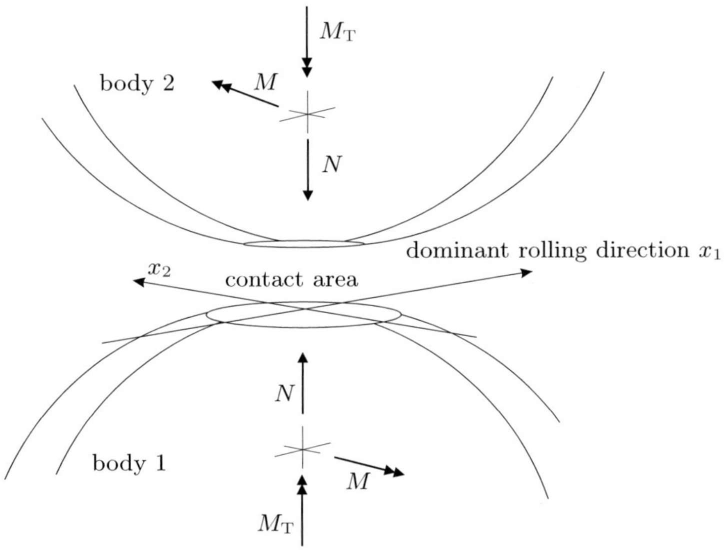

Fig. 1. While undergoing a normal force N, a moment M, and a rotational moment M_T both bodies move, but stay in contact. The common tangential plane contains a coordinate system, orientated at the dominant rolling direction.

The field of displacements and its derivations change first of all in the common rolling direction x_1 (see fig. 1). This leads to an approximation of eq. (3) and by definition of

$$ \begin{aligned} \text{creep } \mathbf{c} &:= {}^{1}\dot{\mathbf{x}} - {}^{2}\dot{\mathbf{x}} \text{ and} \\ \text{strain } \boldsymbol{\epsilon} &:= \frac{\partial\, ({}^{1}\mathbf{u} - {}^{2}\mathbf{u})}{\partial x_1} \end{aligned} \tag{4} $$

we get the equation of simplified kinematics

$$\mathbf{c} = \mathbf{s} + v_c \boldsymbol{\epsilon}. \tag{5}$$

v_c is the relative velocity at the contact point, assumed as constant. For the precise definition and derivation of eq. (5) see [4, p. 11].

The relationship between tangential stress $\mathbf{q}$ and strain is calculated neglecting the curvatures at the contact point. In other words both bodies are seen as half-spaces, which is the material-filled space on one side of a plane. This is necessary to limit the number of different influence coefficients, that are stored in memory. In the general case of arbitrary surfaces the coefficients depend on the location of stress and strain. Then the number of all possible combinations is of quadratic order regarding number of points. With half spaces influence coefficients have the same magnitude for identical distances and therefore, the number of possible combinations reduces to linear order regarding number of points. Taking as an example a 500×500-grid the required memory decreases from 500 GB to 2 MB.

The displacement-field of a half-space under surface stress is a classical result of *Boussinesq* [5, 6]. It leads to a linear relationship between strain and tangential stress. Because details are not used in the following, it is represented by a linear operator $\mathbf{L}$:

$$\boldsymbol{\epsilon} = \mathbf{L}[\mathbf{q}]. \tag{6}$$

With a localized form of *Coulomb*'s friction law the contact area C splits into a stick area A and a slip area S . Then the friction coefficient μ and pressure p determine the

$$\text{stress limit } g = \mu p. \tag{7}$$

The complementary boundary conditions follow with

$$\mathbf{x} \in \mathrm{A}: \quad \|\mathbf{s}(\mathbf{x})\| = 0, \tag{8a}$$

$$g(\mathbf{x}) - \|\mathbf{q}(\mathbf{x})\| > 0, \tag{8b}$$

$$\mathbf{x} \in \mathrm{S}: \quad g(\mathbf{x}) \frac{\mathbf{s}(\mathbf{x})}{\|\mathbf{s}(\mathbf{x})\|} + \mathbf{q}(\mathbf{x}) = 0 \text{ and} \tag{8c}$$

$$\|\mathbf{s}(\mathbf{x})\| > 0. \tag{8d}$$

The kinematic eq. (5) can be used to eliminate the slip $\mathbf{s}$ from the boundary conditions. Then the equations (6) and (8) form a complete set of equations to find $\mathbf{q}$.

2 Discretization

2.1 Introduction

The constitutive eq. (6) contains an integral, that is approximated by a summation with one value per cell. Therefore, a normalized stress $(Q_\alpha)_{i'j'}$ represents the constant value in direction $\alpha \in \{1, 2\}$ of a *cell*, named after their center

$$(i'j') \in \{1, \cdots, m'\} \times \{1, \cdots, n'\}. \tag{9}$$

On the other hand, the set of slip $(S_\alpha)_{ij}$ and relative velocity $(C_\alpha)_{ij}$ show values in direction $\alpha \in \{1, 2\}$ at a *point*, indexed by

$$(ij) \in \{1, \cdots, m\} \times \{1, \cdots, n\}. \tag{10}$$

The difference becomes meaningful, when grid-transformations are performed on the stresses. In the following indices designate elements of sets and not coordinates of a tensor.

The kinematic equation with normalized stress and creep becomes

$$(S_\alpha)_{ij} = (C_\alpha)_{ij} - \sum_\beta \sum_{(i'j')} (V_\alpha^\beta)_{ij}^{i'j'} (Q_\beta)_{i'j'}, \quad (ij) \in \mathrm{C}. \tag{11}$$

The influence coefficient $(V_\alpha^\beta)_{ij}^{i'j'}$ shows the strain $\partial u_\alpha / \partial x_1$ at point (ij) due to stress of magnitude one at $(i'j')$ in direction β. The analytical computation of the influence coefficient demands identical grids for displacements and stresses. While the distinction of cells and nodes by apostrophized indices is sometimes omitted for brevity, the values still have a different meaning.

2.2 The Discrete Boundary Conditions

It is possible, that eq. (8c) is applied to the stick area during the numerical process of solving. Then the diminishing slip in the denominator can cause problems. The discretized form of eq. (8c) expresses the opposite direction of slip and stress; furthermore it assigns the

$$\text{magnitude of stress } Q_{ij} := \sqrt{\sum_\alpha (Q_\alpha)_{ij}^2} \tag{12}$$

to the stress limit G_{ij}. Now the first part is reformulated: Slip and stress must only be parallel and an additional loop controls opposite direction. The discretized eq. (8) is written with help of a nonlinear function $(F_\iota)_{ij}\,[(Q_\alpha)_{kl}]$. It is defined by

$$(ij) \in \mathrm{A}: \quad (F_\iota)_{ij} := (S_\iota)_{ij}, \tag{13a}$$

$$(ij) \in \mathrm{S}: \quad (F_1)_{ij} := G_{ij} - Q_{ij} \quad \text{and} \tag{13b}$$

$$(F_2)_{ij} := (S_1)_{ij}(Q_2)_{ij} - (S_2)_{ij}(Q_1)_{ij}. \tag{13c}$$

Then the following equation must be solved for $(Q_\alpha)_{kl}$:

$$(F_\iota)_{ij}\,[(Q_\alpha)_{kl}] = 0 \tag{14a}$$

under the conditions

$$(ij) \in \mathrm{A}: \quad G_{ij} - Q_{ij} > 0 \quad \text{and} \tag{14b}$$

$$(ij) \in \mathrm{S}: \quad \sum_\alpha (Q_\alpha)_{ij}(S_\alpha)_{ij} > 0. \tag{14c}$$

3 The Multigrid Approach

3.1 Overview

For the contact of rolling bodies exist different numerical approaches. One of them is the method of finite elements, that can be integrated in standard engineering software. Problems with fine resolutions are handled by adaptive mesh refinement ([7], ch. 12) or multigrid methods [8], which are both subject to actual research. The disadvantage of FEM is the complete discretization of contacting bodies, which is not necessary for all problems. If bodies forces are neglected, stationary rolling can be formulated in surface integrals, which leads to boundary elements. In this case the discretization can be restricted to the potential contact area, and therefore takes into account the problem's local character. The combination with half-spaces in the computation of strain leads to a numerically efficient procedure.

Large systems of linear equations are in general solved with iterative methods. Unfortunately many problems of interest converge slowly. An analysis of the mapping's spectrum can show, that only errors with small wavelengths reduce rapidly: A graph of the error-field will be *smoothed* by the iterations. Multigrid algorithms can cope with this problems of convergence velocity by use of grids with different resolution. Mathematical details are described in [9], [10] and [11]. *Venner et. al.* [2] applied a set of multigrid techniques for the contact with separating film of liquid. A preceding work used this combination also for fine discretizations in normal contact[3].

One of the first algorithms for steady state rolling was published by *Kalker* [12]. It is based on maximization of complementary energy and led to the algorithm DUVOROL. Later on, it became part of the program CONTACT, that is still used. CONTACT cannot handle fine resolutions because of a cubic relationship between number of unknowns and calculation time [see 1, p. 175]. This phenomenon limits possible resolutions to approximately 50×50 points, even if an option to use multigrid is offered in the actual version of CONTACT, that seems to originate from an algorithm for linear problems [13]. The multigrid approach in the presented work uses the nonlinear boundary conditions directly, which allows resolutions of 300×300 points.

3.2 Multigrid Elements

The basic structure of the algorithm is the solution of eq. (14a) with two loops, controlling the inequalities (14b) and (14c). The solution of eq. (14a) needs a bundle of numerical instruments at fine discretization. The central method is the multigrid-algorithm for nonlinear problems *Full approximation storage* FAS [see 14, p. 346]. It contains the following elements:

- A smoother for reduction of short wavelength errors,
- a grid-transformation towards finer grids, called prolongation
- a transformation towards coarser grids, called restriction and

– a solver, used on the coarsest grid.

While FAS can reduce the number of iterations to almost $O(nm)$, the calculation of strain still needs $O(n^2m^2)$ operations, because every displacement needs a summation over all points of stress. Furthermore the convergence in the solution of eq. (13a) is threatened by a non-sparse array of influence coefficients. Hence, we need two more numerical methods:

– Multi-level Multi-integration (MLMI) as a fast summation method and
– distributive relaxation for the reduction of interaction between the corrections in the stick area.

Iterative solvers like *Jacobi-* or *Gauss-Seidel* are called smoothers in multigrid context. Because MLMI uses the whole stress array as an argument, the choice of smoothers is limited to the type *Jacobi*.

The described set of methods is a known combination from [2]. The work focuses on lubrication, and therefore, dry contact is only a by-product. Grid-transformations in [2] interpret the variables as values at points. The resulting *vertex*-centered grids are described as "suitable" and "most convenient". The main advantage is seen in the graphical representation of a grid-transformation regarding one point, called a *stencil*. Vertex-centered transformations allow symmetric stencils, but the disadvantage is an incorrect mapping of areas due to the chosen geometry. This is a negligible effect with cohesive contact areas, but it will influence multigrid results with separated, finely structured areas, as they appear with rough or wavy surfaces. Section 2.1 explains in detail the discretization of the constitutive equation and this must lead to a cell-centered geometry (see fig. 2).

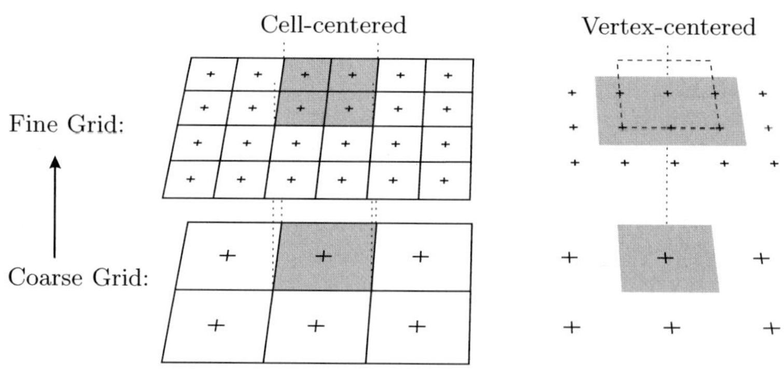

Fig. 2. The comparison between different kinds of grid-transformation. While cell-centered prolongations keep the information about the area, vertex-centered transformations keep the location of points. The shaded zone shows a possible area interpretation after the grid mapping: Size and location change on the right side.

The operators for grid-transformation are taken from a preceding work, that focused on normal contact [3]. A detailed description can be found on page 34ff., where stencils of second order are presented, that satisfy the requirements of FAS. The fast summation by MLMI needs grid-transformations of higher order, because the numerical process replaces an exact calculation. Stencils of 6th order have been developed, which is sufficient, according to [2, p. 166].

In the iterative solution process by FAS the right hand side of eq. (14b) changes. Thus any smoother treats a family of problems

$$(F_\iota)_{ij}[(Q_\alpha)_{kl}] = (h_\iota)_{ij}, \quad (i,j) \in \mathrm{C}. \tag{15}$$

The smoothing-steps for the multigrid solution of eq. (15) differ in stick- and slip area. For fixed stress in the area of slip, the stick area is governed by a set of linear equations. In this case the algorithm improves the stress by a fixed number of *Jacobi*-steps. The calculation is more complex in area of slip, because eq. (14c) is nonlinear. The magnitude of stress is already known by the modified stress limit $G_{ij} - (h_1)_{ij}$, thus it is useful to reformulate $(F_2)_{ij}$ in polar coordinates. With the angle of stress orientation φ_{ij} the remaining problem reads

$$(\hat{F}_2)_{ij}[\varphi_{kl}] = (h_2)_{ij}, \quad (i,j) \in \mathrm{A}. \tag{16}$$

Due to the multigrid formulation, the interest lies in the correction $\Delta\varphi_{ij}$. It can be found by linearization of equation's (16) left hand side, followed by one *Jacobi*-step. This way all appearing summations can be calculated by MLMI, and complexity is lowered to almost linear order, regarding number of unknowns.

The multigrid algorithm FAS is executed a fixed number of cycles, but precision and convergence of the result depend on the initial values. It is advantageous, to use the prolongation of a coarse grid result as estimate. The idea can be applied iteratively, similar to the existing multigrid algorithm *Full Multigrid*. The resulting, complete algorithm was added to an already existing program, named CON3D, that is now able to deal with normal and tangential contact. The computations of chapter 4 are made with help of CON3D.

4 Results

In the tangential contact the qualitative differences between calculations of smooth surfaces with boundary conditions (8a, 8c) and measured results are not completely understood yet [see 15, fig. 17]. Influences in the contact zone like temperature, contamination and roughness change the transmission of force, which cannot be neglected in many cases. CON3D can be used to investigate the influence of surface irregularities, but here the focus lies on the possibility of fine resolutions and lowering of complexity. CON3D has been validated for different load cases and resolutions. The results were correct, but it turned out, that robustness and stability cannot be guaranteed for discretizations higher than 100×100 points. Nevertheless there are computations of disturbed surfaces, that are discretized by approximately 300×300 points (see fig. 3 and 4).

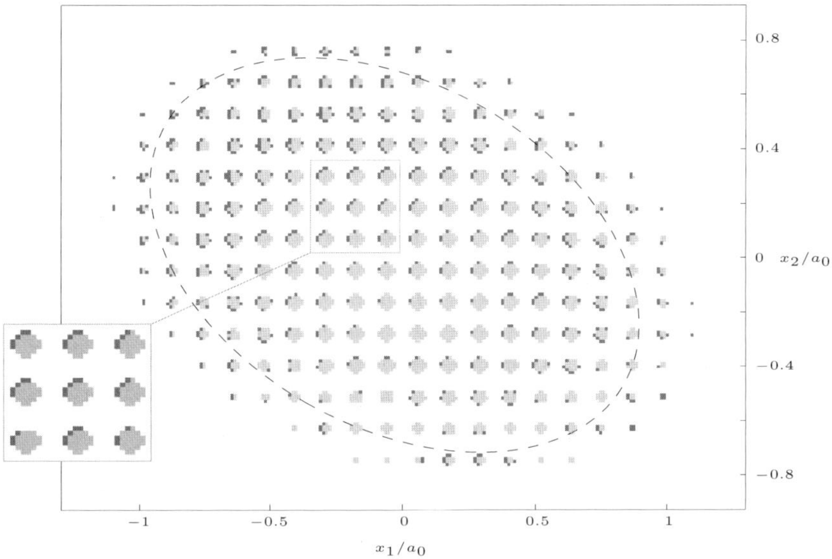

Fig. 3. The separation into stick- (light gray) and slip area (dark gray) during stationary rolling from left to right (160×224 points). The load case is described by a contact pressure of 10 kN, constant friction coefficient $\mu = 0.3$, longitudinal creep $1.9 \cdot 10^{-5}$ %, lateral creep $1.9 \cdot 10^{-6}$ % and rotational creep $1.3 \cdot 10^{-3}$ $\frac{1}{\text{m}}$. The axes are normalized with the greater diameter of *Hertz* ian contact ellipse, shown by a dashed line. One of the surfaces was disturbed by sinusoidal waviness. The magnified detail shows contact islands with area separations similar to the smooth case.

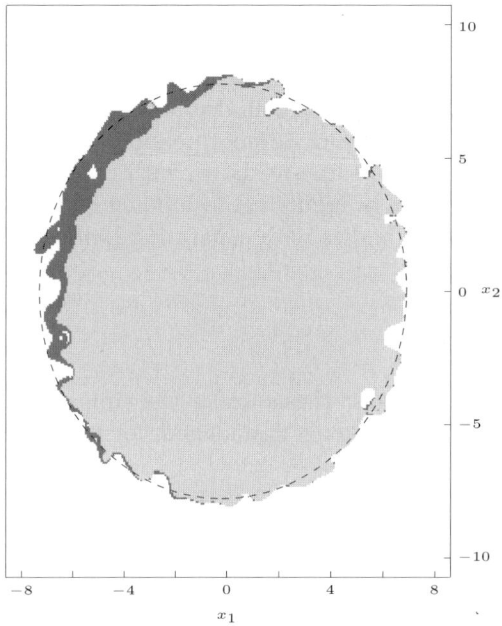

Fig. 4. The separation into stick- (light gray) and slip area (dark gray) in during stationary rolling from left to right (256×320 points). The load case is described by a contact pressure of 100 kN, constant friction coefficient $\mu = 0.3$, longitudinal creep $1.9 \cdot 10^{-4}$ %, and rotational creep $3.7 \cdot 10^{-5} \frac{1}{\mathrm{m}}$. The surface perturbation is simulated roughness, that was smoothed by a cutting frequency in spectral representation. The dashed ellipse shows the contact area in the smooth case

The solution algorithm of CON3D's tangential algorithm for fixed contact set (see ch. 3.2), has theoretically an almost linear complexity regarding the number of unknowns. An example for calculation time, taking area determination into account, is presented in Tab. 1. The complexity of the complete solution can be considered quadratic, in contrast to the cubic dependency in the common algorithm of CONTACT [see 1, p. 175]. With coarse resolutions CONTACT is significantly faster, but this changes at approximately 100×100 points.

Number of elements n	400	576	784	1024	1296	1600
calculation time t [s]	106	222	427	711	1049	1667
$t/n^2 \cdot 10^3$ [s]	0.66	0.67	0.70	0.68	0.63	0.65

Table 1. An evidence for quadratic complexity in a calculation example of pure longitudinal creepage 0.5 % with different resolutions.

5 Summary

The equations, that describe tangential contact during steady rolling, are solvable with a multigrid approach. It decreases complexity and makes the investigation of finely discretized surfaces possible. The algorithm is used in the program CON3D and results are shown with resolutions up to 256×320 points. In the future the set of numerical techniques can be extended to bodies of different materials and bodies with separating liquid.

6 Acknowledgment

This work started as a master thesis under the supervision of Frank Bucher and Klaus Knothe. I would like to thank them for their support and help.

7 Variables

Name	Dim.	Description
μ	-	Friction coefficient
φ_{ij}	-	Orientation of stress at (ij)
$\Delta\varphi_{ij}$	-	Change of stress orientation at (ij)
A	-	Set of cells in stick area
C	-	Set of cells in contact area
$\mathbf{c}$	$\frac{\mathrm{L}}{\mathrm{T}}$	creep
$(C_\alpha)_{ij}$	-	Normalized creep in direction α at (ij)
ϵ	-	Strain
g	$\frac{\mathrm{M}}{\mathrm{T^2L}}$	Stress limit
G_{ij}	-	Normalized stress limit at (ij)
$(h_\iota)_{ij}$	-	Inhomogeneity in direction ι on cell (ij)
i	-	Index in direction x_1
j	-	Index in direction x_2
$(F_\iota)_{ij}$	-	Value of boundary equations in direction ι at (ij)
M	$\frac{\mathrm{ML^2}}{\mathrm{T^2}}$	Moment parallel to tangential area applied at bodies
M_T	$\frac{\mathrm{ML^2}}{\mathrm{T^2}}$	Torsional moment applied at bodies
m	-	Number of cells in direction x_1
m'	-	Number of cells in direction x_1 (stress)
N	$\frac{\mathrm{ML}}{\mathrm{T^2}}$	Normal force applied at bodies
n	-	Number of cells in direction x_2
n'	-	Number of cells in direction x_2 (stress)
$\mathbf{L}$	-	Linear operator in the relation of stress and strain
p	$\frac{\mathrm{M}}{\mathrm{T^2L}}$	Pressure
$\mathbf{q}$	$\frac{\mathrm{M}}{\mathrm{T^2L}}$	Tangential stress on body 1

Name	Dim.	Description
$(Q_\iota)_{i'j'}$	-	Normalized discretized stress at $(i'j')$ in direction ι
$Q_{i'j'}$	-	Magnitude of normalized, discretized stress at $(i'j')$
S	-	Set of cells in slip area
$\mathbf{s}$	$\frac{\mathrm{L}}{\mathrm{T}}$	Slip
$(S_\iota)_{ij}$	-	Normalized, discretized slip in direction ι at (ij)
t	T	Time
${}^j\mathbf{u}$	L	Displacement on body j
v_{C}	$\frac{\mathrm{L}}{\mathrm{T}}$	Contact point velocity
$(V_\alpha^\beta)_{ij}^{i'j'}$	-	Normalized strain $\partial u_\alpha/\partial x_1$ at (ij) due to stress of magnitude one in direction β at $(i'j')$
${}^j\mathbf{x}$	L	Position on rigid body j
x_1	L	Dominant rolling direction
${}^j\mathbf{y}$	L	Position on elastic body j

References

1. Kalker, J. J. *Three-Dimensional Elastic Bodies in Rolling Contact.* Kluver Academic Publishers, (1990).
2. Venner, C. H. and Lubrecht, A. A. *Multilevel methods in lubrication.* Number 37 in Tribology Series. Elsevier, (2000).
3. Bucher, F., Knothe, K., and Lünenschloß, A. *Arch Appl Mech* **73**, 561ff. (2004).
4. Lünenschloß, A., Bucher, F., and Knothe, K. Fortschritt-Berichte Reihe 2, 616, VDI, (2002).
5. Boussinesq, J. *Application des Potentiels à l'Etude de l'Equilibre et du Mouvement des Solides Elastiques.* Gautier-Villars, Paris, (1885).
6. Love, A. E. H. *A Treatise on the Mathematical Theory of Elasticity.* 4th Ed. Oxford, (1944).
7. Wriggers, P. *Computational Contact Mechanics.* John Wiley & Sons, (2002).
8. Kornhuber, R. *Adaptive Monotone Multigrid Methods for Nonlinear Variational Problems.* Teubner, Stuttgart, (1997).
9. Hackbusch, W. *Multi-Grid Methods and Applications.* Springer Verlag, (1985).
10. Brandt, A. *Guide to Multigrid Development*, volume 960 of *Lecture Notes in Math.* Springer Verlag, (1981).
11. Wesseling, P. *J of Comp Physics* **79** (1988).
12. Kalker, J. J. *On the Rolling Contact of Two Elastic Bodies in the Presence of Dry Friction.* PhD thesis, TH Delft, (1967).
13. Vollebregt, E. A. H. Master's thesis, TU Delft, (1992).
14. Brandt, A. *Math Comp* **31**, 333–390 (1977).
15. Knothe, K., Wille, R., and Zastrau, B. W. *Vehicle System Dynamics* **35**, 361–407 (2001).

Rigid and Quasi-Rigid Theories of Granular Media

Sean McNamara

Institut für Computerphysik, Pfaffenwaldring 27, Universität Stuttgart, 70569 Stuttgart, Germany sean@icp.uni-stuttgart.de

Summary. We sketch two different theories for for granular media: a rigid particle theory, where the deformation of grains is neglected, and a quasi-rigid particle theory, where their deformation is taken as infinitesimally small. Both theories are applied to a simple system that can be analytically solved, and results are compared with simulations. In certain situations, self-canceling forces can affect the motion of the particles, in spite of causing no acceleration.

1 Introduction

1.1 Motivation

As science and engineering have become more specialized, researchers have separated themselves into schools that function almost independently of each other. This is often justified, since mastering one subject requires many years of study. If one wishes to be a good chemical engineer, one usually does not have time to become acquainted with the latest work of pure mathematicians. However, it can also happen that specialization hampers progress. The same problem can be studied by two different schools, without either one knowing about the other. Each school has its own perspective, and develops its own terms and notation, so that when the two schools finally discover one another, communication is difficult.

One area where this has happened is the study of granular media. These materials have been studied for a long time in various branches of engineering. Around twenty years ago, physicists entered the field. Due to differences of scientific culture, much relevant work by engineers was unknown and inaccessible to them. This caused some initial controversies [1]. Those debates revolved around continuum theories used by civil engineers (and others) and various newly proposed alternatives.

A similar thing could happen with mechanical engineers, because granular materials can be considered as mechanical systems, i.e., they are a collection of

Peter Eberhard (ed.), Multiscale Problems in Multibody System Contacts, 163–172.

solid bodies that interact through repulsion, friction (and perhaps cohesion). The only difference is that in machines, the bodies have been given their shape an initial position by some intelligent agent so that the system as a whole will carry out a certain task. In granular materials, there is no such agent. Nevertheless, the equations governing the motion of the bodies are the same.

In this paper, therefore, I would like to present a recently developed approach to granular material and its application to a simple example. This theory takes a mechanical approach to granular matter: each grain is considered as a nearly rigid body whose motion is determined by its contacts with its neighbors. If these results are already known, it is hoped that this article will provoke them to complain, thus initiating a series of fruitful exchanges. If they are unknown, it is hoped that these results will be interesting and helpful.

1.2 Synopsis

This paper compares two different approaches to granular media. Both are concerned with a quasi-static granular packing – the applied forces change slowly enough so that the grains can be assumed to be always in static equilibrium. Both approaches have given rise to various theories and and numerical methods.

These two approaches share a common starting point: the assumption that the deformation of the grains is small, but then they differ in how these deformations are treated. In the first approach, the particles are assumed to be infinitely rigid, thus the deformations are set to zero. This approach has yielded the "stress-only" theories of sand piles [5], and the force network ensemble [6] theory of static granular media subjected to an external load. Finally, there is also a numerical method – "contact mechanics" – based on this approach [12].

In the second approach, one takes the particle deformation as a small parameter. The particles deform, but these deformations remain infinitesimally small. This is similar to what is done in continuum mechanics when the strain is taken to be small, and to depend linearly on the displacement field. If this approximation is applied to an assembly of particles, one obtains the "stiffness matrix" theories of granular media [7, 8, 9]. There is also a corresponding numerical method – the "granular element" [10] or "quasi-static method" [11].

In this paper, I will use both approaches to describe a granular "solid". A "granular solid" is an assembly of grains are held together by an external force, where the grains change their neighbors only rarely. The external force changes slowly, so that the packing is always in force equilibrium. The granular media will be described at the microscopic level – that is, each and every grain in the packing will be considered. The grains are assumed to interact with Coulomb friction, but without cohesion.

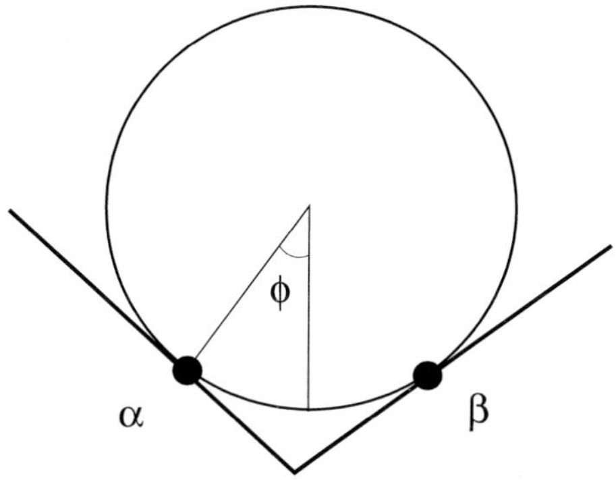

Fig. 1. A disk placed in a circular groove, supported by two contacts α and β. The angle ϕ suffices to specify the geometry.

After sketching the theory for a general configuration of grains, it will be applied to the simple example of a single particle stuck in a triangular shaped groove, as shown in Fig. 1. The disk is pushed into the groove by a downward force and a torque τ is also applied. What is the torque needed to set the disk in motion? Self-canceling forces – combinations of contact forces that cause no acceleration – will also be investigated. What are these forces, and what role do they play in each approach? Can they affect the motion of the disk?

2 Rigid Particles

2.1 Theory

Let us construct a theory for a two-dimensional assembly of infinitely rigid disks subjected to some slowly changing force. Let $i = 1 \ldots N$ number the grains, and $\alpha = 1 \ldots M$ number the contacts between the grains. At each contact α, forces $\boldsymbol{F}_\alpha$ are exerted on each of the two touching particles.

In force equilibrium, the total force and torque exerted on each particle must vanish. Let $\boldsymbol{f}_i$, τ_i be the contact forces exerted on particle i, and let $\boldsymbol{f}_i^{\text{ext}}$ and τ^{ext} be the externally imposed forces. Force equilibrium can now be written

$$\boldsymbol{f}_i = -\boldsymbol{f}_i^{(\text{ext})}, \quad \tau_i = -\tau_i^{(\text{ext})}, \quad i = 1 \ldots N. \tag{1}$$

It is useful to write this equation in vector form. To do so, we gather all the forces into vectors:

$$\mathbf{f}^{(\mathrm{ext})} = \begin{pmatrix} f_{1x}^{(\mathrm{ext})} \\ f_{1y}^{(\mathrm{ext})} \\ \tau_1^{(\mathrm{ext})} \\ \vdots \\ f_{Nx}^{(\mathrm{ext})} \\ f_{Ny}^{(\mathrm{ext})} \\ \tau_N^{(\mathrm{ext})} \end{pmatrix}, \quad \mathbf{f} = \begin{pmatrix} f_{1x} \\ f_{1y} \\ \tau_1 \\ \vdots \\ f_{Nx} \\ f_{Ny} \\ \tau_N \end{pmatrix}, \quad \mathbf{F} = \begin{pmatrix} F_{N,1} \\ F_{T,1} \\ \vdots \\ F_{N,M} \\ F_{T,M} \end{pmatrix}. \tag{2}$$

Note that the force $\boldsymbol{f}_i$ on particle i has been resolved into x and y components, whereas the contact force $\boldsymbol{F}_\alpha$ is resolved into normal (N) and tangential (T) components. Now Eq. (1) becomes simply

$$\mathbf{f} = -\mathbf{f}^{(\mathrm{ext})}. \tag{3}$$

We would like to use this relation to solve for the contact forces. But the $\boldsymbol{f}_i$ and τ_i are sums over these forces. Therefore, $\mathbf{f}$ and $\mathbf{F}$ are linearly related:

$$\mathbf{f} = \mathbf{c}\mathbf{F}, \tag{4}$$

where $\mathbf{c}$ is a $2M \times 3N$ matrix given explicitly elsewhere [2, 3]. Combining Eqs. (3) and (4) gives

$$\mathbf{c}\mathbf{F} = -\mathbf{f}_{\mathrm{ext}}. \tag{5}$$

There are $3N$ equations for $2M$ unknowns. If there are fewer unknowns that equations ($2M < 3N$), the packing is probably not stable, since Eq. (5) will have a solution only when $\mathbf{f}^{(\mathrm{ext})}$ lies within some $2M$-dimensional subset of $\mathbb{R}^{3N}$. If $2M = 3N$, the packing is isostatic, and the contact forces can be uniquely determined from Eq. (5). Assemblies of frictionless particles are often isostatic [7]. For packings of static particles, one has $3N < 2M$. The packing can be stable, but the matrix $\mathbf{c}$ is singular, and its null space contains vectors $\mathbf{F}_0 \neq 0$ such that

$$\mathbf{c}\mathbf{F}_0 = 0. \tag{6}$$

There is no unique solution, because if $\mathbf{F}$ is a solution, so is $\mathbf{F} + a\mathbf{F}_0$. But these undetermined forces are *self-canceling*: they cause no acceleration. One may therefore hope that they have no effect on the motion.

In general, one will have many different, linearly independent self-canceling forces $\mathbf{F}_{0,k}$, all obeying Eq. (6), each with its own amplitude a_k. Bounds can be put on their amplitudes a_k by recalling that the particles interact through Coulomb friction, and without cohesion. Therefore, one must have

$$F_{N,\alpha} > 0, \quad |F_{T,\alpha}| \leq \mu F_{N,\alpha}, \quad \alpha = 1 \ldots M. \tag{7}$$

As shown in the next section, these inequalities can be transformed into inequalities involving the a_k.

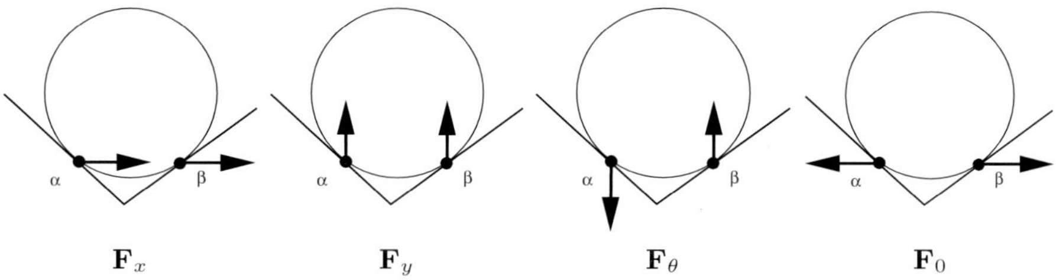

Fig. 2. The meanings of the basis vectors shown in Eq. (9).

2.2 Application to a single particle

Let us apply this theory to the single particle placed in a triangular groove, as shown in Fig. 1. In this situation, the contact matrix is

$$\mathbf{c} = \begin{pmatrix} \sin\phi & \cos\phi & -\sin\phi & \cos\phi \\ \cos\phi & -\sin\phi & \cos\phi & \sin\phi \\ 0 & r & 0 & r \end{pmatrix}. \tag{8}$$

It is convenient to use the following orthogonal basis of $\mathbb{R}^4$:

$$\mathbf{F}_x = \frac{1}{2}\begin{pmatrix} \sin\phi \\ \cos\phi \\ -\sin\phi \\ \cos\phi \end{pmatrix}, \quad \mathbf{F}_y = \frac{1}{2}\begin{pmatrix} \cos\phi \\ -\sin\phi \\ \cos\phi \\ \sin\phi \end{pmatrix},$$
$$\mathbf{F}_\theta = \frac{1}{2}\begin{pmatrix} -\cos\phi \\ \sin\phi \\ \cos\phi \\ \sin\phi \end{pmatrix}, \quad \mathbf{F}_0 = \frac{1}{2}\begin{pmatrix} \sin\phi \\ \cos\phi \\ \sin\phi \\ -\cos\phi \end{pmatrix}. \tag{9}$$

This physical meaning of this basis is shown in Fig. 2. $\mathbf{F}_x$ are forces that cause acceleration in the horizontal direction (and also exert a torque on the particle). $\mathbf{F}_y$ exerts only an upwards force on the particle, but no torque. $\mathbf{F}_\theta$ exerts a torque, but no linear force on the disk. Finally $\mathbf{F}_0$ is the self-canceling set of forces. It exerts neither a force nor a torque.

If $\mathbf{F}$ is written in this basis:

$$\mathbf{F} = a_x\mathbf{F}_x + a_y\mathbf{F}_y + a_\theta\mathbf{F}_\theta + a_0\mathbf{F}_0, \tag{10}$$

then one can verify that

$$\mathbf{cF} = \begin{pmatrix} a_x \\ a_y \\ a_x r\cos\phi + a_\theta r\sin\phi \end{pmatrix} = -\begin{pmatrix} 0 \\ -mg \\ \tau \end{pmatrix} = -\mathbf{f}_{\text{ext}}. \tag{11}$$

This equation enables one to easily construct solutions to Eq. (5), for one has immediately $a_x = 0$, $a_y = mg$, and $a_\theta = -\tau/(r\sin\phi)$. However, a_0

cannot be determined in this way because $\mathbf{F}_0$ is a null eigenvector of $\mathbf{c}$. This is an expression of the indeterminacy in the problem. We put limits on a_0 by requiring Eq. (7) to hold at both α and β. Using Eqs. (9) and (10) to obtain an expression for $\mathbf{F}$, and then Eq. (11) to obtain expressions for a_x, a_y, and a_0, Eq. (7) becomes

$$a_0\left(\mu\tan\phi+1\right)+\left[mg+\frac{\tau}{r\sin\phi}\right]\left(\mu-\tan\phi\right)\geq 0, \tag{12}$$

$$a_0\left(\mu\tan\phi-1\right)+\left[mg+\frac{\tau}{r\sin\phi}\right]\left(\mu+\tan\phi\right)\geq 0, \tag{13}$$

$$a_0\left(\mu\tan\phi-1\right)+\left[mg-\frac{\tau}{r\sin\phi}\right]\left(\mu+\tan\phi\right)\geq 0, \tag{14}$$

$$a_0\left(\mu\tan\phi+1\right)+\left[mg-\frac{\tau}{r\sin\phi}\right]\left(\mu-\tan\phi\right)\geq 0. \tag{15}$$

When $\tan\phi < 1/\mu$, these conditions are sufficient to predict when the disk will move, because it becomes impossible to find a a value of a_0 that satisfies these conditions when τ is larger than a certain value τ_*. However, when $\tan\phi > 1/\mu$, these conditions can always be satisfied for any value of τ, simply by making a_0 large and positive. This suggests that an arbitrarily large torque can be applied to the disk, without causing it to rotate. Is this in fact what happens?

3 Numerical Interlude

Simulations of the system shown in Fig. 1 were done using contact dynamics (CD) [12] and molecular dynamics (MD) [13]. Contact dynamics follows the theory presented in the previous section, and assumes that the particles are infinitely rigid. The contact forces are treated as unknowns, and an iterative procedure is used to calculate them. Since there are many possibilities, the solution that is actually found depends upon the details of the iteration procedure. Molecular dynamics, on the other hand, calculates the forces as a function of the grain positions. When two grains are not in contact, the force between them vanishes. When they first touch, the force becomes non-zero, and it rises rapidly as the particles interpenetrate.

In the simulations, the disk was first placed just in contact with the groove. Then gravity was slowly increased from 0 to a maximum value $g_{\max}$, and then reduced to a second value g_*, and then maintained there. Then the torque τ was linearly increased with time. We investigate the torque τ_* needed to rotate the disk. These applied forces are sketched in the left panel of Fig. 3.

The other three panels of Fig. 3 show a_0 as a function of time for three different angles ϕ and various values of $g_{\max}$. The coefficients a_x, a_y, and a_θ are all determined through Eq. (11), so they are not shown.

Some conclusions that can be drawn from Fig. 3 are

Lateral Impacts on Flexible Beams in Multibody Dynamics Simulations

Jaap Meijaard

School of MMME, The University of Nottingham, University Park, Nottingham NG9 2RD, United Kingdom jaap.meijaard@nottingham.ac.uk

Summary. The motion of a simply supported beam struck by a spherical ball is studied. Time scales are identified, which reveal the relative importance of several effects. The resulting motion is calculated with the aid of fundamental solutions. It is shown that the approach leads to results that agree with those obtained by means of a spatial discretization. A further development into a generally applicable method remains a subject for further investigations.

1 Introduction

This contribution aims at shedding some light on the problem of lateral impacts on beams as can occur in flexible multibody systems. To be specific, a system consisting of a simply supported beam with rectangular cross-section that is struck by a body, the striker, at its centre is considered, see Fig. 1 (*left*). The relevant parameters are: the length of the beam, l; its mass per unit of length, ρA, being the product of the density, ρ, and the area of the cross-section, A; its flexural rigidity, EI, being the product of Young's modulus, E, and the area moment of inertia, $I = Ah^2/12$; the beam height, h; the mass of the striker, m_0; the radius of curvature of its spherical striking face, R; the stiffness coefficient according to the Hertzian contact model, $k_{\rm H} = 4G\sqrt{R}/[3(1-\nu)]$, where G is the shear modulus and ν is Poisson's ratio; and the initial velocity of the striker, v_0. The coordinate along the beam axis is x, with its zero at the beam centre, and the time is denoted by t. Owing to the symmetry, only the part of the beam for which $x \geq 0$ need be considered. If the lateral deflections of the beam, $u = u(t, x)$, remain small, the only nonlinearity that needs consideration arises from the contact between the striker and the beam. The motion of the beam is described by the equations

$$\rho A \frac{\partial^2 u}{\partial t^2} + EI \frac{\partial^4 u}{\partial x^4} = 0\,, \quad 2EI \frac{\partial^3 u(t,0)}{\partial x^3} = f_0(t)\,, \quad m_0 \frac{{\rm d}^2 u_0}{{\rm d}t^2} = -f_0(t)\,, \quad (1)$$

where f_0 is the compressive contact force and u_0 is the displacement of the striker, measured from the point where it just touches the beam. The bound-

Peter Eberhard (ed.), Multiscale Problems in Multibody System Contacts, 173–182.

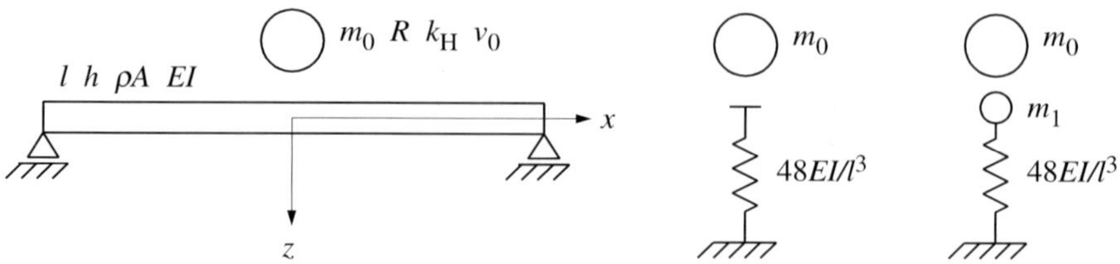

Fig. 1. Simply supported beam laterally hit at its centre by a body (*left*) and two approximate models (*right*)

ary conditions at $x = l/2$ are that the displacement and its second spatial derivative, representing the curvature, are zero.

The study of problems of this kind has already a fairly long history, see for instance [1] for a limited account. If the mass of the striker is large in comparison with the mass of the beam, several approximations can be made, Fig. 1 (*right*). The simplest approximation is to assume that the beam is massless, in which case it can be represented by a spring, which was used by Thomas Young in 1807. Eaton Hodgkinson in 1833–35 proposed to model the beam as a spring with a rigid body of mass m_1 attached to its end, on which the striker impacts without direct rebound. He proposed to use half the mass of the beam, $m_1 = \rho Al/2$, which can be considered a lumped mass approach. Homersham Cox in 1849 proposed to use $m_1 = 17\rho Al/35$, which he derived by assuming that the deflection curve of the beam was as in the static case, an approximation that can be called a consistent mass approach.

Saint-Venant in 1853–4 [2] proposed a more exact solution. He assumed that the rigid striker remains in contact with the beam, at least until the maximal deflection is reached. The motion is expanded in eigenmodes, $\phi_i(x)$, whose participation factors are derived from the initial conditions at the moment of impact. The lateral displacement of the beam can then be represented as

$$u(t,x) = \sum_{i=1}^{\infty} u_i \phi_i(x) \sin \omega_i t \,, \quad \text{with} \quad u_i = \frac{m_0 v_0 \phi_i(0)}{m_i \omega_i} \,, \tag{2}$$

where ω_i are the circular eigenfrequencies and m_i is the modal mass. Because $u_i = O(1/i^3)$, the amplitude of the shear force for each mode is of the order of unity, and the resulting series may not converge everywhere and many sign changes may occur. So the assumption that the striker remains in contact with the beam is not valid.

The possibility of multiple impacts was considered by Timoshenko [3]. The striker was a spherical ball and the contact between the ball and the beam was modelled by Hertzian theory. The motion of the beam was expanded in a series of eigenmodes,

$$u(t,x) = \sum_{i \text{ odd}} u_i(t) \cos \frac{i\pi x}{l} \,, \tag{3}$$

while the displacement of the ball is u_0. With the compression, contact force, modal masses and eigenfrequencies,

$$\delta = u_0 - \sum_{i \text{ odd}} u_i \,, \quad f_0 = k_{\mathrm{H}} \delta^{3/2} \,, \quad m_i = \frac{1}{2} \rho A l \,, \quad \omega_i = \frac{\pi^2 i^2}{l^2} \sqrt{\frac{EI}{\rho A}} \,, \tag{4}$$

the equations of motion in terms of the participation factors and the displacement of the ball are

$$m_i \ddot{u}_i + m_i \omega_i^2 u_i = f_0 \,, \quad m_0 \ddot{u}_0 = -f_0 \,. \tag{5}$$

The results indicated that multiple impacts could occur, which has been validated by experiments [4, 5]. The influence of the shear deformation according to the Timoshenko beam model has been investigated in [6, 7]. The apparent paradox that the shear force for each mode is zero at the centre of the beam is explained by noting that the point force is approximated by a finite series, so the force is spatially spread out. A disadvantage of this method is that for an accurate determination of the bending stresses, a quite large number of modes have to be used.

Fundamental solutions for a beam of infinite length, where the displacement, the rotation, the lateral force or the moment at the end is prescribed, have been derived by Boussinesq [8]. These have been applied in [9] to calculate the bending stresses in a long beam centrally hit by a striker (note the error of a factor 2 in the parameter α in [9]).

2 Scaling and Fundamental Solutions

Seven characteristic time scales can be discerned in the problem, which are defined in Table 1. The six independent ratios are the dimensionless parameters that govern the problem. The first two time scales, t_1 and t_2, represent the time elastic waves need to traverse the height of the beam and the diameter of the striker, respectively. The time scale t_3 represents the time needed for a shear wave to travel from the centre of the beam to the supports. The scale t_4 is a characteristic time for the impact if the beam would be supported at its centre. The scale t_5 is a characteristic impact time for a rigid striker with a beam of infinite length, where $\mu = m_0/(\rho A l)$ is the mass ratio. The scale t_6 is the reciprocal of the fundamental circular frequency of the beam. Finally, t_7 is a time scale for the rigid-body motion.

Two examples, which were also used in [3], are considered. Common parameter values for these are $k_{\mathrm{s}} = 0.84$, $\rho = 7960\,\mathrm{kg/m^3}$, $E = 215.82\,\mathrm{GPa}$, $G = 88.29\,\mathrm{GPa}$, $h = 0.01\,\mathrm{m}$, $A = 1 \times 10^{-4}\,\mathrm{m}^2$ and $v_0 = 0.01\,\mathrm{m/s}$. In the first example, $l = 0.1535\,\mathrm{m}$, $R = 0.01\,\mathrm{m}$, $m_0 = 0.033342\,\mathrm{kg}$, $k_{\mathrm{H}} = 1.5696 \times 10^{10}\,\mathrm{N/m^{3/2}}$ and $\mu = 0.27289$, while in the second example, $l = 0.307\,\mathrm{m}$, $R = 0.02\,\mathrm{m}$, $m_0 = 0.26674\,\mathrm{kg}$, $k_{\mathrm{H}} = 2.2198 \times 10^{10}\,\mathrm{N/m^{3/2}}$ and $\mu = 1.0915$. The numerical

Table 1. Characteristic time scales with example values in ms

definition	example 1	example 2
$t_1 = h\sqrt{\rho/E}$	0.00192	0.00192
$t_2 = 2R\sqrt{\rho/E}$	0.00384	0.00768
$t_3 = (l/2)\sqrt{\rho/(k_\mathrm{s}G)}$	0.0251	0.0503
$t_4 = (m_0/k_\mathrm{H})^{2/5}v_0^{-1/5}$	0.0538	0.108
$t_5 = \mu^2 l^2\sqrt{\rho A/EI}$	0.117	7.47
$t_6 = (l/\pi)^2\sqrt{\rho A/EI}$	0.159	0.635
$t_7 = l/v_0$	15350	30700

values for the time scales for these two examples are shown in Table 1. It can be seen that t_1 and t_2 are much smaller than t_4, so the effects of transverse wave propagation can be neglected, as was done in the model in the Introduction. The next four time scales do not differ much in the first example, so all of them are expected to play a role, where also the finite shear wave propagation time is important, which is not included in the model. For the second example, t_5 is relatively large, so this can be considered a case of a short beam, although the physical length of the beam is larger. For these examples, the rigid-body motion is unimportant.

Fundamental solutions are solutions for an infinite beam to specific simple prescribed forces or displacements at a point. For the case under consideration, either a prescribed displacement at the centre if the contact stiffness is large or a prescribed force if the contact stiffness is taken into account are of interest. The input can be an impulse, a step or a ramp, from any of which the others can be found by simple integrations and differentiations. The fundamental solutions have a self-similar form. The procedure for determining them is based on [10].

Because the linearized motion occurs only in a single direction, the length scales along the beam and transverse to the beam can be considered to be independent, and the four basic dimensions are time T, mass M, longitudinal length L_x and transverse length L_z. The dimension of a quantity is indicated by square brackets. The relevant parameters for the beam have the dimensions

$$[x] = L_x\,, \quad [t] = T\,, \quad [\rho A] = ML_x^{-1}\,, \quad [EI] = T^{-2}ML_x^3\,, \quad [u] = L_z\,, \tag{6}$$

from which a dimensionless length parameter,

$$\xi = \frac{x}{\sqrt{t}}\sqrt[4]{\frac{\rho A}{EI}}\,, \tag{7}$$

is obtained. For a ramp prescribed displacement, which corresponds to an impulse input for the acceleration, with a velocity v_0, $[v_0] = T^{-1}L_z$, the lateral displacement can be made dimensionless with $v_0 t$, so the solution has the form

$$u(t,x) = v_0 t \, u_v(\xi) \; , \tag{8}$$

where u_v is a dimensionless function of the dimensionless parameter ξ, which characterizes the fundamental solution. By substituting this into the partial differential equation (1), the ordinary differential equation

$$u_v'''' + \frac{1}{4}\xi^2 u_v'' - \frac{1}{4}\xi u_v' = 0 \tag{9}$$

is obtained, where a prime denotes a derivative with respect to ξ. The solution which approaches zero at infinity and has unit value and zero slope at zero is

$$u_v = 1 + \xi^2\Big[C\Big(\frac{\xi}{\sqrt{2\pi}}\Big) - \frac{1}{2}\Big] - \frac{2\xi}{\sqrt{2\pi}}\sin\frac{\xi^2}{4} - 2S\Big(\frac{\xi}{\sqrt{2\pi}}\Big) \; , \tag{10}$$

where $C(z)$ and $S(z)$ are the Fresnel integrals [11],

$$C(z) = \int_0^z \cos\frac{\pi}{2}\zeta^2 \mathrm{d}\zeta \; , \quad S(z) = \int_0^z \sin\frac{\pi}{2}\zeta^2 \mathrm{d}\zeta \; . \tag{11}$$

This solution describes the initial phase of the motion if a rigid striker hits the beam. The bending moment and shear force are

$$M_v = \sqrt{EI\rho A}\, v_0 \Big[2C\Big(\frac{\xi}{\sqrt{2\pi}}\Big) - 1\Big] \; , \tag{12}$$

$$D_v = (EI)^{1/4}(\rho A)^{3/4} t^{-1/2} v_0 \frac{2}{\sqrt{2\pi}} \cos\frac{\xi^2}{4} \; . \tag{13}$$

The scaled displacement, bending moment and shear force are shown in Fig. 2. Note that for large values of ξ, the shear force does not decay, while its oscillations become more closely spaced, so for a beam of finite length, the contact may be lost as soon as the first reflections from the supported ends arrive, which is immediately in an Euler–Bernoulli beam. This is a further explanation for the limitation of Saint-Venant's solution (2) and shows that in general, it is necessary either to include a contact stiffness or to take the finite propagation time into account.

For calculating the response if there is a finite contact stiffness, the response to a force impulse is more useful. For an impulse with strength I_P, $[I_P] = T^{-1} M L_z$, the response has the form

$$u = I_P (EI)^{-1/4} (\rho A)^{-3/4} \sqrt{t}\, u_I(\xi) \; , \tag{14}$$

where $u_I(\xi)$ is a dimensionless function for this case. By substituting this in (1), the differential equation

$$u_I'''' + \frac{1}{4}\xi^2 u_I'' + \frac{1}{4}\xi u_I' - \frac{1}{4} u_I = 0 \tag{15}$$

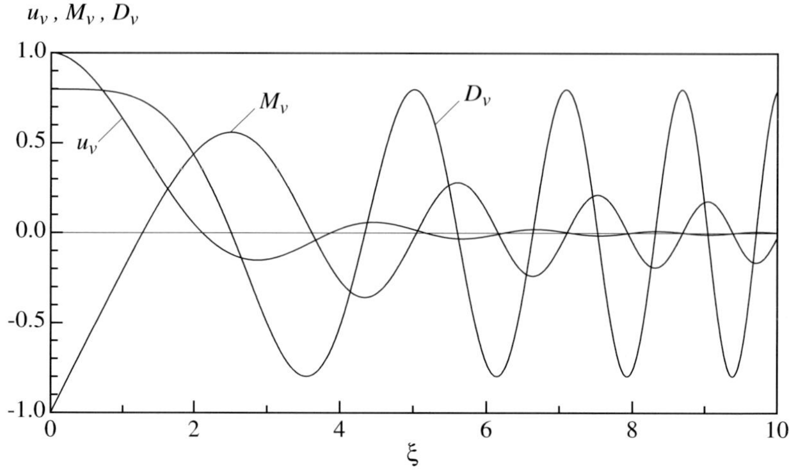

Fig. 2. Scaled displacement, bending moment and shear force for an infinite beam subject to a prescribed ramp displacement input

is obtained, which can be integrated once to yield

$$u_I''' + \frac{1}{4}\xi^2 u_I' - \frac{1}{4}\xi u_I = 0 \; . \tag{16}$$

The constant of integration is zero, because the shear force at $x = 0$ is zero. The equation has the same form as (9) and the required solution which has zero value at infinity and zero slope at $x = 0$ is

$$u_I = \frac{\xi}{2}\left[S\Big(\frac{\xi}{\sqrt{2\pi}}\Big) - C\Big(\frac{\xi}{\sqrt{2\pi}}\Big)\right] + \frac{1}{\sqrt{2\pi}}\left[\cos\frac{\xi^2}{4} + \sin\frac{\xi^2}{4}\right] \; . \tag{17}$$

The velocity is

$$v = \frac{\partial u}{\partial t} = I_P(EI)^{-1/4}(\rho A)^{-3/4}\frac{u_I - u_I'}{2\sqrt{t}} = I_P(EI)^{-1/4}(\rho A)^{-3/4}\frac{v_I}{\sqrt{t}} \; , \tag{18}$$

where

$$v_I = \frac{1}{2\sqrt{2\pi}}\left[\cos\frac{\xi^2}{4} + \sin\frac{\xi^2}{4}\right] \; . \tag{19}$$

The multiplicative constant follows from the impulse balance, which requires that the impulse of the beam for $t > 0$ is equal to I_P. The scaled displacement and velocity are shown in Fig. 3. Note that for large values of ξ, the velocity does not approach zero, but its integral has a finite value. On the other hand, $u_I = O(\xi^{-2})$ for large values of ξ, so it decays rather fast.

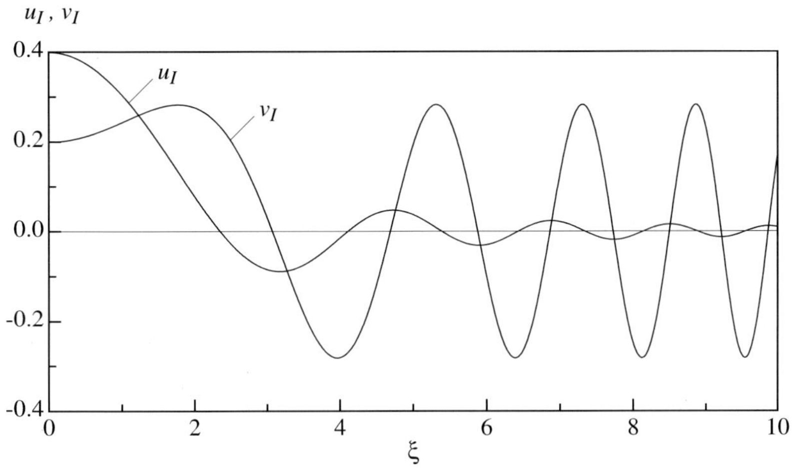

Fig. 3. Scaled displacement and velocity for an infinite beam subject to a force impulse

3 Solution Procedure

The solution of a problem is achieved in a step-by-step way, as in an ordinary numerical integration of equations of motion. For a given time with known history of the contact force, the displacement at the centre of the beam is

$$\begin{aligned} u(t,0) &= \alpha \int_0^t f_0(\bar{t})\sqrt{t-\bar{t}}\,\mathrm{d}\bar{t} \\ &+ 2\sum_{k=1}^{\infty}(-1)^k \int_0^t u_I\left(\frac{kl}{\sqrt{t-\bar{t}}}\sqrt[4]{\frac{\rho A}{EI}}\right) f_0(\bar{t})\sqrt{t-\bar{t}}\,\mathrm{d}\bar{t}\,, \end{aligned} \tag{20}$$

where

$$\alpha = u_I(0) = \frac{1}{\sqrt{2\pi}}(EI)^{-1/4}(\rho A)^{-3/4}\,. \tag{21}$$

A discretized version of (20) is based on the midpoint rule for the numerical integration; only a finite number of terms from the infinite sum need to be included. The Fresnel integrals are evaluated by the procedure from [12]; several other algorithms are readily available as subroutines, all of which are based on approximations by rational functions and asymptotic expansions. The advancing of the solution is by a version of the central difference scheme proposed by the author [13]. In this scheme, first an approximation for the displacements at the midpoint is made by an explicit Euler step. Then with these displacements and the original velocities, the force is evaluated at the midpoint and the accelerations are determined; finally the solution is advanced by assuming that this acceleration is constant over the time step. A modification here is that only the motion of the striker is directly integrated, while the motion of the beam is determined by (20).

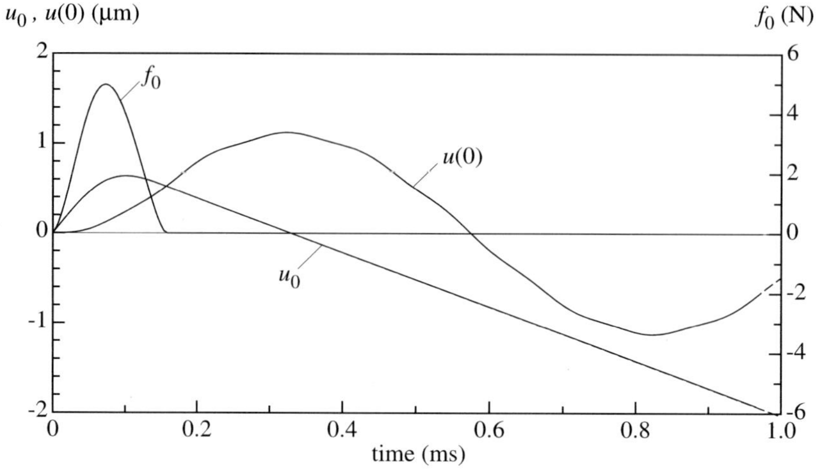

Fig. 4. Displacement of the striker, u_0, deflection of the beam at its centre, $u(0)$, and the contact force, f_0, for the first example; the solid lines are calculated with fundamental solutions and the dashed lines with a modal expansion

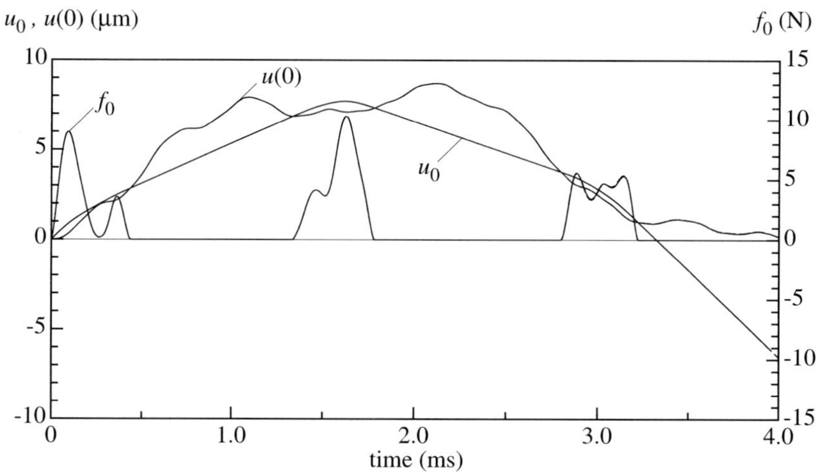

Fig. 5. Results as in the previous figure for the second example

4 Application examples

As an application, the two examples defined in Sect. 2 are used. The solution is determined in two ways, first by using a modal expansion of the motion of the beam with 10 modes, and then by the use of fundamental solutions as proposed here. The results for the displacement of the ball, the deflection of the beam at its centre and the contact force are shown in Figs. 4 and 5. It can be seen that the results calculated with both methods almost coincide. In the first example, there is a single impact. The coefficient of restitution for this case is 0.302, and the maximal deflection of the beam is 1.124 μm. Cox's approximation gives the values 0.360 and 0.709 μm, which is especially bad for the displacement. In the second example, there are three impacts. In [3], the third impact was missed, because the calculation was prematurely terminated. The coefficient of restitution for this case is 0.970 for the modal approximation and 0.963 for the presented solution method , and the maximal deflection of the beam is 8.74 μm. The coefficient of restitution is unusually high: this is atypical; if the third impact would not have happened, which can be so for slightly changed parameter values, the coefficient would have been 0.351. Cox's approximation gives 0.692, and 7.87 μm for the maximal displacement.

In the calculations with the modal approximations, the time step for the intergation method is limited by stability requirements. No such limitation exists with the use of fundamental solutions, and the step size is determined by the required accuracy. For longer time intervals, the evaluation of the convolution integrals becomes more elaborate. If the impact take place over a small time interval, as in the first example, the present method has certainly advantages.

5 Conclusions

Considering time scales reveals which effects are important in an impact problem. Here, it appears that the shear deformation and rotary inertia effects may not be neglected in the example problems.

The fundamental solutions for the lateral vibrations of beams seem to be little known and hardly used. They assist in obtaining an understanding of the impact phenomena and yield scaling relations. The application to impact phenomena seems appropriate, because the duration of the contact is small, which keeps the convolution integrals manageable.

The examples given show that the traditional discretization with vibration modes and the use of fundamental solutions lead to the same results. It showed that the restitution coefficient is determined by the details of the impact and depends critically on the precise parameters.

Before the presented method can be used in general flexible multibody dynamics simulations, more work has to be done. Switching between models is necessary, because between impacts a different model is used than during the impact. More general boundary conditions and beams with non-constant cross-section as well as the influence of shear deformability need to be considered. Damping of the vibrations generated by the impact, especially of those with high frequencies, needs to be considered as well.

References

1. Timoshenko SP (1953) History of strength of materials, with a brief account of the history of theory of elasticity and theory of structures. McGraw-Hill Book Company, New York
2. Saint-Venant AJCB de (1857) Mémoire sur l'impulsion transversale et la résistance vive des barres élastiques appuyées aux extrémités. Comptes Rendus Hebdomadaires des Séances de l'Académie des Sciences 45:204–208
3. Timoschenko S (1913) Zur Frage nach der Wirkung eines Stoßes auf einen Balken. Zeitschrift für Mathematik und Physik 62:198–209
4. Mason HL (1936) Impact on beams. Journal of Applied Mechanics (ASME) 3:55–61
5. Arnold RN (1937) Impact stresses in a freely supported beam. Proceedings of the Institution of Mechanical Engineers 137:217–281, plates 1–4
6. Christopherson DG (1951) Effect of shear in transverse impact on beams. Proceedings of the Institution of Mechanical Engineers 165:176–188
7. Dengler MA, Goland M (1952) Transverse impact of long beams, including rotary inertia and shear effects. In: Sternberg E, et al. (eds.) Proceedings of the first U.S. National Congress of Applied Mechanics. American Society of Mechanical Engineers, New York, pp. 179–186
8. Boussinesq, J (1885) Application des potentiels à l'étude de l'équilibre et du mouvement des solides élastiques, principalement au calcul des déformations et des pressions que produisent, dans ces solides, des efforts quelconques exercés sur une petite partie de leur surface ou leur intérieur; mémoire suivi de notes étendues sur divers points de physique mathématique et d'analyse. Gauthier-Villars, Paris
9. Schwieger H (1965) Die maximale mechanische Beanspruchung beim zentralen Balkenquerstoß. Zeitschrift für angewandte Mathematik und Mechanik 45:499–506
10. Barenblatt GI (2003) Scaling. Cambridge University Press, Cambridge
11. Abramowitz M, Stegun IA (eds.) (1965) Handbook of mathematical functions, with formulas, graphs, and mathematical tables. Dover Publications, New York
12. Boersma J (1960) Computation of Fresnel integrals. Mathematics of Computation 14:380
13. Meijaard JP (1991) A comparison of numerical integration methods with a view to fast simulation of mechanical dynamical systems. In: Haug EJ, Deyo RC (eds.) Real-time integration methods for mechanical system simulation. Springer-Verlag, Berlin and New York, pp. 329–343

Contact Indentations Determined by a Direct Super Finite Element Approach

Pauli Pedersen

Dept. of Mechanical Engineering, Solid Mechanics, Technical University of Denmark
Nils Koppels Allé, Building 404, 2800 Kgs.Lyngby, Denmark.
pauli@mek.dtu.dk

1 Introduction and contact parameters

For more general introduction to analytical contact analysis, see [1] and [2], and for more general introduction to numerical contact analysis, see [3]. In this paper we present something in-between analytical solutions and advanced numerical solutions; a direct numerical analysis of elastic contact without incrementation and iteration. As a short introduction the main parameters of analytical contact analysis are presented.

We assume two elastic bodies (termed 1 and 2) in contact. The *total contact force* is F and the force distribution is described by the *contact pressure distribution* $p = p(s)$, where s gives the position at the contact surface Γ, i.e.,

$$F = \int_{\Gamma} p d\Gamma \tag{1}$$

The *contact size* is characterized by a single parameter a, say the extension of contact from the origin to a, i.e., the contact domain is specified by $0 \leq s \leq a$, with s as the distance from the origin. Part of the solution to a contact problem is to determine a and $p(s)$ for a specified total force F. As we shall see the inverse problem of determining F and $p(s)$ for a given contact size a is more directly solved. The contact pressure must be non-negative, i.e., $p(s) \geq 0$ everywhere. In this paper friction is neglected.

The solution is depending on the geometry of the undeformed surfaces. For circular geometry with two surface radii R_1 and R_2, the mean value of curvature $(\frac{1}{b} = \frac{1}{2}(\frac{1}{R_1} + \frac{1}{R_2}))$ is the relevant parameter. For the presented analytical solution we thus define the *geometry parameter*

$$b := 2R_1R_2/(R_1 + R_2) \tag{2}$$

and for the specific cases of equal radii, or for a flat surface we have

Peter Eberhard (ed.), Multiscale Problems in Multibody System Contacts, 183–193.

$$b = R \quad \text{for} \quad R = R_1 = R_2 \qquad \text{and} \qquad b = 2R \quad \text{for} \quad R_1 = \infty \tag{3}$$

For conforming surfaces we insert $R_1 = -R_1$ in (2) and get

$$b = 2R_1R_2/(R_1 - R_2) \tag{4}$$

and the important conforming case of $R_1 = R_2 + \Delta R$ with $\Delta R/R_2 << 1$ gives

$$b = 2R^2/\Delta R \qquad \text{for} \qquad R = R_2 \simeq R_1 = R_2 + \Delta R \tag{5}$$

From a numerical point of view the mutual geometry of the contact surfaces is described by a gap function (here gap vector $\{g\}$)

$$\{g\} := \{h_1\} - \{h_2\} \tag{6}$$

In "classical" contact we normally have $\{g\} \geq \{0\}$, say with initially only one contact point (one zero component of $\{g\}$), while for shrink fit contact we normally have $\{g\} \leq \{0\}$, and $-\{g\}$ is then named interference.

The materials of the bodies in contact are also of vital importance. In analytical solutions the materials are described with two *material flexibilities* m_1 and m_2, that for plane strain problems and 3D problems are defined as

$$m_1 := (1 - \nu_1^2)/E_1 \qquad \text{and} \qquad m_2 := (1 - \nu_2^2)/E_2 \tag{7}$$

where E_i, ν_i are the modulus of elasticity and the Poisson's ratio for body i. For non-isotropic materials a description by a single parameter is not possible. For a rigid surface i we get $m_i = 0$.

The parameter of major importance for the present paper is *the indentation*, that in the present paper is denoted e_0. In the literature this parameter is mostly denoted by the prefix δ and it is also named penetration, compliance and normal approach. The parameter describes the uniform, mutual compressive translation of the two bodies in contact; the translation that is needed to obtain non-negative contact pressure. For this important quantity we can in the literature find non-equal expressions, so let us restrict to proportionality $e_0 \propto a^2/b$ as stated in [4]

2 Hertz solutions

The classical Hertz solutions ([5]) are presented and derived in [6] and in [2] with a special section on conforming surfaces, including results from [7]. The Hertz solutions are presented in explicit analytical forms, and the basic assumption for the Hertz solutions is that they rely on the elasticity solution for a point force, acting on an infinite half space. The primary results give the size of the contact domain a as a function of the total force F (with material flexibilities and geometry as parameters). For the 3D case the result is

$$a = \left(\tfrac{3}{8}(m_1 + m_2)bF\right)^{1/3} \tag{8}$$

and for the 2D case the result is

$$a = \sqrt{\tfrac{2}{\pi}(m_1 + m_2)b\tfrac{F}{t}} \tag{9}$$

where t is the thickness/length in the third dimension.

The pressure distribution $p(s)$ is found elliptical and by integration then given as a function of the total force F

$$p = p_0\sqrt{1 - (\tfrac{s}{a})^2} \tag{10}$$

with the amplitude for 3D problems

$$p_0 = \frac{3}{2\pi}(\frac{8}{3(m_1 + m_2)b})^{2/3}F^{1/3} \tag{11}$$

and for the 2D case the amplitude is

$$p_0 = (\frac{2}{\pi t(m_1 + m_2)b})^{1/2}F^{1/2} \tag{12}$$

For the variety of problems solved numerically during the present study we find rather good agreement with these simple analytical results. However, for the determination of the indentation, numerical calculation is needed in most cases. A fruitful combination of analytical and numerical methods therefore seems a good approach.

3 The suggested direct method, using super finite element

The essential point of the applied finite element direct method is that the contacting bodies are analyzed individually. Starting with body 1, then for the contact nodal degrees of freedom by condensation we have

$$[S_1]\{D_1\} = \{A\} \quad \Rightarrow \quad \{D_1\} = [S_1]^{-1}\{A\} \tag{13}$$

where $[S_1]$ is the super element stiffness matrix (in the actual cases of order equal to the number of nodes on the contact surface), $\{D_1\}$ contains the contact nodal displacements in the direction of mutual compression, and $\{A\}$ contains the corresponding contact nodal forces. The practical determination of the super element stiffness matrix $[S_1]$ is described in [8] and [9]. (The solution in terms of the nodal displacements $\{D_1\}$ for given nodal loads $\{A\}$ will give the same result as in the calculation with the full stiffness matrix (say of order 10000 - 100000)). Our goal is to determine $\{A\}$, knowing $\{D_1\}$

only relative to the corresponding displacement $\{D_2\}$ of the other body at the contact surface.

For body 2 with only the contact nodal degrees of freedom we have a similar super element description

$$[S_2]\{D_2\} = -\{A\} \qquad \Rightarrow \qquad \{D_2\} = -[S_2]^{-1}\{A\} \tag{14}$$

Note that the equilibrium of the nodal actions is fulfilled by assuming the same nodal forces $\{A\}$ in (13) and (14) except for the sign.

The positions in the force direction of the contact nodes of the two bodies before displacement are given by functions (here vectors) $\{h_1\}$ and $\{h_2\}$, and the corresponding displacements by $\{D_1\}$ and $\{D_2\}$. The initial gap function (vector) $\{g\}$ is defined in (6) and the gap vector $\{\tilde{g}\}$ after displacements is

$$\{\tilde{g}\} = (\{h_1\} + \{D_1\}) - (\{h_2\} + \{D_2\}) = \{g\} + (\{D_1\} - \{D_2\}) \tag{15}$$

Substituting (13) and (14) into (15) and demanding the resulting gap vector $\{\tilde{g}\}$ to be zero we get the corresponding force vector $\{A_g\}$

$$\{0\} = \{g\} + ([S_1]^{-1} + [S_2]^{-1})\{A_g\} \Rightarrow \{A_g\} = -\left([S_1]^{-1} + [S_2]^{-1}\right)^{-1}\{g\} \tag{16}$$

In general we cannot expect a meaningful solution ($\{A_g\} \geq \{0\}$) corresponding to any prescribed gap vector $\{g\}$.

However, with an additional uniform translational indentation e_0 in the direction of mutual compression we obtain non-negative contact forces. This scalar indentation e_0 is the mutual uniform impression of the two bodies. The displacement vector from the additional indentation is $e_0\{1\}$, where the one vector $\{1\}$ has the value 1 for all components. We determine the value of the indentation e_0 such that the minimum value in the resulting force vector $\{A_r\}$ is zero. Let the force vector resulting only from the uniform indentation vector $\{1\}$ be $\{A_u\}$, calculated as in (16)

$$\{A_u\} = \left([S_1]^{-1} + [S_2]^{-1}\right)^{-1}\{1\} \tag{17}$$

then the indentation e_0 is determined by the condition

$$\{A_g\} + e_0\{A_u\} \geq \{0\} \qquad \Rightarrow \qquad e_0 \tag{18}$$

i.e., the indentation e_0 is determined by equating the minimum value in the vector $\{A_g\} + e_0\{A_u\}$ to zero. The resulting force vector $\{A_r\}$ (contact pressure described in the finite element model) is then

$$\{A_r\} = \{A_g\} + e_0\{A_u\} \tag{19}$$

Having determined the mutual contact forces $\{A_r\}$, the two bodies in contact can be analyzed individually, and all the detailed informations from a normal finite element analysis are available. It may be criticized that the problem is solved inversely, assuming the contact size a and determining the total contact force F. However computer times are a few seconds and parameter studies are therefore easily performed.

The matrix $[K] = ([S_1]^{-1}+[S_2]^{-1})^{-1}$ is a stiffness matrix, which is positive definite and strictly diagonal dominant. From this follows for each row (or column) in $[K]$ that $\sum_{j=1}^{n} K_{ij} > 0$ for all i; therefore $\{A_u\} > \{0\}$, and we can always find an indentation e_0. Thus the contact problem has a solution with only positive contact pressure.

4 Solid and Hollow spheres in contact

By specific examples we compare results from Hertz analytical solutions with the obtained results from the finite element solutions. Single sphere problems illustrate when the analytical determined indentation is not correct.

4.1 A single sphere

The first specific example to be presented is a problem where symmetry arguments can be applied to analyze the axisymmetric domain shown to the right in Figure 1, with the appropriate boundary conditions.

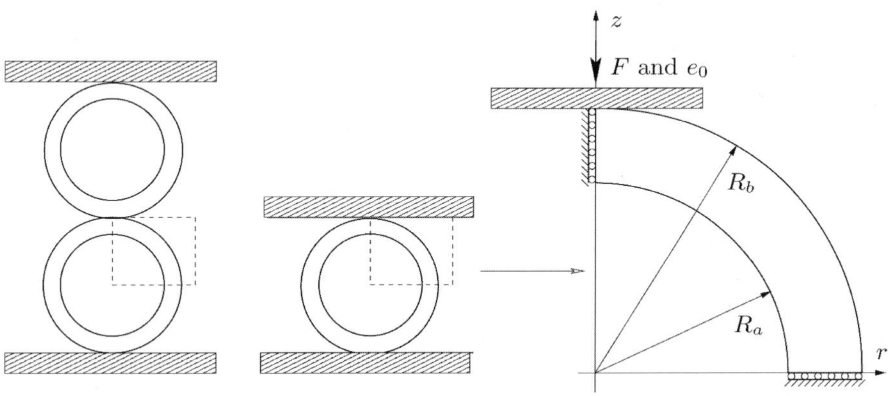

Fig. 1. Model for a hollow sphere in contact with a similar sphere, or between two rigid plane plates. By symmetry only a quarter of one sphere is modeled. For solid spheres we have $R_a = 0$.

For a solid sphere Table 1 lists, for four values of total contact force, the main quantities from the Hertz solutions and from the finite element solution. For all four cases of total load we note good agreement between the FE and Hertz values for the contact extension a. Treating only half the sphere in the

FE model we must multiply e_0 by 2 before comparing to the Hertz value δ (say from [10]). For the solid sphere we note good agreement also for the indentation.

Table 1. Results with solid single sphere. Comparison of results from finite element procedure with Hertz solutions using the same total forces F.

Quantity	case I	case II	case III	case IV
Force F in kN	148.	695.	1900.	4040.
FE a in mm	7.87	13.14	18.43	23.74
Hertz a_H in mm	7.96	13.34	18.65	23.98
FE $2e_0$ in mm	0.060	0.174	0.344	0.568
Hertz δ in mm	0.063	0.178	0.348	0.575

Table 2. Results with hollow single sphere. Comparison of results from finite element procedure with Hertz solutions using the same total forces F.

Quantity	case I	case II	case III	case IV
Force F in kN	46.	252.	700.	1525.
FE a in mm	5.9	9.9	13.8	17.8
Hertz a_H in mm	5.4	9.5	13.7	17.3
FE $2e_0$ in mm	0.042	0.182	0.44	0.88
Hertz δ in mm	0.029	0.090	0.179	0.30

For a hollow sphere with $R_a = 0.9R_b$ Table 2 lists resulting values, similar to the results for the solid sphere. For all four cases of total load we again note good agreement between the FE and Hertz values for the contact extension a. For this hollow sphere there is a disagreement in values for indentation which is not unexpected, because of the larger flexibility of the hollow sphere as compared to the solid sphere. The analytical results (same as for the solid sphere) do not include a parameter for inner radius of the hollow sphere. An attempt to physically explain the result that $2e_0 > \delta$ may be as follows: the bending tends to increase the original curvature at the contact surface, and this implies a further indentation before a straight deformed contact surface is obtained.

4.2 Two sphere in outer contact

As an example where the contact surface is not plane, Figure 2 shows two different elastic spheres in contact. The chosen size parameters are for the smaller upper hollow sphere: $b_1 = 20$ mm and $a_1 = 10$ mm and for the larger, lower hollow sphere: $b_2 = 40$ mm and $a_2 = 30$ mm. Both spheres are made of the same material with modulus of elasticity $E = 2.0 \cdot 10^{11}$ Pa and

Poisson's ratio $\nu = 0.3$. We analyze two different cases of support conditions, corresponding to the angle for the fixed supports (bold lines in Figure 2) being $\phi = 45^o$ and $\phi = 5^o$. This change in boundary conditions resulted in a change in compliance from indentation $e_0 = 1.41$ mm to $e_0 = 2.76$ mm, and this difference in compliance cannot be explained by Hertz analysis.

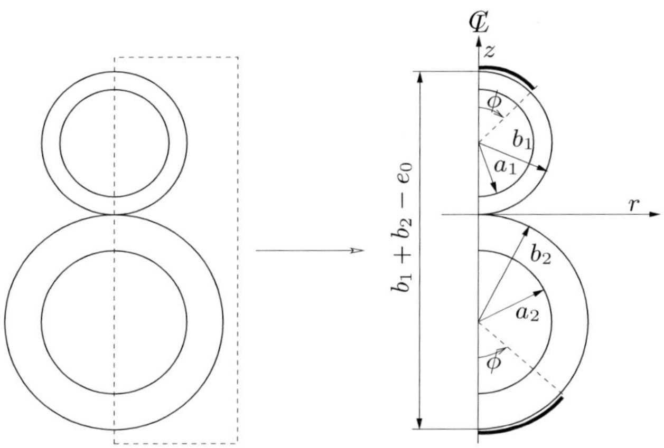

Fig. 2. Model for two elastic hollow spheres in contact. The indentation measure is e_0, outer radii dimensions b_1, b_2, and hole radii dimensions a_1, a_2. The domains of fixed supports described by the angle ϕ.

The relation between the extension of the contact area a and the total contact force F agrees well with the Hertz solution. For both support cases we have chosen $a = 3.75$ mm, resulting for both support cases in a total force $F = 4.96 \cdot 10^5$ N. For this force the Hertz solution according to [10] gives $a = 3.5$ mm, which in fact is a rather large contact area for the actual dimensions. We conclude that good agreement is obtained for the size of the contact area.

5 Shrink fit analysis and design

Shrink fit or interference fit is an important connection in a number of mechanical parts, say to obtain a prestress (residual stress) state or simply to connect different parts in a reliable and non expensive manner. A thick cylinder that must sustain high internal pressure is an important example, and so are shaft disk connections that must transmit high torsional and/or axial loads. Stress concentration is important for the fatigue of the shafts as described in [10] with reference given to the early research of [11]. It is recognized that the design of the shrink fit components is an important issue, but still no final answer is given to the problem.

5.1 Classical 2D analytical analysis and its limitations

For the two dimensional classical models, we have analytical results that follow from a formulation which gives the Euler differential equation. We first show such a two dimensional specific solution and then evaluate these results relative to an axisymmetric finite element solution. Using a three dimensional, axisymmetric finite element model we validate the accuracy of this solution which is based on the plane stress assumption. The shown explicit solution is independent of Poisson's ratio ν, but with $\nu > 0$ the inner part expands in the axial direction due to the shrink fit pressure, while the outer part shrinks in the axial direction, so ν must have an influence. Also the stress distribution will not be constant in the axial direction z, and especially at axial boundaries the modifications to the analytical solutions may be important. The model in Figure 3 is discussed in detail in [8]

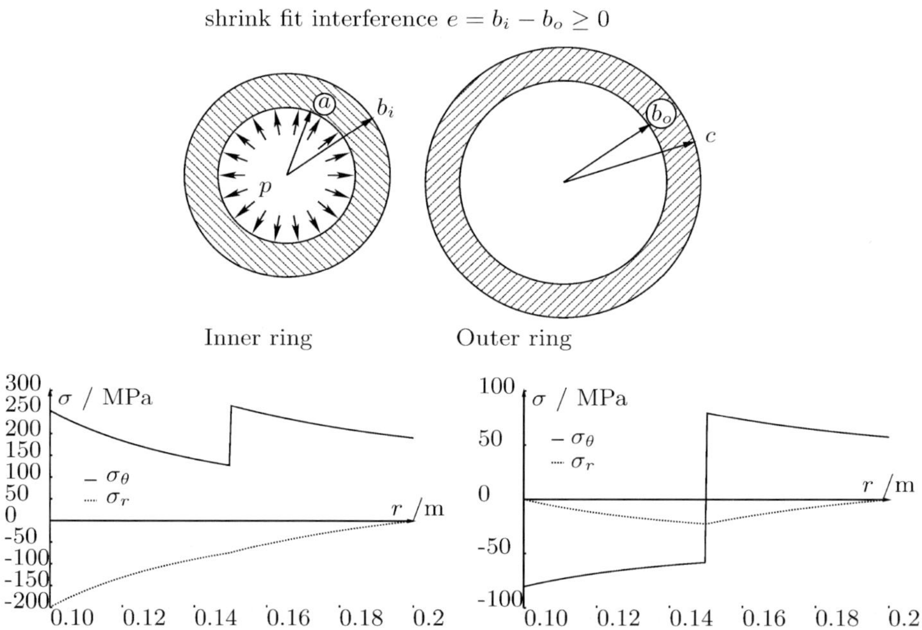

Fig. 3. Shrink fit model from [12] with internal pressure p and shrink fit interference e. The resulting analytical circumferential stresses σ_θ and radial stresses σ_r are shown to the left with and to the right without internal pressure.

5.2 Design of shrink fit surfaces

We turn to the design problem, where the initial contact surface may be curved, and ask for the shape, which will return a prescribed distribution of contact pressure.

We suggest to postulate the shrink fit pressure and then derive the corresponding shrink fit interference; in the axisymmetric case , the interference function $e = e(z)$. In reality this is a simple linear and direct analysis. Two resulting interference functions are shown in Figure 4, for detail see [8].

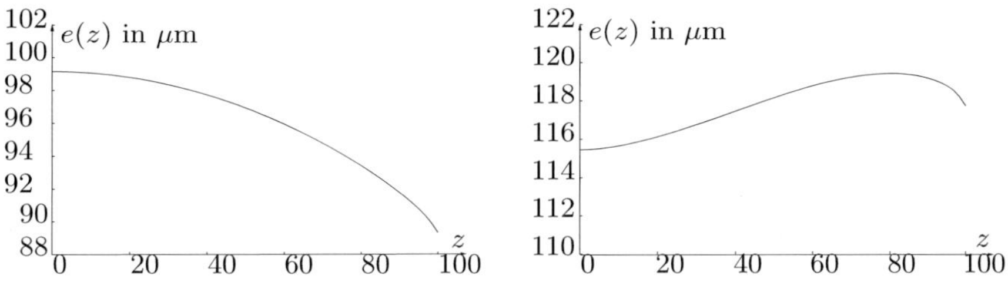

Fig. 4. Interference function $e = e(z)$ from analysis with two different pressure distributions. The coordinate z is the axial position from axial symmetry measured in mm, and the shrink fit interference function $e = e(z)$ is measured in μm.

6 Orthotropic 2D disc to pin contact

A recent study, [13], is restricted to two dimensional problems of conforming cylindrical contact with focus on the orthotropic disc-pin contact where the hole is in an orthotropic disc,i.e., in a non-isotropic material. Especially the indentation is badly estimated by the analytical methods and therefore deserves special attention. Results from a number of parameter studies of the influence of clearance is presented, and from this follows that almost linear relations are found. It is concluded that a simple analytical Hertz formula is useful, but it can-not give detailed information.

Inserting (5) in the Hertz solution (9) and solving for the material flexibilities, we get

$$(m_d + m_p) = \frac{\pi}{4}\frac{1}{F}\frac{ta^2 \Delta R}{R^2} \tag{20}$$

with index d for the disc material and index p for the pin material. A possible approach is to estimate $(m_d + m_p)$ from this when all other quantities are known for a specific case and then assume independence of the actual case. In order to test this approach we have analyzed fifteen different cases by the method described in Section 3, i.e., prescribe t, a, ΔR, R and determine F. Table 3 shows the individual estimated values of $(m_d + m_p)$.

The almost constant estimated values of $(m_d + m_p)$ as a function of the clearance ΔR, more or less confirms the linear dependence in (20). However, the influence of the contact size parameter a is more complicated than shown in (20), and the determined total flexibilities vary in the range 3.4 to 4.0 $(10^{-11}\ \mathrm{Pa}^{-1})$.

Table 3. Estimated total contact flexibility $(m_d + m_p)$ from 15 different cases, analyzed by the super element direct method described in Section 3.

	$\Delta R = 20\mu m$	$\Delta R = 40\mu m$	$\Delta R = 60\mu m$	$\Delta R = 80\mu m$	$\Delta R = 100\mu m$
size a in mm	Flexibility sum $m_d + m_p$ in Pa^{-1}:				
$a = 0.5277$	$3.88 \cdot 10^{-11}$	$3.92 \cdot 10^{-11}$	$3.96 \cdot 10^{-11}$	$4.00 \cdot 10^{-11}$	$4.04 \cdot 10^{-11}$
$a = 1.0181$	$3.58 \cdot 10^{-11}$	$3.62 \cdot 10^{-11}$	$3.65 \cdot 10^{-11}$	$3.68 \cdot 10^{-11}$	$3.72 \cdot 10^{-11}$
$a = 1.4512$	$3.40 \cdot 10^{-11}$	$3.43 \cdot 10^{-11}$	$3.46 \cdot 10^{-11}$	$3.48 \cdot 10^{-11}$	$3.51 \cdot 10^{-11}$

The almost linear dependence of contact forces as well as of indentations upon the clearance is clear in Figures 5, and this may be stated as a nice practical result that is not intuitively clear. However, the a^2 dependence has some discrepancy, illustrated in Figures 5 by the separated curves.

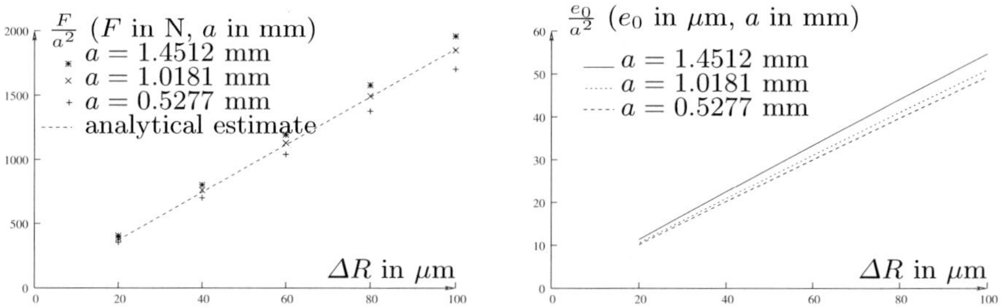

Fig. 5. To the left finite element determined forces as a function of clearance for three different sizes of contact surface, together with analytical result based on estimated total material flexibility. To the right finite element determined indentation as a function of clearance for three different sizes of contact surface. Analytical result for the indentations e_0 are not available.

7 Conclusion

Contact problems with two flexible bodies are involved problems, and they are difficult to solve even with the finite element method. A direct method based on a super element procedure is described and documented by examples. The procedure is tested with classical examples for which analytical results based on Hertz solutions are available. For the size of the contact area and the contact pressure distribution as a function of total contact force this confirms the accuracy of the Hertz solutions.

However, disagreement in contact compliance/indentation is found for problems where the total model is flexible, and thus the assumption of an infinite domain for the Hertz solutions is not satisfied. For these cases it is found that the needed indentation, to satisfy the condition of non-negative contact pressure, is larger than the indentation determined by formulas found in the literature. Thus the actual contact compliance must be determined numerically, say, by the suggested finite element procedure.

An alternative practical procedure for some contact problems may be as follows. Assume that the Hertz solution for pressure distribution and size of contact area, as a function of given total contact force, can be used. Then we analyze the contact bodies individually with a FE program. With verified contact condition, the FE results will give the correct indentation. If contact conditions are violated, iterations may be needed, but no direct couplings in the FE calculations are needed. This procedure has been used to verify the solutions obtained by the direct analysis.

References

1. G. M. L. Gladwell. *Contact problems in the classical theory of elasticity.* Sijthoff and Noordhoff, Alphen aan den Rijn, The Netherlands, 1980. 716 pages.
2. K. L. Johnson. *Contact mechanics.* Cambridge University Press, Cambridge, UK, 1987. 452 pages.
3. P. Wriggers. *Computational contact mechanics.* Wiley, Chichester, UK, 2002. 441 pages.
4. T. A. Stolarski and S. Tobe. *Rolling Contacts.* Professional Engineering Publishing Ltd., London, UK, 2000. 450 pages.
5. H. Hertz. Über die berührung fester Körper. *Journal für die reine und angewandte Mathematik*, 92:156–171, 1882.
6. S. Timoshenko and J. N. Goodier. *Theory of elasticity.* McGraw-Hill/Kogakusha, Tokyo, Japan, second edition, 1951. 506 pages.
7. A. Persson. *On the stress distribution of cylindrical elastic bodies in contact.* Chalmers Tekniska, Goteborg, Sweden, 1964. Ph.D. thesis.
8. P. Pedersen. On shrink fit analysis and design. *Comput Mech* , 37(2):121–130, 2006.
9. P. Pedersen. A direct analysis of elastic contact. *Comput Mech* , 37(1):221–231, 2006.
10. R. L. Norton. *Machine design - an integrated approach.* Prentice-Hall, New Jersey 07458, USA, second edition, 2000. 1078 pages.
11. R. E. Peterson and A. M. Wahl. Fatigue of shafts at fitted members, with a related photoelastic analysis. *ASME J. Appl. Mech.*, 57:A1–A11, 1935.
12. P. P. Benham, R. J. Crawford, and C. G. Armstrong. *Mechanics of engineering materials.* Longman, Harlow, Essex, England, second edition, 1996. 627 p.
13. P. Pedersen. On the influence of clearance in orthotropic disc-pin contacts. *Composite Structures*, 2006. to appear.

Dynamic Analysis of Intermittent-Motion Mechanisms Through the Combined Use of Gauss Principle and Logical Functions

Ettore Pennestrì, Pier Paolo Valentini, and Leonardo Vita

Università di Roma Tor Vergata
Dipartimento di Ingegneria Meccanica, via del Politecnico, 1 00133 Roma, Italy
pennestri@mec.uniroma2.it

1 Introduction

Intermittent-motion mechanisms play an important role in modern technology. For instance they are key elements of many automatic machines. Scientific literature records different modelling analyses of this kind of mechanism (e.g. [1, 2]). Due to the widespread use of such devices, their analysis and design taking into account impact phenomena appears to be significant.

The dynamic simulation of intermittent motion involves several issues. For example the presence of impact and sudden changes in velocity leads to the requirement of including transient mechanics. Several mathematical models of impact could be found in most of the investigations which deal with mechanism clearances. However in ratchet mechanisms the impact is an usual event whose presence does not depend on the clearances.

The main contributions presented in this paper are:

- an extension of the dynamic formulation proposed by Udwadia and Kalaba [3] to the analysis of impacts;
- an engineering model for the analysis of the impact phenomena in ratchet mechanisms;
- a procedure of optimal design of ratchet mechanism.

2 Brief summary of the Gauss-Udwadia-Kalaba dynamic formulation

The main advantages of this formulation concern with the possibility of reducing the equations of motion to a system of ordinary differential equations (ODE), even in presence of redundant constraints or sudden topology changes. The numerical efficiency of this formulation has been discussed in the following bibliographical references [4, 7].

Peter Eberhard (ed.), Multiscale Problems in Multibody System Contacts, 195–204.

Let us denote with

- $\{F\}$ the vector of external generalized forces;
- $[M]$ the mass matrix;
- $[\Psi_q]$ the Jacobian of constraints equations;
- $\{q\}$ the vector of generalized coordinates;
- $\{\gamma\} = -\left([\Psi_q]\{\dot{q}\}\right)_q \{\dot{q}\} - 2[\Psi_{qt}]\{\dot{q}\} - \{\Psi_{tt}\}$;
- the upperscript $+$ denotes the operation of pseudoinverse of a matrix.

When a redundant set of coordinates is used, the following system of differential-algebraic system of equations (DAE) is obtained

$$\begin{bmatrix} M & \Psi_q^T \\ \Psi_q & 0 \end{bmatrix} \begin{Bmatrix} \ddot{q} \\ \lambda \end{Bmatrix} = \begin{Bmatrix} F \\ \gamma \end{Bmatrix} \tag{1}$$

If we let

$$[M]^{-1} = [M]^{-\frac{1}{2}}[M]^{-\frac{1}{2}} \ , \quad [D] = [\Psi_q][M]^{-\frac{1}{2}} \ , \quad \{\ddot{q}_f\} = [M]^{-1}\{F\} \tag{2}$$

one can demonstrate [3, 4, 7] that

$$\{\ddot{q}\} = \{\ddot{q}_f\} + [M]^{-\frac{1}{2}}[D]^+\left(\{\gamma\} - [\Psi_q]\{\ddot{q}_f\}\right) . \tag{3}$$

3 Multibody dynamics simulation with impact

Let us assume that the impact is at time $t = t_I \in [t_1 \ t_2]$. The variation of speed $\{\Delta\dot{q}\} = \{\dot{q}(t_I^+)\} - \{\dot{q}(t_I^-)\}$ is computed by solving the system

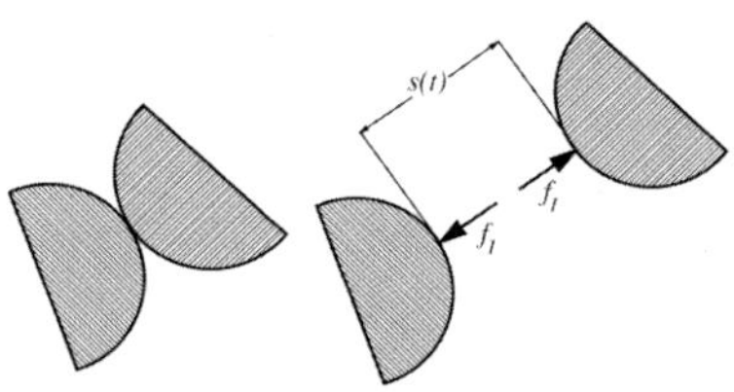

Fig. 1. Impact between bodies

$$\begin{bmatrix} M & \Psi_q^T & \left\{\frac{\partial s}{\partial\{q\}}\right\}^T \\ \Psi_q & 0 & 0 \\ \left\{\frac{\partial s}{\partial\{q\}}\right\} & 0 & 0 \end{bmatrix} \begin{Bmatrix} \Delta\dot{q} \\ \lambda_I \\ -p \end{Bmatrix} = \begin{Bmatrix} 0 \\ 0 \\ -(1+e)\left\{\frac{\partial s}{\partial\{q\}}\right\}\{\dot{q}(t_I^-)\} \end{Bmatrix} . \tag{4}$$

where, as shown in Figure 1, s is the distance between the impacting bodies, e the coefficient of restitution. System (4) can be solved efficiently taking into account the partitioned nature of its coefficient matrix

$$[A] = \begin{bmatrix} A_{11} & A_{12} \\ A_{21} & A_{22} \end{bmatrix} , \tag{5}$$

where $[A_{11}] = [M]$, $[A_{12}] = \left[\Psi_q^T \middle| \left\{ \frac{\partial s}{\partial \{q\}} \right\}^T \right]$, $[A_{22}] = [0]$, $[A_{21}] = [A_{12}]^T$.

Through inversion one obtains

$$[A]^{-1} = \begin{bmatrix} M^{-1} - M^{-1}A_{12}\left(A_{21}M^{-1}A_{12}\right)^{-1} A_{21}M^{-1} & M^{-1}A_{12}\left(A_{21}M^{-1}A_{12}\right)^{-1} \\ \left(A_{21}M^{-1}A_{12}\right)^{-1} A_{21}M^{-1} & -\left(A_{21}M^{-1}A_{12}\right)^{-1} \end{bmatrix} .$$

Therefore, from system (4), after the introduction of the equalities

$$\{v\} = \left\{ 0 \middle| - (1+e) \left\{ \frac{\partial s}{\partial \{q\}} \right\} \left\{ \dot{q}\left(t_I^-\right) \right\} \right\}^T , \tag{6}$$

and $[D_1] = [A_{21}] [M]^{-\frac{1}{2}}$ follows

$$\{\Delta\dot{q}\} = [M]^{-\frac{1}{2}} [D_1]^+ \{v\} . \tag{7}$$

By means of this original equation, the velocity jump $\{\Delta\dot{q}\}$ due to impact is computed. It shares the features of the dynamic formulation expressed by (3). In particular, (7) can be effectively used for the case of mechanisms with redundant constraints and varying kinematic structure, as in the case of intermittent motion mechanisms.

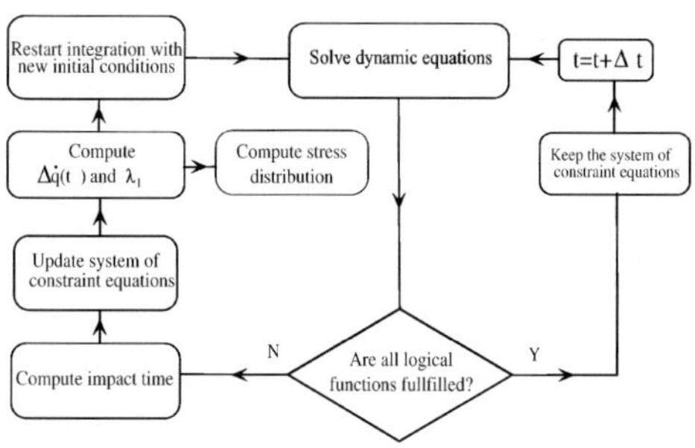

Fig. 2. Flow-chart of the computational procedure for the analysis of intermittent mechanisms

The flow chart shown in Figure 2 summarizes the main steps of the procedure to be adopted in the case of mechanisms with varying kinematic structure. The logical functions are used to monitor such variation.

4 A method for computing stress due to impact

In this section a methodology for computing stress due to anelastic impact is presented. According to W. Goldsmith [5] the transverse collision of a rigid striker with a beam could be treated as an equivalent mass–spring–damper system. The dynamic deflection curve $w(x,t)$ of the beam is assumed geometrically similar to the modal shape $u(x)$, therefore

$$w(x,t) = u(x)\,\eta(t) \; . \tag{8}$$

The values of the equivalent beam mass m and stiffness k follow from the well known conditions

$$\frac{1}{2}m\dot{\eta}^2 = \frac{1}{2}\rho \int_0^l A(x)\left(\frac{\partial w}{\partial t}\right)^2 dx \; , \qquad \frac{1}{2}k\eta^2 = \frac{1}{2}E\int_0^l I_n(x)\left(\frac{\partial^2 w}{\partial x^2}\right)^2 dx \; , \tag{9}$$

where $A(x)$ is the area of the cross section, $I_n(x)$ the moment of inertia, E the Young modulus and ρ the mass density and x coordinate along the beam axis.

The differential equation of motion of the equivalent system in terms of displacement $\eta(t)$ is

$$m\ddot{\eta}(t) + c\dot{\eta}(t) + k\eta(t) = 0 \; , \tag{10}$$

with $c = 2\zeta\sqrt{km}$ damping coefficient. The initial conditions to be imposed are $\eta(0) = 0$ and $\dot{\eta}(0) = v_I$. These represent the kinematic conditions of the beam at impact time. In fact the position instantaneously is not changed while a sudden variation of its velocity is registered. The value of initial velocity v_I follows from (7).

For the complete definition of the dynamic deflection curve, as reported in (8), the modal shape curve $u(x)$ needs to be evaluated. The Rayleigh-Ritz method has been applied to compute the first and second natural frequencies and their relative modal shapes. For this purpose a static deflection curve, consistent with boundary conditions of the beam, is expressed in polynomial form:

$$u(x) = \sum_{i=0}^{n} a_i x^i \; . \tag{11}$$

The coefficients a_i $(i = 0, \ldots, n)$ are computed solving the linear system of equations

$$\frac{\partial V}{\partial a_i} - \Lambda^2 \frac{\partial T}{\partial a_i} = 0 \;\; (i = 0 \ldots n) \; . \tag{12}$$

where V and T represent the potential and kinetic energy of the beam, respectively.

Finally, the time varying expression of the bending moment M_b and the related bending stress σ_b along the beam are computed, respectively, by means of the following equations

$$M_b = -EI_n(x)\frac{\partial^2 w}{\partial x^2} \; , \qquad \sigma_b = \frac{M_b}{W_b} \; . \tag{13}$$

where W_b is the section modulus.

The Figure 3 reports the flow-chart summarizing the steps of the stress recovery methodology used for the optimization of the ratchet mechanism.

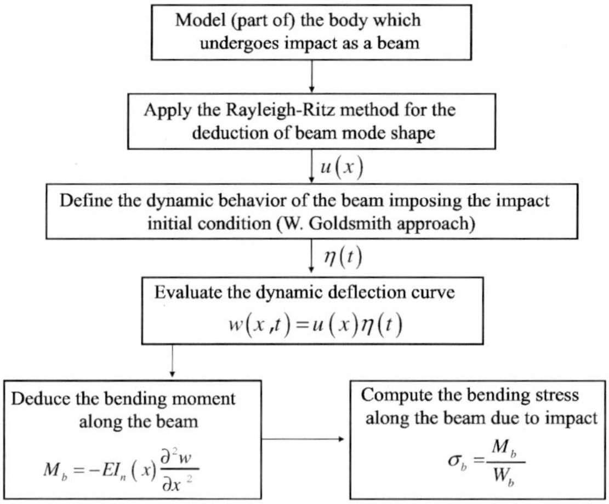

Fig. 3. Beam stress recovery method flow-chart

5 Optimal design of a ratchet mechanism

Ratchets belong to the class of intermittent motion mechanisms. Although the component bodies are subjected to impact, their design is usually based on the hypothesis of static application of loads [6]. Thus, it seems appropriate to apply the previously discussed theory to the optimal design of the toothed wheel. The main geometry of the system is depicted in Figure 4.

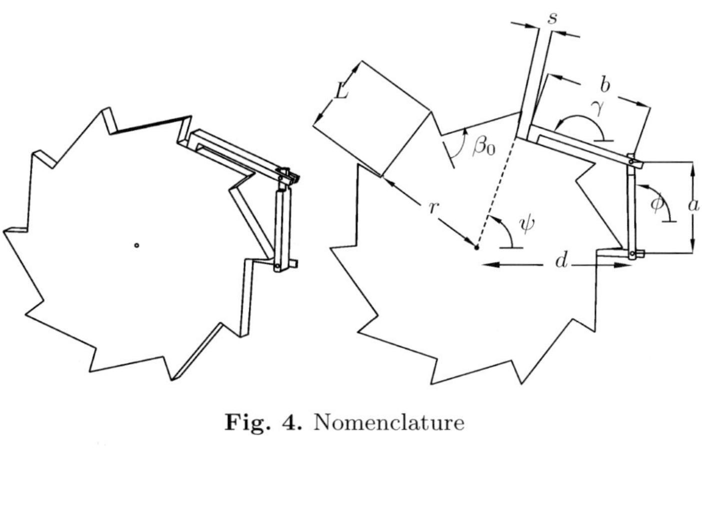

Fig. 4. Nomenclature

The dimensions and the inertial properties of the links are summarized in Table 1[1] and Table 2. The initial configuration of the system is set at $\phi = 77°$,

Table 1. Geometric dimensions (See Figure 4)

$d = 0.130$ m	Distance between the fixed hinges
$a = 0.110$ m	Length of the crank
$b = 0.100$ m	Length of the follower
$r = 0.100$ m	Inner radius of the ratchet wheel

Table 2. Inertial properties

Link	**Mass** [kg]	**Inertia** [kgmm^2]
Crank	0.1000	0.0900
Follower	0.0976	0.0853
Ratchet wheel	5.6264	32.7948

$\gamma = 190°$, $\psi = 75°$ and all the links are initially at rest. A driving torque of 1 Nmm acts on the crank. A resisting torque is applied on the ratchet wheel by means of a torsional spring with stiffness 15 Nmm/rad. The impact is assumed anelastic (*i.e.* restitution coefficient e=0) and friction neglected.

The dynamic analysis of the system is executed by means of Gauss-Udwadia–Kalaba formulation and stabilized by Baumgarte criterion with gain parameters α=10, β=1. These values of Baumgarte's coefficients are properly chosen in order to limit the introduction of extraneous frequencies in the time response.

The impact instant t_I is evaluated by monitoring the distance s. When $s = 0$, an event function stops the ODE solver. Then the variation of velocity $\{\Delta\dot{q}\}$ is computed by means of the expression (7). Afterwards the ODE solver restarts the numerical integration with new initial conditions on velocities.

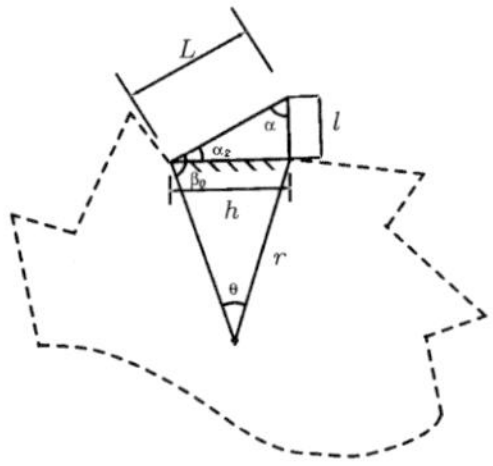

Fig. 5. Ratchet wheel tooth

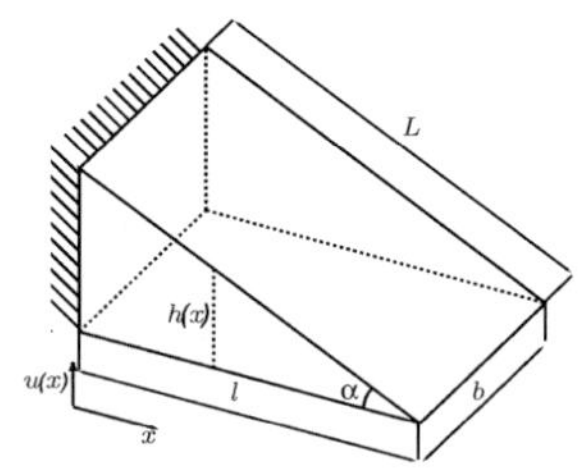

Fig. 6. Cantilever beam model

Once the dynamic analysis of the system is completed, the stress analysis on the ratchet wheel tooth subjected to impact is performed. For our purposes and with reference to Figure 5, the tooth has been modelled as a cantilever beam with its fixed end corresponding to the inner side of the tooth. The

[1] One hinge is the revolute joint between the frame and the ratchet wheel, the other one is the revolute joint between the crank and the frame.

length h depends on teeth number of the ratchet wheel through the angle θ. The length L of the sloping side, which represents the path of the pawl tip, has been arbitrarily set proportional to h by means of a coefficient $c_{prop} = 1.115$. This will ensure realistic tooth proportions for different teeth number configurations. The length l of the beam is function of two parameters, which have been varied (see section Case studies). These parameters are the number nt of teeth, the angle β_0 between the inner radius of the ratchet wheel and L. In order to simplify the analysis without accuracy loss, the shape of the beam has been always considered as a right–angle triangle. In fact, two different geometries are possible during the analysis according to the angle $\alpha^* = \alpha + \alpha_2$. The configuration for $\alpha^* < 90°$ is the one depicted on the left hand side of Figure 7, otherwise it is the one on the right hand side. The shape of the beam considered neglects the grey part in both sides of Figure 7. This simplification is supported by observing that the grey part has a stiffening effect on the adjacent portion of the beam and thus can be regarded as rigid without introducing a great error. Following the stress recovery methodology discussed in

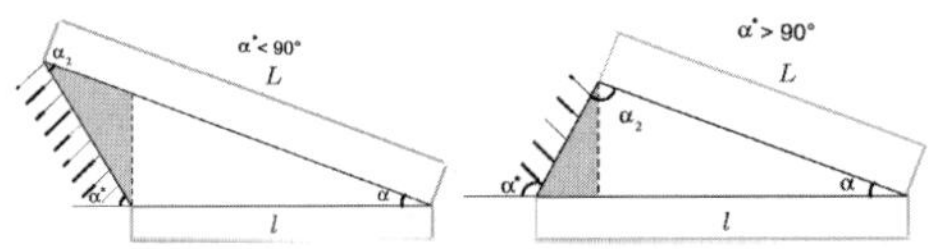

Fig. 7. Shapes of ratchet wheel tooth

the previous section, a modal analysis has been performed. The first two natural frequencies and the relative modal shapes have been computed. However, the contribution of the second modal shape on the beam dynamic behavior has a little effect. For this purpose the deflection law introduced (see Fig. 6 for nomenclature) is $u(x) = a_1 \frac{x^2}{l^2} + a_2 \frac{x^3}{l^3}$.

Since,

$$A(x) = b(l - x)\tan\alpha \ , \qquad I_n(x) = \frac{b}{12}(l - x)^3 \tan^3\alpha \ . \tag{14}$$

by means of the Rayleigh-Ritz method the first two natural frequencies of the cantilever beam are obtained

$$\Lambda_1 = 1.535 \frac{\sqrt{\rho E}\tan\alpha}{\rho l} \ , \qquad \Lambda_2 = 4.994 \frac{\sqrt{\rho E}\tan\alpha}{\rho l} \ . \tag{15}$$

Now solving system (12) once substituting Λ_1 and then Λ_2, one obtains the values of coefficients a_1 and a_2 for the first and the second beam modal shape, $u_1(x)$ and $u_2(x)$, respectively.

After computing the first two modal shapes, the expression of the dynamic deflection w of the beam is obtained from

$$w = u_1(x)\,\eta_1(t) + u_2(x)\,\eta_2(t) \ . \tag{16}$$

where $\eta_1(t)$ and $\eta_2(t)$ are yet to be determined. For this reason, equations (9) and (10) have been solved once with u_1 and then with u_2, respectively. For both of cases the damping factor of the system has been set equal to $\zeta = 0.03$ which corresponds to the damping factor of steel. Identical initial conditions have been imposed. It is important to impose an initial velocity consistent with the cantilever beam boundary condition. In fact the actual configuration of tooth is such that its inner part, (*i.e.* the one regarded as fixed), has an initial velocity due to the angular velocity of the ratchet wheel at impact time. Modeling the tooth as a cantilever beam may not seem appropriate. However, a possible way to mimic the actual behaviour of the tooth is to consider the *relative* velocity between the free-end and the supposed fixed-end. Thus, the value of v_I has to be the difference between the velocity of the free end and the velocity of the end supposed fixed in the model. Finally, the time varying expression of the bending moment and stress along the cantilever beam is computed by means of equations (13), with $W_b = \frac{1}{6}b\left(l-x\right)^2\tan^2\alpha$ section modulus.

5.1 Case Studies

Two main parameters have been varied during the analysis in order to investigate their influence on the stress distribution along the ratchet tooth. The first one is the number of teeth nt which has been varied in the interval $10 \div 25$ with increments of 5. The second parameter is the angle β_0 (see Figure 5) which varies between $85° \div 105°$ with step 5. In Figure 8 the changing in ratchet wheel shape with the number of teeth is shown ($\beta_0 = 95°$). The other three configurations concerning the angle β_0 display the same behavior reported in Figure 8.

In Figure 9 the relative tooth shapes are presented. Tooth thickness has been set equal to $b = 0.02$ m. For all the case studies, the input parameters of the dynamic analysis are kept constant (*e.g.* input torque, initial configurations, etc.). The output parameters monitored are the impact time t_I, the angular velocity of ratchet wheel after impact $\omega\left(t_I^+\right)$ and the maximum stress σ_{max} registered in the tooth.

Moreover in Figure 10 the comparison between the bending stress along the tooth is reported. In each plot are compared the bending stress curves for the same number of teeth and different values of β_0. The tooth length has been normalized in the range $[0,1]$. For all the case studies presented, the bending stress attains its maximum at about 40% of the length of the tooth from the fixed end. A chart of the maximum stress as function of the number of teeth and the slope of the tooth is presented in Figure 11. Once known the values of varying parameters, it is possible to predict the maximum bending stress which will occur. Generally, the maximum stress value reduces by increasing the number of teeth but increases with β_0.

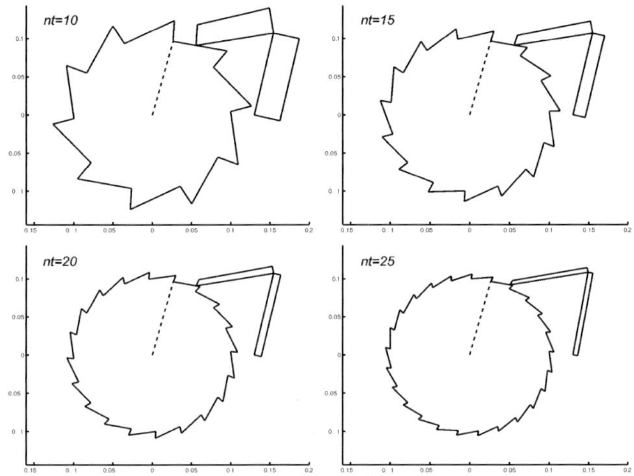

Fig. 8. Ratchet wheel shape for different number of teeth ($\beta_0 = 95°$)

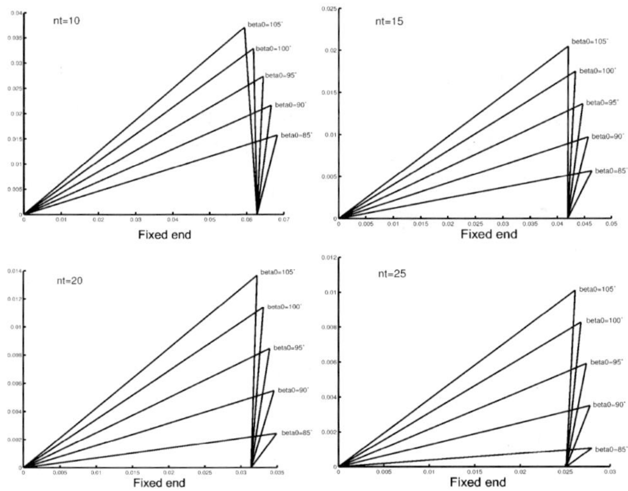

Fig. 9. Shape of ratchet wheel tooth for different configurations

References

1. Bickford JH (1972) Mechanisms for Intermittent Motion. Industrial Press, New York
2. Lee T W, Wang AC (1983) On the Dynamics of Intermittent-Motion Mechanisms. Part 2: Geneva Mechanism, Ratchets, and Escapements. ASME J. of Mech., Transm. and Autom. in Design, September, Vol. 105, pp. 541-551
3. Udwadia F, Kalaba R (1996) Analytical Dynamics a New Approach. Cambridge University Press, Cambridge

4. Ara Arabyan, Fei Wu (1998) An Improved Formulation for Constrained Mechanical Systems, *Multibody System Dynamics*, vol.2, pp.49-69
5. Goldsmith W (2001) Impact. Dover Publications Inc.
6. Lingaiah K. (1994) Machine Design Data Handbook. McGraw-Hill Inc, Chap. 25.
7. de Falco D., Pennestrì E., Vita L. (2005) The Udwadia-Kalaba Formulation: A Report on its Numerical Efficiency in Multibody Dynamics Efficiency and on its Teaching Effectiveness, Eccomas Thematic Conference, Madrid 2005.

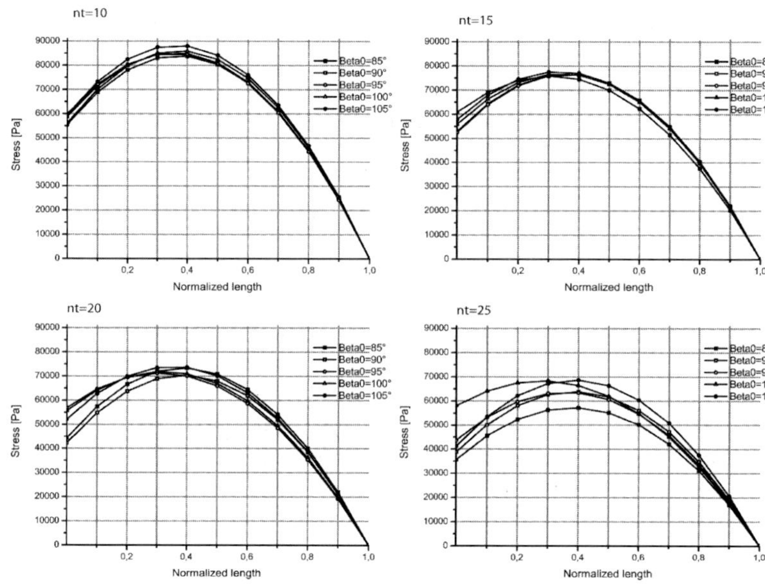

Fig. 10. Stress along the tooth

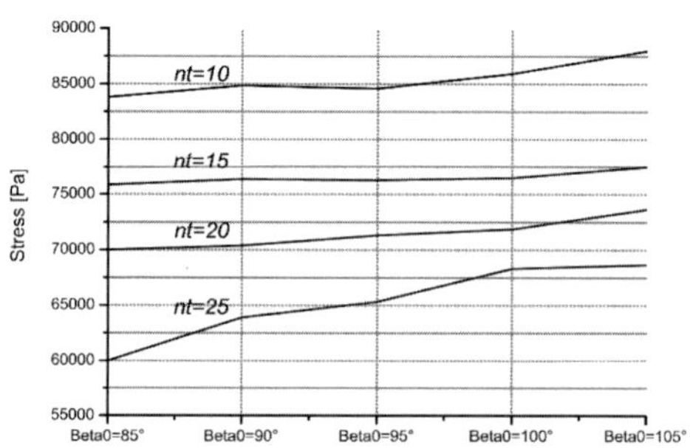

Fig. 11. Maximum bending stress

Multibody Modeling of Pantographs for Pantograph-Catenary Interaction

Frederico Grases Rauter[1,2], João Pombo[2], Jorge Ambrósio[2], Manuel Pereira[2]

[1] SNCF – Direction de l'Innovation et de la Recherche, 45 Rue de Londres, 75008 Paris, France frederico.grases@sncf.fr
[2] IDMEC - -Instituto Superior Técnico, Technical University of Lisbon, Av. Rovisco Pais, 1049-001 Lisboa, Portugal {jpombo,jorge,mpereira}@dem.ist.utl.pt

Summary. In the great majority of railway networks the electrical power is provided to the locomotives by the pantograph-catenary system. From the mechanical point of view, the single most important feature of this system consists in the quality of the contact between the contact wire(s) of the catenary and the contact strips of the pantograph. Therefore not only the correct modeling of the catenary and of the pantograph must be achieved but also a suitable contact model to describe the interaction between the two systems must be devised. The work proposed here aims at enhancing the understanding of the dynamic behavior of the pantograph and of the interaction phenomena in the pantograph-catenary system. The catenary system is described by a detailed finite element model of the complete subsystem while the pantograph system is described by a detailed multibody model. The dynamics of each one of these models requires the use of different time integration algorithms. In particular the dynamics of the finite element model of the catenary uses a Newmark type of integration algorithm while the multibody model uses a Gear integration algorithm, which is variable order and variable time step. Therefore, an extra difficulty that arises in study of the complete catenary-pantograph interaction concerns the need for the co-simulation of finite element and multibody models. As the gluing element between the two models is the contact model, it is through the representation of the contact and of the integration schemes applied for the finite and multibody models that the co-simulation is carried on. The work presented here proposes an integrated methodology to represent the contact between the finite element and multibody models based on a continuous contact force model that takes into account the co-simulation requirements of the integration algorithms used for each subsystem model.

1 Introduction

The pantograph-catenary system is still today the only reliable form of collecting electric energy for running trains. This system should ideally run with relatively low contact forces, in order to minimize wear and damage of the contacting elements of the system, and no contact loss should be observed, so that the power supply would be constant and no electric arching would be observed.

Peter Eberhard (ed.), Multiscale Problems in Multibody System Contacts, 205–226.

Unfortunately such ideal state is impossible to achieve and the best that can be done is to design the catenary and pantograph systems, with particular relevance to their interface, in order to minimize contact loss and to maintain the contact forces within an acceptable operational envelope.

Due to the increasing importance of the catenary pantograph interface with the developments in high-speed trains a large number of works have been developed to study this system [1-5]. Gardou [1] presents a rather simple model for the catenary, using 2D finite elements, where all nonlinear effects are neglected. The single model of catenary analyzed is excited by a lumped mass model of a pantograph. Jensen [2] presents a very detailed study on the wave propagation problem on the catenary and a 2D model for the catenary-pantograph dynamics. In both references [1] and [2] not only the representation of the contact forces is not discussed but also no reference is made on how the integration algorithms are able to handle the contact loss and impact between registration strip and contact wire. Recently Veitl and Arnold [3] proposed a co-simulation strategy between the code PROSA, where a catenary is described by the finite difference method and the SIMPACK commercial multibody code used to simulate the pantograph. All models involved in this work are 3D but the catenaries are modeled are restricted, and therefore, the models and programs can hardly be used for different catenary systems. Labergi [4] presents a very thorough description of the pantograph catenary system that includes a 2D model for the catenary based on the finite element method, and a pantograph model based on a multibody approach. The contact is treated by unilateral constraints. In the work by Seo et al. [5] the catenary contact wire is modeled by a finite elements based on the absolute nodal coordinate formulation and the pantograph is a full 3D multibody model. The contact is represented by a kinematic constraint between contact wire and registration strip and no loss of contact is represented. None of the models used has been validated in this study.

Although major efforts have been done to provide reliable models of pantographs and catenaries and a very good insight on the dynamics of the pantograph catenary interaction has been offered by the studies referred currently there are no accepted general numerical tools designed to simulate the pantograph-catenary system in nominal, operational, and deteriorated conditions. Here it is understood that operating conditions must into account the wear effects and the deteriorated conditions include extreme climatic conditions, material defects or mechanical problems. However, several important efforts have been reported to understand the mechanisms of wear in catenaries and the effect of defect conditions on the dynamics of the complete system [6-8]. In addition the capability to analyze real three-dimensional models of the catenary and pantograph is very limited and no validated models exist for either of the subsystems. The European project EUROPAC project (European Optimized Pantograph Catenary Interface) aims at developing general software that is able to overcome the shortcomings just listed and in the process to develop accurate models for different types of catenaries and pantographs. The software and models are to be used not only for maintenance and design but also to appraise the interoperability between different systems by verifying compliance with standards on the mean contact force value, standard deviation, uplift at supports or maximum deflection of the catenary. The work now presented addresses the choice of the methods used to develop and simulate the pantograph models and to interface them with the code used for catenary analysis in a co-simulation environment.

2 Multibody Dynamics

In this work the methods presented to describe multi-body systems are based on the use of Cartesian coordinates, which lead to a set of Differential-Algebraic equations that need to be solved. Due to the very large simulation periods it very important that appropriate numerical procedures are used to integrate the equations of motion. In particular, the numerical issues that result from the use of this Cartesian coordinates, such as the existence of redundant constraints, the possibility of achieving singular positions or the drift of the constraint position and velocity equations that lead to violation of such equations, are also solved.

2.1 Multibody Equations of Motion

A typical multibody model is defined as a collection of rigid or flexible bodies that have their relative motion constrained by kinematic joints and is acted upon by external forces. The forces applied over the system components may be the result of springs, dampers, actuators or external applied forces describing gravitational, contact/impact or other forces. A wide variety of mechanical systems can be modeled as the schematic system represented in Figure 1.

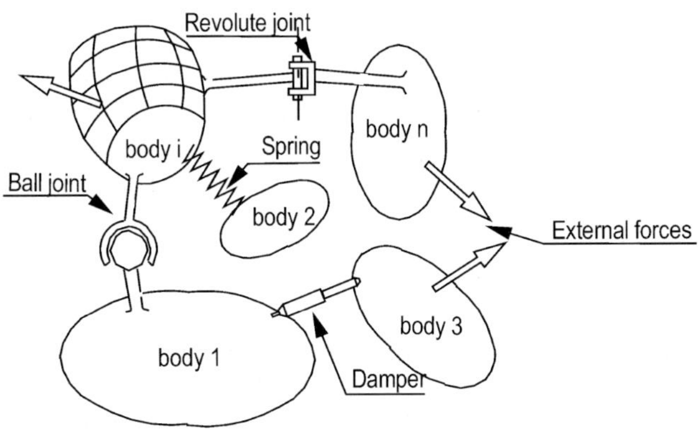

Fig. 1. Generic multibody system.

Let the configuration of the multibody system be described by n Cartesian coordinates $\mathbf{q}$, and a set of m algebraic kinematic independent holonomic constraints $\mathbf{\Phi}$ be written in a compact form as

$$\mathbf{\Phi}(\mathbf{q},t)=\mathbf{0} \tag{1}$$

Differentiating equation (1) with respect to time yields the velocity constraint equation. After a second differentiation with respect to time the acceleration constraint equation is obtained

$$\mathbf{\Phi}_{\mathrm{q}}\dot{\mathbf{q}}=\boldsymbol{\upsilon} \tag{2}$$

$$\mathbf{\Phi}_{\mathrm{q}}\ddot{\mathbf{q}}=\boldsymbol{\gamma} \tag{3}$$

where $\mathbf{\Phi}_{\mathrm{q}}$ is the Jacobian matrix of the constraint equations, $\boldsymbol{\upsilon}$ is the right side of velocity equations, and $\boldsymbol{\gamma}$ is the right side of acceleration equations, which contains the terms that are exclusively function of velocity, position and time.

The equations of motion for a constrained multibody system of rigid bodies are written as [9]

$$\mathbf{M}\ddot{\mathbf{q}}=\mathbf{g}+\mathbf{g}^{(c)} \tag{4}$$

where $\mathbf{M}$ is the system mass matrix, $\ddot{\mathbf{q}}$ is the vector that contains the state accelerations, $\mathbf{g}$ is the generalized force vector, which contains all external forces and moments, and $\mathbf{g}^{(c)}$ is the vector of constraint reaction equations. The joint reaction forces can be expressed in terms of the Jacobian matrix of the constraint equations and the vector of Lagrange multipliers as [9]

$$\mathbf{g}^{(c)}=-\mathbf{\Phi}_{\mathrm{q}}^{T}\boldsymbol{\lambda} \tag{5}$$

where $\boldsymbol{\lambda}$ is the vector that contains m unknown Lagrange multipliers associated with m holonomic constraints. Substitution of equation (5) in equation (4) yields

$$\mathbf{M}\ddot{\mathbf{q}}+\mathbf{\Phi}_{\mathrm{q}}^{T}\boldsymbol{\lambda}=\mathbf{g} \tag{6}$$

A unique solution is obtained when the constraint equations are considered simultaneously with the differential equations of motion and a proper set of initial conditions [9]. Therefore, equation (3) is appended to equation (6), yielding a system of differential algebraic equations to be solved for $\ddot{\mathbf{q}}$ and $\boldsymbol{\lambda}$. This system is

$$\begin{bmatrix}\mathbf{M} & \mathbf{\Phi}_{\mathrm{q}}^{T}\\ \mathbf{\Phi}_{\mathrm{q}} & \mathbf{0}\end{bmatrix}\begin{bmatrix}\ddot{\mathbf{q}}_{r}\\ \boldsymbol{\lambda}\end{bmatrix}=\begin{bmatrix}\mathbf{g}\\ \boldsymbol{\gamma}\end{bmatrix} \tag{7}$$

In each integration time step, the accelerations vector, $\ddot{\mathbf{q}}$, together with velocities vector, $\dot{\mathbf{q}}$, are integrated in order to obtain the system velocities and positions at the next time step. This procedure is repeated up to final time will be reached.

The set of differential algebraic equations of motion (7) does not use explicitly the position and velocity equations associated to the kinematic constraints, equations (1) and (2), respectively. Consequently, for moderate or long time simulations, the original constraint equations are rapidly violated due to the integration process. Thus, in order to stabilize or keep under control the constraints violation, equation (7) is solved by using the Baumgarte Stabilization Method [10] or the Augmented Lagrangean formulation [11], and the integration process is performed using a predictor – corrector algorithm with variable step and order [12]. Furthermore, due to the long time simulations typically required for pantograph-catenary interaction analysis, it is also necessary to implement constraint violations correction methods. The Coordinate Partition method is used for the purpose [9, 13].

2.2 Solution of the Equations of Motion

The forward dynamic analysis of a multibody system requires that the initial conditions of the system, i.e. the position vector $\mathbf{q}^{0}$ and the velocity vector $\dot{\mathbf{q}}^{0}$, are given. With this information equation (5) is assembled and solved for the unknown accelerations, which are in turn integrated in time together with the velocities. Although a constraint violations stabilization method is used it is not possible to proceed with a thorough correction of the numerical errors that occur typically in long running analysis periods. The only solid numerical procedure that always guarantees the fulfillment of the constraint equations is the use of the Coordinate Partition Method to correct the position and velocity constraint equations at any time that the violations exceed a prescribed acceptable tolerance. The coordinate partitioning method is based on the fact that the n coordinates $\mathbf{q}$ are not independent [9, 13]. The method partitions $\mathbf{q}$ into a set of m independent coordinates $\mathbf{v}$ and a set of k dependent coordinates $\mathbf{u}$. Only the independent coordinates are integrated and for the positions and velocities of the dependent coordinates are solved afterwards

$$\boldsymbol{\Phi}(\mathbf{u},\mathbf{v}) = \mathbf{0} \tag{8}$$

$$\boldsymbol{\Phi}_{\mathrm{u}}\dot{\mathbf{u}} = -\boldsymbol{\Phi}_{\mathrm{v}}\dot{\mathbf{v}} \tag{9}$$

The process, schematically shown in Figure 2, starts with an automatic partition of the coordinates using L-U matrix factorization of the Jacobian matrix followed by the correction of the initial conditions. This is one of the key issues is the selection of the independent coordinates, which are here assumed to be maintained as independent throughout the complete simulation process. Due to the type of system that is aimed to be modeled here, which only have moderate relative rotations between system components, the invariance of the independent coordinate set and the redundant constraint set is an acceptable assumption.

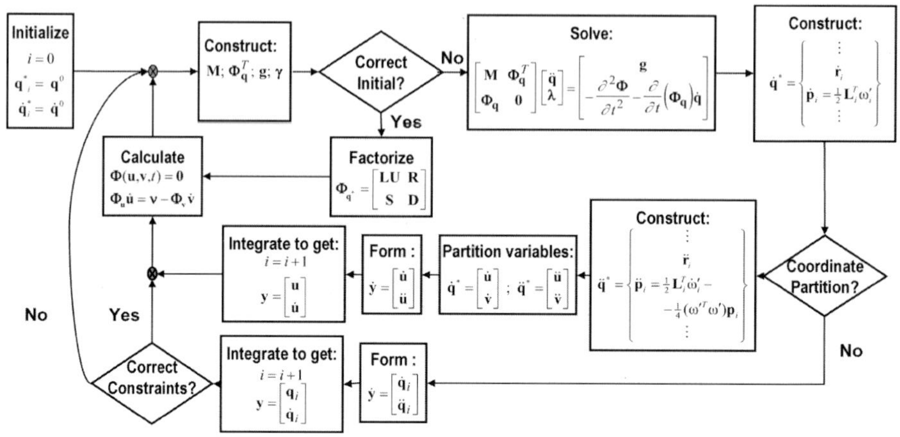

Fig. 2. Flowchart representing the forward dynamic analysis of a multibody system with a numerical strategy for constraint violation correction.

The pantograph-catenary system is characterized by an intermittence of the contact between the contact wire of the catenary and the registration strip of the pantograph. Ideally not only the contact force between these elements should be low but also no loss of contact should be observed. In reality it is impossible to achieve any of these objectives. The numerical methods used for the dynamic simulation must be able to represent the loss and start of contact. This fact puts particular restrictions on the numerical integration algorithms for both pantograph and catenary with particular emphasis on the time step size selection.

3 Pantograph-Catenary System

3.1 Multibody Modeling Issues for the Pantograph

In general, a pantograph consists of a collection of bodies and mechanical elements, as depicted in Figure 3, attached to a railway carbody that is moving along the track Two modeling strategies can be used to define the reference motion of the pantograph: define the kinematics of the pantograph base; define the motion of the railway vehicle and simply fix the pantograph in the top of the train carbody. Once the kinematics of the pantograph is compatible with the geometries of the catenary and railway both strategies are acceptable.

Although the moving components of the pantograph have some level of flexibility in this model they are represented as rigid bodies connected by perfect kinematic joints. These kinematic joints are responsible to control the relative motion between the different pantograph elements. Deformable elements, such as springs and dampers, are used to model relevant internal forces that represent interactions between rigid bodies of the system. In existent pantographs a good number of non-linear elements differ from standard kinematic and compliance elements used in traditional mechanical systems, as those exemplified in Figure 4.

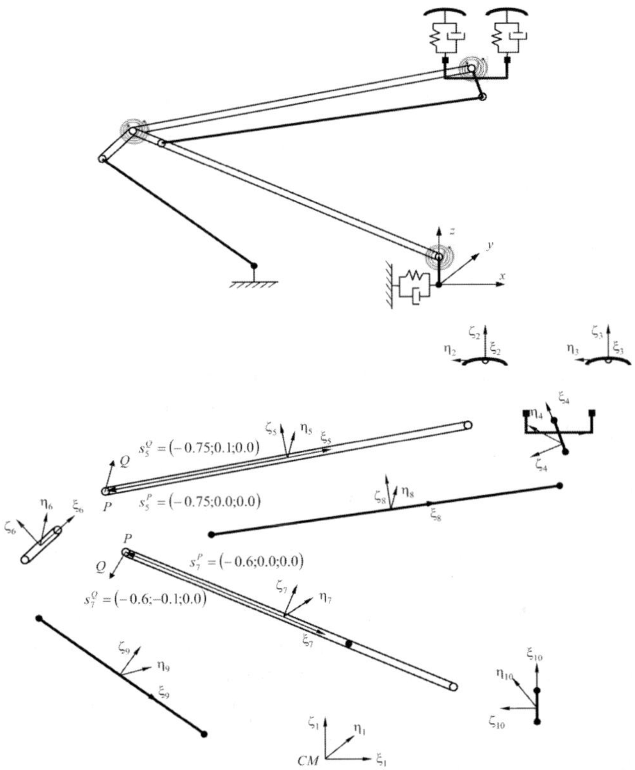

Fig. 3. Graphical representation of a planar pantograph.

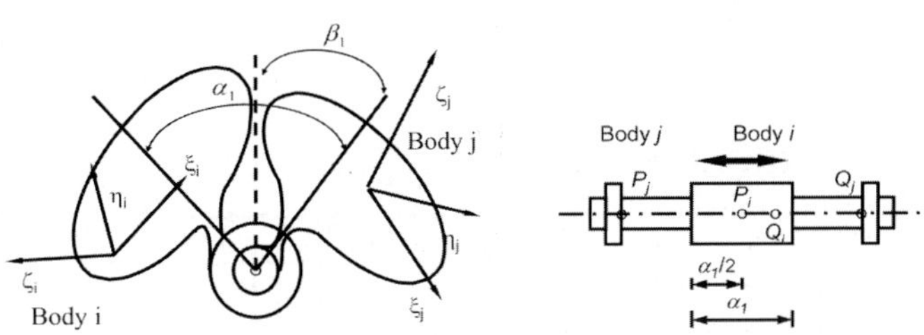

Fig. 4. Revolute and translational joints operational limits used for the pantograph model.

There are different types of force elements that do not have a linear type behavior in the pantograph model. Among these, some relevant force elements, represented in Figure 5, include friction effects besides the elastic forces and the damping.

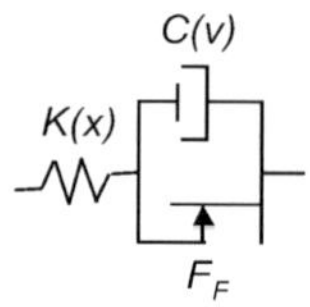

Fig. 5. Model for the low-frequency damper system.

In order to study the pantograph-catenary contact, the motion of the carbody must be defined. For this purpose, a set of control points that are representative of the trajectory of the CM of the carbody and the roll angle at each point must be defined. With such nodal points, the computational tool uses a piecewise cubic interpolation scheme to parameterize the carbody trajectory and orientation as a function of the traveled distance. Then, a prescribed motion constraint is used in order to enforce the carbody to follow the prescribed trajectory with the inclination defined by the roll angle, as depicted in Figure 6. Details of the implementation of this general curve following constraint can be fond in the work by Pombo and Ambrósio [14] and are not repeated here.

3.2 Finite Element Model of Catenaries

The second part of the electric collecting system is the catenary. These systems do not exhibit large displacements, i.e., large rotations, and therefore, they are typically modeled by using linear finite elements. It is not the objective of this work to discuss in detail models for the catenaries but simply to highlight some of the features of these models. Catenaries are complex periodical structures, as that represented in Figure 7. Examples of typical structural elements involved in the catenary model are the contact, stitch and messenger wires, droppers and registration arms. Depending on the catenary systems there are other elements that may have to be considered. In any case, the contact wire is the responsible for the contact between catenary and pantograph and it is the correct modeling of its 'foundations' that is the essential feature of the correct catenary modeling.

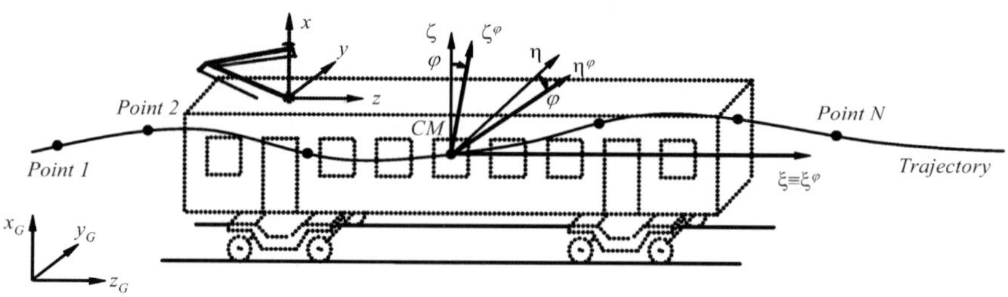

Fig. 6. Prescribed motion constraint of the carbody.

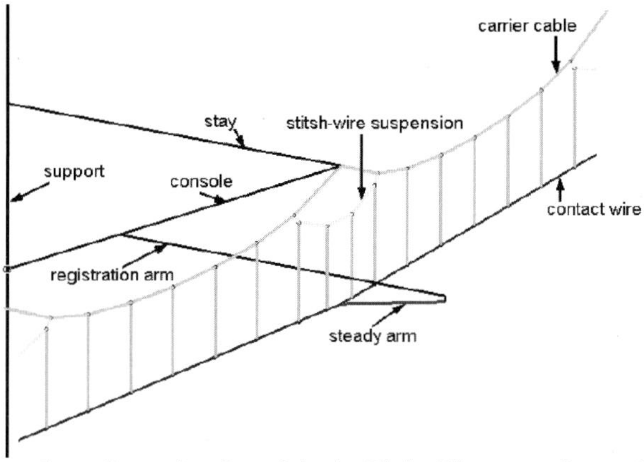

Fig. 7. Catenary three-dimensional mesh in the Finite Element software OSCAR (permission by SNCF).

At this point it must be referred that there is a large variety of catenary systems used worldwide. Even in a single European country there are different types of catenaries in use with different particularities in their construction. For the purpose of this work it is assumed that the model of a contact wire on a flexible foundation is used and modeled by linear finite beam elements.

4 Pantograph-Catenary Interaction Analysis

The main objective of this work is to present a general methodology to analyze the dynamic behavior of the pantograph-catenary system. Therefore it is of crucial importance to be able to consider each subsystem independently modeled, by different methodologies, and analyzed using different computer codes. The use of co-simulation paradigms is, therefore, of fundamental importance in this work. The contact forces model that represents the pantograph-catenary interaction is the driving element of the co-simulation environment. The second element of importance in the co-simulation is the compatibility between the numerical integration strategies used in FE and MB codes.

4.1 Contact Force Model

The pantograph and catenary systems are related by to the interaction between the registration strips of the pantograph and the contact wires of the catenary. The contact force due to pantograph-catenary interaction, regarding present operating conditions and pantograph and catenary technology, is characterized by a high-frequency oscillating force with high relative amplitude. Railway industry measurement data shows that reasonable values for the contact force are, for a train running at approximately 80 m/s: a mean value of 200N oscillating between 400N and 100N. Loss of contact in particular points of the catenary may also occur. Therefore impact effects must be included in the model.

The contact model used here is based on the Hunt and Crossley work [15] and it has been proposed by Lankarani and Nikravesh [16] that developed a contact force model with hysteresis damping for impact in multibody systems. In this work, the Hertzian type contact force including internal damping can be written as [16]

$$F_N = K\delta^n \left[1 + \frac{3(1-e^2)}{4} \frac{\dot{\delta}}{\dot{\delta}^{(-)}} \right] \tag{10}$$

where K is the generalized stiffness contact, e is the restitution coefficient, $\dot{\delta}$ is the relative penetration velocity and $\dot{\delta}^{(-)}$ is the relative impact velocity. The proportionality factor K is obtained from the Hertz contact theory as the external contact between two cylinders with perpendicular axis.

4.2 Co-simulation Between the Pantograph and Catenary Modeling Codes

The analysis of the pantograph-catenary interaction is done through two independent codes, the pantograph code, developed using the multibody methodology reported in this work, and the catenary code, developed using a finite element methodology. Both of these programs can work as stand-alone. The structure of their communication scheme is shown in Figure 8. The MB code provides the FE code with the position and velocity of the pantograph's registration strip. The FEM code calculates the contact force, using the contact model represented by equation (10), and its application points in the pantograph and catenary, using a geometric interference model. These forces are applied to the catenary, in the finite element code, and to the pantograph model, in the MB code. Each code integrates separately the equations of motion of each sub-system based on the shared force information.

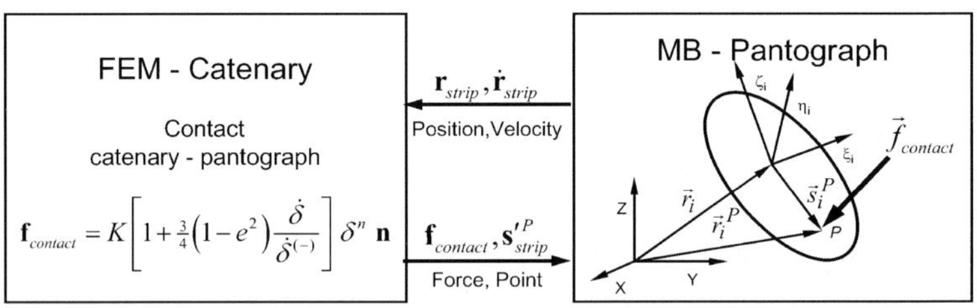

Fig. 8. Scheme for the communication between the FEM and MB codes.

The typical numerical integration algorithms used by FEM codes are Newmark family algorithms [17]. Therefore the FEM code needs a prediction of the positions and velocities not only of the catenary but also of the pantograph in a forthcoming time before advancing to a new time step. A predicted contact force is calculated and, using the finite element method equations of equilibrium, the catenary accelerations are computed for the new time-step. The calculated acceleration values are used to correct the initially predicted positions and velocities of the catenary. On the other hand the MB code uses a Gear multi-step multi-order integration algorithm

[18,19]. To proceed with the dynamic analysis, the MB code needs information about the positions and velocities of the pantograph components and also the contact force and its application point coordinates at different time instants during the integration time period, and not only at its start and end. The compatibility between the two integration algorithms imposes that the state variables of the two subsystems are readily available during the integration time but also that a reliable prediction of the contact forces is also available at any given time step. Several strategies can be envisaged to tackle this co-simulation problem such as the gluing algorithms proposed by Hulbert [20] or the co-simulation procedures suggested by Schiehlen [21]. However, the problem that it is tackled in this work is characterized by a non neglectable frequency of loss of contact between the two systems, almost a problem of intermittent contact, and therefore, a different approach is proposed.

4.3 Communication Synchronization Procedure

The key of the synchronization procedure between the MB and FE codes is the time integration, which must be such that it is ensured the correct dynamic analysis of the pantograph-catenary system, including the loss and regain of contact. Let it be assumed that the FE integration code is of the Newmark family and has a constant time step. Moreover, let it be assumed that the time step of the FE is small enough not only to assure the stability of the integration of the catenary but also to be able to capture all starts of contact between the pantograph registration strip and the contact wire of the catenary. The only restriction that is made to the integration algorithm of the multibody code is that its time step cannot exceed the time step of the FE code. Finally let it be assumed that both codes can start independently from each other, i.e., the catenary FE model and the pantograph MB model include the initial conditions for the start of the analysis expressed in terms of the initial positions and velocities of all components of the respective systems.

4.4 Multibody Communication Interface

The MB code requires the knowledge of the history of the contact forces between the catenary and pantograph and of the location of their points of application in the pantograph during any integration time step being attempted. Let it be assumed that at some instant in the middle of the time step, eventually even past its end, such forces and application points are known. Then, a table with the values of the forces and application points can be constructed. By interpolation of this table the equilibrium equations of the MB system can be formulated at any time and the integration algorithm can estimate the state variables. The integration algorithms typically used for multibody codes are variable step and variable order algorithms, such as those presented by Petzold [12], Shampine and Gordon [18] and Gear [19], to name a few. Besides internal control of the time step size these algorithms need to predict the state variables at different instants of the integration time step before calculating the state variables at the end of the time step. Therefore, the existence of a table of forces that can be interpolated is fundamental to avoid that the user has to mess with the details of the integration algorithm of the MB code.

The communication procedure implemented for the MB code consists on the updating of the table, assumed to be available at every instant, as follows:

1. Enter the process when the forces on the MB system are being evaluated
2. If the MB integrator is predicting the state variables terminate this communication and return to the MB code because the table with the forces cannot be updated. Otherwise, the integrator is correcting the state variables. Therefore, continue to next step in the communication procedure.
3. Inquire if the data generated by the MB code has already been used by the FE code. If the data still exists then halt the MB code until the FE code uses it and catches up with the current time of the MB analysis. When the data ceases to exist the MB communication can proceed to the next step.
4. If the current time in the MB integration of the pantograph is larger than the final time for which there is data available in the table with the forces proceed to the next step to update the table. Otherwise, the table with the forces can still be interpolated and the communication finishes being the control transferred back to the MB code.
5. Generate data for the FE code with the information on the position and velocity of the pantograph registration strip (see Figure 8).
6. Check if there is data available from the FE code to the MB code with the information required to update the table of the forces on the pantograph. If such data does not exist wait until it is available. When the data from the FE to the MB code exists, proceed to the next step.
7. Update the table with the new forces and application points locations on the pantographs and time at which they have been calculated ($t_{catenary}$).
8. Erase the data from the FE to the MB code indicating in this form that the analysis can proceed on both codes.
9. Return the control of the analysis of the pantograph to the MB code.

Nowhere in the communication procedure outlined it is implied what kind of integration algorithm is used for the FE catenary analysis, provided that it is a fixed time step integrator. Even this condition can be relaxed but it is not usual that FE dynamic analysis is performed with variable time step algorithms.

4.5 Finite Elements Communication Interface

The FE code used here for the dynamic analysis of the catenary has been developed by SNCF to study models of different catenaries. It uses a Newmark algorithm to integrate the FE equations of motion. For what follows it is important to review the features of these algorithms. At any given time step the algorithm proceeds by first predicting the displacements and velocities for the new time step by using the information of the last completed time step as

$$\begin{aligned}\tilde{\mathbf{d}}_{t+\Delta t} &= \mathbf{d}_t + \Delta t\,\mathbf{v}_t + \frac{\Delta t^2}{2}(1-2\beta)\mathbf{a}_t \\ \tilde{\mathbf{v}}_{t+\Delta t} &= \mathbf{v}_t + \Delta t(1-\gamma)\mathbf{a}_t\end{aligned} \tag{11}$$

Based on the position and velocity predictions for the FE mesh and on the pantograph predicted position and velocity in the MB code the contact forces are evaluated for $t+\Delta t$ and the FE mesh accelerations are calculated from the equilibrium equation

$$\left(\mathbf{M}+\gamma\,\Delta t\,\mathbf{C}+\beta\,\Delta t^2\mathbf{K}\right)\mathbf{a}_{t+\Delta t}=\mathbf{f}_{t+\Delta t}-\mathbf{C}\tilde{\mathbf{v}}_{t+\Delta t}-\mathbf{K}\tilde{\mathbf{d}}_{t+\Delta t} \tag{12}$$

Then, with the acceleration $\mathbf{a}_{t+\Delta t}$ the positions and velocities of the finite elements at time $t+\Delta t$ are corrected by

$$\begin{aligned}\mathbf{d}_{t+\Delta t}&=\tilde{\mathbf{d}}_{t+\Delta t}+\beta\,\Delta t^2\mathbf{a}_{t+\Delta t}\\ \mathbf{v}_{t+\Delta t}&=\tilde{\mathbf{v}}_{t+\Delta t}+\gamma\,\Delta t\,\mathbf{a}_{t+\Delta t}\end{aligned} \tag{13}$$

A difficulty is that the positions and velocities of the pantographs may not be available at time $t+\Delta t$ when they are required to calculate the contact forces included in vector $\mathbf{f}_{t+\Delta t}$. Therefore, a prediction on these positions must also to be made at a given time step based on the most updated information obtained from the MB code.

In what the interfacing between MB and FE is concerned, it is within this code that the contact forces and their application points are calculated. The calculation of the contact forces requires that the FE code has access to the positions and velocities of the registration strips of the pantographs, provided by the MB code. In turn, the FE code makes available to the MB code the contact forces and application points required to update the table of forces. The communication procedure implemented for the FE code consists on the following steps:

1. Enter the communication procedure when the contact forces between the catenary and pantographs need to be calculated.
2. If the data provided by the MB code to the FE code does not yet exist wait until it is available and then proceed to next step.
3. Access the MB data getting the time at which it was generated in the first place, i.e., get $t_{pantograph}$.
4. If the $t_{pantograph}$, i.e., the time at which the most updated information on the pantograph is available, is less than the current FE integration time, i.e., $t_{catenary}$, erase it and go back to step 2. Otherwise, proceed to the next step
5. Get the position and velocity of the pantograph registration strip from the data provided by the MB code to the FE code and erase the original information.
6. Calculate $D_{time} = t_{catenary} + \Delta t - t_{pantograph}$.
7. Calculate the predicted positions and velocities of the pantographs for $t+\Delta t$

$$\begin{aligned}\tilde{\mathbf{r}}_{t+\Delta t}&=\mathbf{r}_{t\_pantograph}+D_{time}\,\dot{\mathbf{r}}_{t\_pantograph}+\frac{\Delta t^2}{2}\left(1-2\beta\right)\ddot{\mathbf{r}}_{t\_pantograph}\\ \tilde{\dot{\mathbf{r}}}_{t+\Delta t}&=\dot{\mathbf{r}}_{t\_pantograph}+D_{time}\left(1-\gamma\right)\ddot{\mathbf{r}}_{t\_pantograph}\end{aligned} \tag{14}$$

8. With the predicted positions and velocities of the catenary, evaluated by equation (11), and of the pantograph, by equation (14), calculate the contact points and the respective forces using equation (10).
9. If the data provided by the FE to the MB code at time $t = t_{catenary}$ exists wait until it is erased by the MB code. If the data does not exist or has been erased by the MB code, create a new set of data for $t+\Delta t$ including the current integration time for the catenary $t_{catenary} = t+\Delta t$, the contact force $f_{contact}$ and the the application point of the force in the pantograph $a_{contact}$.
10. Return to the FE code from the communication procedure.

The communication procedures for both MB and FE codes are described in Figure 9. Also in the same figure possible time stepping sequences are pictured for the both codes. Again, it is worth emphasizing a key issue in the co-simulation procedure described is that the MB time step must always be smaller or equal to the FE time step.

The effective transference of data between the two codes can be achieved by file exchange or by access to a common memory block by both programs. The first strategy is easier to implement because only standard operations with files are involved and it is not required to address the operating systems under which the codes are running. However, the access by file sharing may be slow. Another alternative is to have the codes sharing a common block of memory, which is only feasible if both codes are running in the same computer and provided that the operating system allows such sharing. A third alternative is to use communications via TCP/IP. These issues address knowledge on the computer sciences that is out of the scope of this work and are not be discussed here any further.

5 Application Example

The computational simulation described in this work aims at modeling the behavior of a real pantograph and catenary system used in French high speed lines. It is of interest to be able to compare the numerical results not only with regard to experimental data but also to numerical results obtained using commercial or built-in codes. In this way it is possible to validate the communication procedure and implicitly the contact force model. In this work the experimental data used is provided by SNCF and corresponds to a measurement run using a commercial French high speed line with the train running at a speed of approximately 80 $m \cdot s^{-1}$. To compare the results calculated in this work with the experimental and numerical results obtained with other codes a lumped mass model of the pantograph, that has a dynamic response equivalent to the complex multibody pantograph model described in Figure 4, is used.

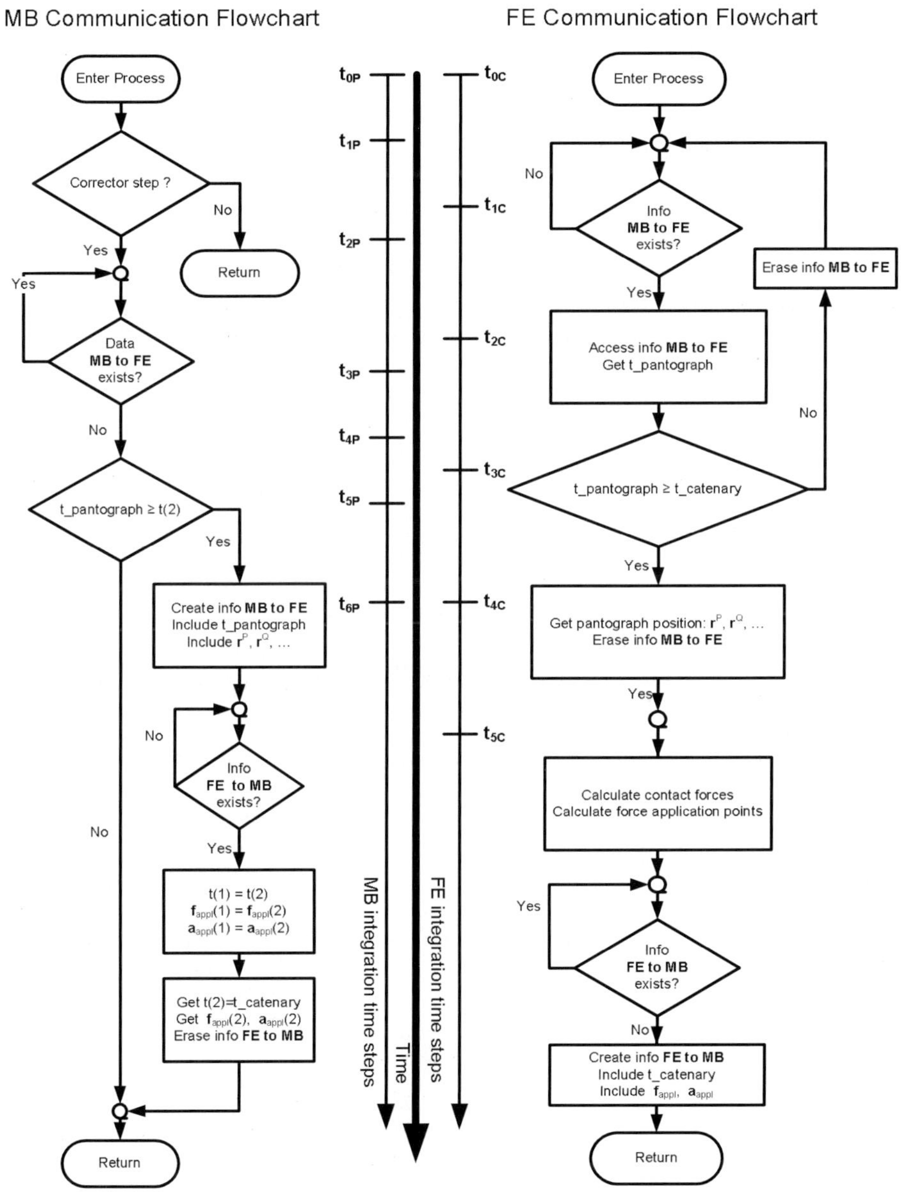

Fig. 9. Flowchart of the communication procedure between the MB and FE codes.

5.1 Pantograph Model

The pantograph model used in this work corresponds to a SNCF three level lumped-mass pantograph model described in Figure 10. The bow of the pantograph and suspension are modeled by the top mass and spring-damper element, respectively. A local reference frame (ξ,η,ζ) is rigidly attached to the center of mass of each body in such a way that the axes are aligned with the principal inertia directions of the rigid bodies. The initial conditions for the bodies of the system are given by the location of the center of mass and by the orientation of the local reference frames (ξ,η,ζ) with respect to the pantograph coordinate system (x,y,z). The relevant data required for the bodies is presented in Tables 1 and 2. The system is composed by two subsystems: the carbody (Subsystem 1) and the pantograph (Subsystem 2). Table 1 presents the characteristics of the carbody and in Table 2 the properties of the remaining bodies in the pantograph. In both cases the data described for the models should not be deemed as exact. The attachment of the pantograph to the carbody is described by a parent-child joint. The relevant information used to assemble the kinematic joints is presented in Tables 3 and 4.

Table 1. Properties and initial conditions for each rigid body of the carbody (subsystem 1)

ID	Rigid Body	Mass (Kg)	Inertia Properties (Kg.m^2) $I_{\xi\xi}$ / $I_{\eta\eta}$ / $I_{\zeta\zeta}$	Initial Position (m) x_0 / y_0 / z_0	Initial Orientation e_1 / e_2 / e_3
1	Carbody	11160	14953/ 225365/ 224994	-5.5/ 0.0 /-1.2	0/ 0/ 0

Table 2. Properties and initial conditions for each rigid body of the pantograph (subsystem 2)

ID	Rigid Body	Mass (Kg)	Inertia Properties (Kg.m^2) $I_{\xi\xi}$ / $I_{\eta\eta}$ / $I_{\zeta\zeta}$	Initial Position (m) x_0 / y_0 / z_0	Initial Orientation e_1 / e_2 / e_3
2	Low Mass	4.8	- / - / -	0.0/ 0.0/ 0.5	0/ 0/ 0
3	Mid Mass	4.63	- / - / -	0.00 / 0.00 / 0.80	0/ 0/ 0
10	Top Mass	8.5	- / - / -	0.0/ 0.0/ 1.0	0/ 0/ 0

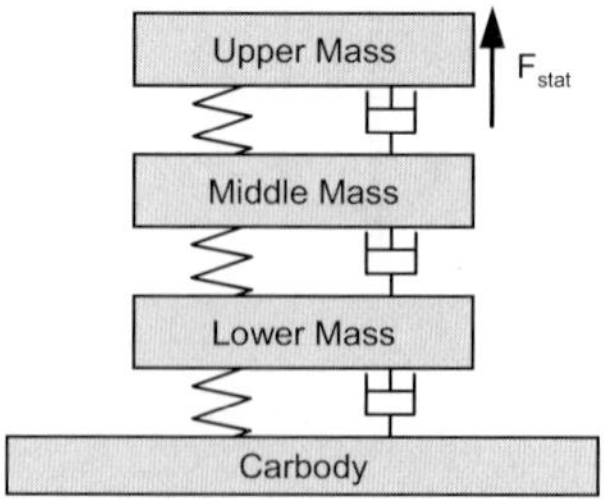

Fig. 10. Graphical representation of the pantograph model used.

Table 3. Kinematic joints used between pantograph and carbody (subsystems 1 and 2)

ID	Kinematic Joint	Connected Bodies		End Stroke Stop		Attach Points Local Coordinates (m)	
						Body *i*	Body *j*
		i	*j*	α	β	$\xi_i/\eta_i/\zeta$	$\xi_j/\eta_j/\zeta_j$
1	Parent-Child	-	-	-	-	-5.5/ 0.0/ 1.2	0/ 0/ 0
				-	-	-	0/ 0/ 0

Table 4. Kinematic joints used in the multibody model of the carbody (subsystem 1)

ID	Kinematic Joint	Connected Bodies		End Stroke Stop		Attach Points Local Coordinates (m)	
						Body *i*	Body *j*
		i	*j*	α	β	$\xi_i/\eta_i/\zeta$	$\xi_j/\eta_j/\zeta_j$
1	Ground Body	1	-	-	-	-	-

Spring-damper-actuator elements are used to model the forces transmitted among the rigid bodies that compose the pantograph system. The characteristics of linear force elements are presented in Table5. In Table 6 it is defined the static force conditions upon each mass in order to respect the static equilibrium conditions with the catenary.

5.2 Catenary Finite Element Model

The model of the catenary used in this work is described in Figure 7 and corresponds to the Paris-Lyon French high speed railway track. All the structural elements that compose the real catenary are modeled including the contact, messenger and stitch wires, droppers, registration and steady arms. The catenary static state (displacements, forces), which is a very sensitive parameter, is computed using a non-linear strategy since the wires' stiffness depends on their actual displacements.

Table 5. Characteristics of the spring-damper elements used in the model of the pantograph

ID	Spring Stiffness (N/m)	Undef. Length (m)	Damping Coeffic. (N.s/m)	Actuator Force (N)	Connect Bodies		Attach Points Local Coordinates (m)	
							Body *i*	Body *j*
					i	*j*	$\xi_i/\eta_i/\zeta_i$	$\xi_j/\eta_j/\zeta_j$
1	1	0.5	32	0	1	2	-5.5/ 0.0/ 1.2	0.0/0.0/ 0.0
2	5400	0.3	5	0	2	3	0.0/ 0.0/ 0.0	0.0/ 0.0/ 0.0
3	6045	0.2	10	0	3	4	0.0/ 0.0/ 0.0	0.0/ 0.0/ 0.0

Table 6. Static force conditions of each rigid body of the pantograph (subsystem 2)

ID	Rigid Body	Static Force (N)
2	Lower mass	0.0
3	Middle mass	0.0
4	Top mass	200.0

5.3 Results

The experimental and calculated contact forces in the pantograph-catenary interface are depicted by Figure 11. The results show a satisfactory correlation between the numerical and experimental data for the contact force, according to the EN50318 standard (filtered at 20 Hz). The contact force peaks happens at the time that the pantograph passes under the catenary supports, identified in Figure 13 as the maximum and minimum peaks of the displacement of the contact point in the transversal direction (stagger). The filtered results include the effects of the structural elements of the catenary, due to the travel of the pantograph, but do not include the effects of singular defects, for instance. The experimental data is obtained at a speed of approximately 80 $m s^{-1}$ over a commercial high-speed catenary in operating conditions, and therefore subjected to climatic defects, wear and other general defects. Such effects were not included in the catenary model which explains some of the discrepancies between numerical and experimental results. It can also be seen that the correlation between the numerical and experimental contact force increases as the pantograph moves forward. This is due to the fact that the pantograph is raised at the beginning of the simulation to its operating position and tends to reach its equilibrium position in time. Nevertheless it is to be noted that the frequency of the numerical contact forces is similar to the one observed experimentally, even in the stages corresponding to the raising of the pantograph. Regardless of these differences the results show that the contact model developed and the co-simulation methodology applied are suitable to describe the pantograph-catenary contact.

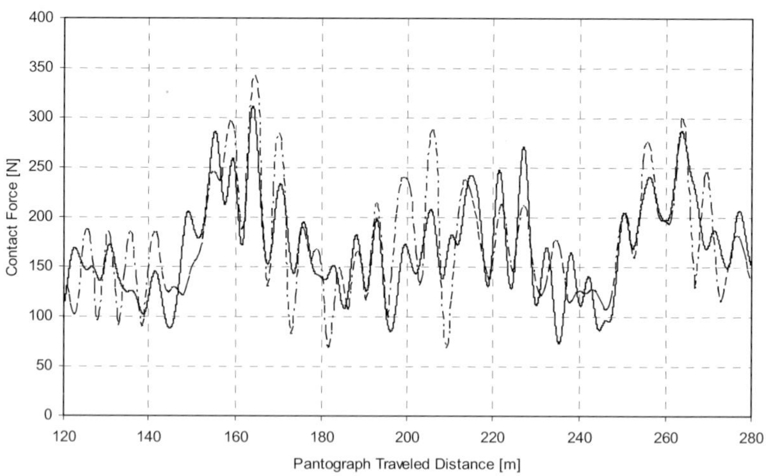

Fig. 11. Numerical results (dash lines) and experimental measurements (solid line) for a train speed of approximately 80 $m s^{-1}$.

The numerical results of the co-simulation between the FEM and the MB codes are displayed together with those obtained with the OSCAR code and a built-in pantograph model in Figure 12. The vertical motion of the masses of the pantograph lumped-mass model is presented being the similarity of results a sign that both models are equivalent. The motion of the pantograph follows the geometry of the catenary. Each peak of the vertical displacement coincides with the registration strip passing under a support of the catenary as seen in Figure 14.

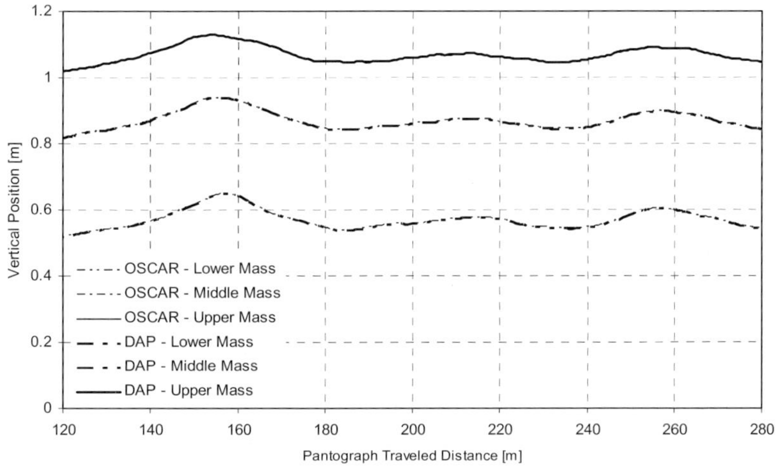

Fig. 12. Vertical position of lumped-mass pantograph model bodies.

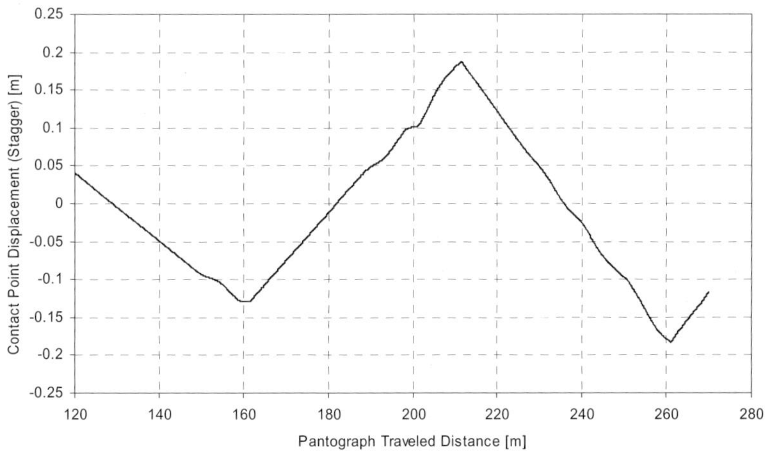

Fig. 13. Contact point stagger position along the registration strip.

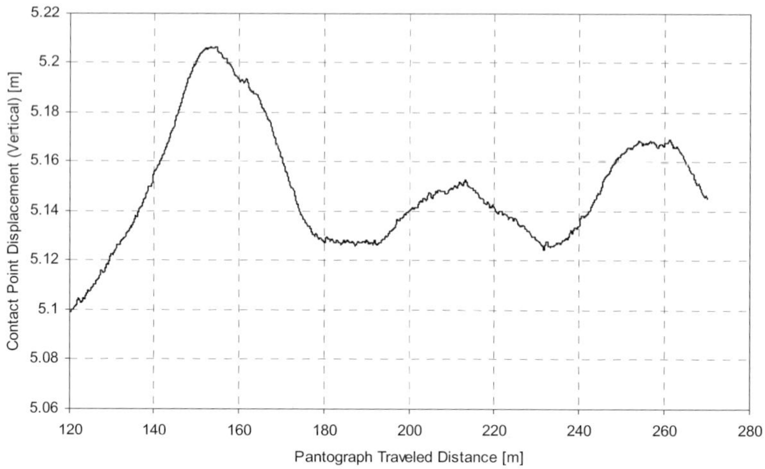

Fig. 14. Contact point vertical position with respect to the ground reference.

6 Conclusions

This work presents a co-simulation methodology to be applied to the dynamic analysis of the pantograph-catenary system. The pantograph system is modeled using a multibody code while the catenary system is modeled using an independent finite element code. The communication between the two codes is done through the contact forces. The only limitation on the communication procedure proposed is that the maximum value for the integration time-step of the multibody code must be that of the finite element code. Due to the pantograph-catenary contact with the possibility of loss of contact requires that the finite element time-step has a maximum value allowed that does not constitute a major drawback of the procedure as it is not expected that the multibody code would have an integration time-step much higher than that of the FE code. It is shown in this work that the contact model and the co-simulation methodology represent valid approaches to the analysis of the interaction of the pantograph-catenary interaction, not presenting a heavy overload on the calculations. Furthermore the proposed methodology allows the use of codes to simulate each system independently enabling the enhancement of the efficiency of the analysis of the catenary and pantograph dynamics. Generic fully three-dimensional models of the pantograph may be implemented without increasing the complexity of the code development. The results show a good correlation between measured and calculated contact forces showing in this form the suitability of the approaches proposed to analyze the problem of the pantograph-catenary interaction.

Acknowledgement

The work presented has been developed in the framework of the European funded project EUROPAC (European Optimized Pantograph Catenary Interface, contract nº STP4-CT-2005-012440) having as partners Société Nationale des Chemins de Fer Français (Leader), Alstom Transport, ARTTIC, Banverket, Ceské dráhy akciová společnost, Deutsche Bahn, Faiveley Transport, Mer Mec SpA, Politecnico di Milano, Réseau Ferré de France, Rete Ferroviara Italiana, Trenitalia SpA, Union Internationale des Chemins de Fer, Kungliga Tekniska Högskolan. The collaboration of SNCF, Faiveley Transport and Politécnico di Milano is specially acknowledged. The support of Fundação para a Ciência e Tecnologia (FCT) through the grant SFRH/BD/18848/2004 is also gratefully acknowleged.

References

1. Gardou M (1984) Etude du Comportement Dynamique de l'Ensemble Pantographe-Caténaire (Study of the Dynamic Behavior of the Pantograph-Catenary) (in French). Diploma Thesis, Conservatoire National des Arts et Metiers, Paris, France
2. Jensen C N (1997) Nonlinear Systems with Discrete and Discontinuous Elements. Ph. D. Thesis, Technical University of Denmark, Lyngby, Denmark
3. Veitl A and Arnold M (1999) Coupled Simulations of Multibody Systems and Elastic Structures. In: Ambrósio and Schiehlen (eds) Proceedings of EUROMECH Colloquium 404 Advances in Computational Multibody Dynamics, Lisbon, Portugal, September 20-23, 635-644.
4. Labergri F (2000) Modélisation du Comportement Dynamique du Système Pantographe-Caténaire (Model for the Dynamic Behavior of the System Pantograph-Catenary) (in French). Ph.D. Thesis, Ecole Central de Lyon, Lyon, France
5. Seo J-H, et al. (2004) Large Deformation Analysis of the Pantograph/Catenary Systems. Technical Report #MBS04-7-UIC, Department of Mechanical Engineering, University of Illinois at Chicago, Chicago, Illinois
6. Collina A and Bruni S (2002) Numerical Simulation of Pantograph-Overhead Equipment Interaction. Vehicle System Dynamics, 38(4), 261-291.
7. Collina A, Melzi S and Facchinetti A. (2002) On the Prediction of Wear of Contact Wire in OHE Lines: a Proposed Model. Vehicle System Dynamics Supplement, 37, 579-592.
8. Resta F, Collina A and Fossati F (2001) Actively Controlled Pantograph: and Application. In: (eds) Proceedings of the IEEE/ASME International Conference on Advanced Intelligent Mechatronics, Como, Italy, July 8-12, 243-248.
9. Nikravesh P E (1988) Computer-Aided Analysis of Mechanical Systems. Prentice-Hall, Englewood Cliffs, New Jersey.

10. Baumgarte J (1972) Stabilization of Constraints and Integrals of Motion in Dynamical Systems. Computer Methods in Apllied Mechanics and Engineering, 1, 1-16.
11. Jalón J and Bayo E (1994) Kinematic and Dynamic Simulation of Multibody Systems: The Real-Time Challenge. Springer-Verlag, New York, New York
12. Petzold L (1994) Computational challenges in mechanical systems simulation. In: Pereira and Ambrósio (eds) Computer-Aided Analysis of Rigid and Flexible Mechanical Systems. Kluwer Academic Publishers, Dordrecht, The Netherlands, 483-499.
13. Augusta Neto M and Ambrósio J (2003) Stabilization Methods for the Integration of Differential-Algebraic Equations in the Presence of Redundant Constraints. Multibody Systems Dynamics, 10(1), 81-105.
14. Pombo J and Ambrósio J (2003) General Spatial Curve Joint for Rail Guided Vehicles Kinematics and Dynamics, Multibody Systems Dynamics, 9(3), 237-264.
15. Hunt K H and Crossley F R (1975) Coefficient of Restitution Interpreted as Damping in Vibroimpact. Journal of Applied Mechanics, 7, 440-445.
16. Lankarani H M and Nikravesh P E (1990) A Contact Force Model with Hysteresis Damping for Impact Analysis of Multibody Systems. AMSE Journal of Mechanical Design, 112, 369-376.
17. Newmark N M (1959) A method of Computation for Structural Dynamics. ASCE J. of the Engineering Mechanics Division, 85(EM 3) 67-94.
18. Shampine L and Gordon M (1975) Computer Solution of Ordinary Differential Equations: The Initial Value Problem. Freeman, San Francisco, California.
19. Gear C W (1971) Simultaneous Numerical Solution of Differential-Algebraic Equations. IEEE Transactions on Circuit Theory, 18(1), 89-95.
20. Hulbert G, et al. (2005) Gluing for Dynamic Simulation of Distributed Mechanical Systems. In: Ambrósio (eds) Advances on Computational Multibody Systems. Springer, , Dordrecht, The Netherlands, 69-94.
21. Kubler R and Schiehlen W (2000) Modular Simulation in Multibody System Dynamics. Multibody System Dynamics, 4, 107-127, 2000.

Advances in Discrete Time Transfer Matrix Method of Multibody System

Xiaoting Rui, Bin He, Laifeng Yun, Guoping Wang, Fufeng Yang, and Yuqi Lu

Institute of Power Engineering of Nanjing University of Science and Technology, Nanjing, P. R. China, 210094 ruixt@mail.njust.edu.cn

Summary. A new method of multibody system dynamics, named discrete time transfer matrix method of multibody system (MS-DT-TMM) is developed and widely used in engineering field in recent years. When using this method for the problems of multibody system dynamics, the global dynamics equations of the system are not needed and the orders of involved system matrices are always small. The advance in MS-DT-TMM in recent years has been studied in this paper as following: 1 Discrete time transfer matrix method of rigid-flexible multibody system is developed to study rigid-flexible multibody system dynamics. 2 Mixed method of MS-DT-TMM and finite element method is developed to study rigid-flexible multibody system dynamics. 3 Mixed method of MS-DT-TMM and multibody system dynamics method is developed to study rigid-flexible multibody system dynamics. 4 Riccati discrete time transfer matrix method of multibody system is developed to increase the computation speed and the computation stability. 5 Discrete time transfer matrix method of controlled multibody system is developed to study dynamics of controlled multibody system. Several corresponding numerical examples are done to validate these proposed methods. All this implies that the proposed methods provide a powerful tool for the study of multibody system dynamics.

1 Introduction

Lots of methods for multibody system dynamics have been studied by many authors in theory and computational method [1, 2]. In general, almost all methods of the multibody system dynamics have the same characteristics as follows: 1. It is necessary to develop global dynamics equations of the system. If the system structure is changed, generally, corresponding global dynamics equations must be deduced again. 2. The orders of involved system matrices increase with the increase of the number of the degrees of freedom of the system; hence the orders of matrices involved in global dynamics equations are rather high for complex multibody systems.

Rui etc (1989, 2000) developed transfer matrix method of multibody system (MS-TMM) for vibration analysis of linear multibody system by developing new transfer matrices and orthogonal property of multibody system [3-5]. Rui etc (2005) developed transfer matrix method for 2-dimension system for vibration analysis of 2-dimension system by developing new transfer matrices [6]. Kumar and Sankar

Peter Eberhard (ed.), Multiscale Problems in Multibody System Contacts, 227–241.

(1986) developed discrete time transfer matrix method (DT-TMM) for structure dynamics of time variant system by combining the transfer matrix method with the numerical integration procedure [7]. Rui etc (1999, 2005) developed discrete time transfer matrix method of multibody system (MS-DT-TMM) [8-10] for the study of multibody system dynamics. This method combines and expands the advantages of the TMM, DT-TMM, MS-TMM and the numerical integration procedure. Using the method, dynamics of multi-rigid-body system and dynamics of multi-flexible-body systems, including chain multibody system, branched multibody system, close-looped multibody system, network multibody system, have been studied in detail. MS-DT-TMM presents a new method for multibody system dynamics and has been used widely in engineering [8-13]. There are five characteristics of MS-DT-TMM: (1) It does not require globe dynamical equations of the system and simplifies the solution procedure. (2) It has high computationnal speed, because the system matrices are always small irrespective of the size of a system. 3 It avoids the difficulties caused by developing globe dynamical equations of the system and by computing too high order matrices. (4) It provides maximum flexibility in modeling various configurations of multibody systems, by creating library of transfer matrices and assembling them easily, and by introducing any suitable numerical integration scheme. (5) It can be combined with various dynamics methods to solve complicated problems. The advance in the MS-DT-TMM in recent years includes: (1) Discrete time transfer matrix method of rigid-flexible multibody system is developed to study rigid-flexible multibody system dynamics. (2) Mixed method of discrete time transfer matrix method of multibody systems and finite element method (MS-DT-TMM-FEM) is developed to combine DT-TMM-MS with FEM firstly to study rigid-flexible multibody system dynamics. (3) Mixed method of discrete time transfer matrix method of multibody systems and multibody systems dynamics method (MS-DT-TMM-MS-MSDM) is developed to combine DT-TMM-MS with ordinary method of multibody dynamics firstly to study the rigid-flexible multi body system. (4) Riccati discrete time transfer matrix method of multibody system (MS-R-DT-TMM) is developed to introduce Riccati transformation in MS-DT-TMM to decreased the order of the system matrix involved and to increased the computational speed and the computational stability. (5) Discrete time transfer matrix method of controlled multibody system (CMS-DT-TMM) is developed, to develop transfer matrices of controlled elements to study controlled multibody system dynamics using MS- DT-TMM.

When using the proposed methods, the global dynamics equations of the system are not needed and the orders of involved system matrices are always low. Several corresponding numerical examples have been done to validate the proposed methods. Due to the limit of the paper space, these numerical examples of multibody systems studied using the proposed methods are omitted. All this implies that MS-DT-TMM is a powerful tool for the study of multibody system dynamics.

2 General Theorems and Steps of MS-DT-TMM

2.1 Coordinate System and Sign Convention, Dynamics equations and Discretization

Use the right-handed inertial Cartesian coordinate system *oxyz* as the reference system, use θ_1, y and z as the position coordinates of involved points, use space-three-angles 1-2-3, which was defined in the reference [14], as the orientation angles of involved bodies, that is, by rotating about x, y and z axis directions successively in the same coordinate system, then we get θ_1, θ_2 and θ_3 [14].

Sign convention: sign convention is the same with [8-10]. A complex multibody system is divided into a certain number of subsystems, which can be represented by various elements including bodies (rigid bodies, elastic bodies, lumped masses etc.) and hinges (joints, ball-and-socket, pins, linear springs, rotary springs, linear dampers and rotary dampers, etc.). It should be pointed out that the positions of the bodies and hinges are considered equivalent in transfer equations and transfer matrices. So, the dynamics equations of every body and hinge should be developed relatively to the inertial reference system respectively.

2.2 Linearization of Dynamics Equations of Elements

According to numerical integration procedures, the motion parameters of multibody system $\ddot{\xi}$ and $\dot{\xi}$ at the time instant t_i are expressed as the linear function of ξ in form

$$\ddot{\xi}(t_i) = \chi_1 \xi(t_i) + \chi_{2,\xi} \tag{1}$$

$$\dot{\xi}(t_i) = \chi_3 \xi(t_i) + \chi_{4,\xi} \tag{2}$$

where, the variable ξ may represent column matrix of the positions coordinates x, y, z or the orientation angles $\theta_1, \theta_2, \theta_3$ respectively; $\ddot{\xi}$ and $\dot{\xi}$ represent one order derivative and two order derivative of ξ with respect to time, that is, corresponding column matrices of acceleration and velocity, or corresponding column matrices of angular acceleration and angular velocity for planar motion, at the same time instant t_i. The quantities χ_1, $\chi_{2,\xi}$, χ_3 and $\chi_{4,\xi}$, will have different expressions for different numerical integration procedures. There are many accurate and commonly available numerical integration procedures that can be chosen, please see reference [7]. For example, if Newmark-β method [9] is used, then we obtained

$$\chi_1 = \frac{1}{\beta \Delta T^2} \boldsymbol{I}_k, \quad \chi_{2,\xi} = -\frac{1}{\beta \Delta T^2}\left[\xi(t_{i-1}) + \dot{\xi}(t_{i-1})\Delta T + (\tfrac{1}{2} - \beta)\ddot{\xi}(t_{i-1})\Delta T^2\right] \tag{3}$$

$$\chi_3 = \gamma \chi_1 \Delta T, \quad \chi_{4,\xi} = \gamma \chi_{2,\xi} \Delta T + \dot{\xi}(t_{i-1}) + (1-\gamma)\ddot{\xi}(t_{i-1})\Delta T \tag{4}$$

Where, time step $\Delta T = t_i - t_{i-1}$, β and γ are the coefficients of Newmark-β method. Bold capital symbol $\boldsymbol{I}_k$ is the unit matrix, its subscripts k denotes the order of the unit matrix and equals to three for a system moving in space or equals to two for a system moving in plane.

Using Taylor expansion theorem, at time instant t_i direction cosine matrix $\boldsymbol{A}(t_i)$ can be approximately expressed with respect to t_{i-1} by the truncated Taylor series of order 3, that is

$$\begin{aligned}\boldsymbol{A}(t_i) = {} & \boldsymbol{A}(t_{i-1})\tilde{\boldsymbol{T}}_1(t_{i-1})\theta_1(t_i) + \boldsymbol{A}(t_{i-1})\tilde{\boldsymbol{T}}_2(t_{i-1})\theta_2(t_i) + \boldsymbol{A}(t_{i-1})\tilde{\boldsymbol{T}}_3(t_{i-1})\theta_3(t_i) \\ & + \boldsymbol{\Phi}(t_{i-1}) + \boldsymbol{B}o(\Delta T^2)\end{aligned} \tag{5}$$

Where, $c_i = \cos\theta_i$, $s_i = \sin\theta_i$, ($i = 1,2,3$), the meaning of other elements in Eq. (5) can be found in document [9]. Eq. (5) will be used in the expressions of position coordinates of corresponding points.

In the items of product of the trigonometric functions and the elements of state vectors, the trigonometric functions at time t_i are expanded with respect to t_{i-1} using the truncated Taylor series of order 3, that is

$$\begin{aligned}\sin\theta(t_i) &= \sin[\theta(t_{i-1}) + \Delta\theta] = \bar{s} + o(\Delta T^2) \\ \cos\theta(t_i) &= \cos[\theta(t_{i-1}) + \Delta\theta] = \bar{c} + o(\Delta T^2)\end{aligned} \tag{6}$$

where

$$\bar{s} \overset{\Delta}{=} \sin\theta(t_{i-1})\left\{1 - \frac{1}{2}\left[\dot{\theta}(t_{i-1})\Delta T\right]^2\right\} + \cos\theta(t_{i-1})\left[\dot{\theta}(t_{i-1})\Delta T + \frac{1}{2}\ddot{\theta}(t_{i-1})\Delta T^2\right] \tag{7}$$

$$\bar{c} \overset{\Delta}{=} \cos\theta(t_{i-1})\left\{1 - \frac{1}{2}\left[\dot{\theta}(t_{i-1})\Delta T\right]^2\right\} - \sin\theta(t_{i-1})\left[\dot{\theta}(t_{i-1})\Delta T + \frac{1}{2}\ddot{\theta}(t_{i-1})\Delta T^2\right] \tag{8}$$

$$\begin{aligned}\sin\theta(t_i) &= \cos\theta(t_{i-1})\theta(t_i) + \bar{S} + o(\Delta T^2) \\ \cos\theta(t_i) &= -\sin\theta(t_{i-1})\theta(t_i) + \bar{C} + o(\Delta T^2)\end{aligned} \tag{9}$$

$$\bar{S} = \sin\theta(t_{i-1}) - \theta(t_{i-1})\cos\theta(t_{i-1}) - \frac{1}{2}\sin\theta(t_{i-1})[\dot{\theta}(t_{i-1})\Delta T]^2 \tag{10}$$

$$\bar{C} = \cos\theta(t_{i-1}) + \theta(t_{i-1})\sin\theta(t_{i-1}) - \frac{1}{2}\cos\theta(t_{i-1})[\dot{\theta}(t_{i-1})\Delta T]^2 \tag{11}$$

It can be proved that

$$\begin{aligned}\begin{bmatrix}\dot{\theta}_1^2 \\ \dot{\theta}_2^2 \\ \dot{\theta}_3^2\end{bmatrix} &= \boldsymbol{\kappa}_1\left(\chi_3\begin{bmatrix}\theta_1 \\ \theta_2 \\ \theta_3\end{bmatrix}_{t_i} + \boldsymbol{\kappa}_3\right) + \begin{bmatrix}\ddot{\theta}_1^2 \\ \ddot{\theta}_2^2 \\ \ddot{\theta}_3^2\end{bmatrix}_{t_{i-1}}\Delta T^2 + \begin{bmatrix}o(\Delta T^2) \\ o(\Delta T^2) \\ o(\Delta T^2)\end{bmatrix} \\ \begin{bmatrix}\dot{\theta}_1\dot{\theta}_2 \\ \dot{\theta}_2\dot{\theta}_3 \\ \dot{\theta}_1\dot{\theta}_3\end{bmatrix} &= \boldsymbol{\kappa}_2\left(\chi_3\begin{bmatrix}\theta_1 \\ \theta_2 \\ \theta_3\end{bmatrix}_{t_i} + \boldsymbol{\kappa}_3\right) + \begin{bmatrix}\ddot{\theta}_1\ddot{\theta}_2 \\ \ddot{\theta}_2\ddot{\theta}_3 \\ \ddot{\theta}_1\ddot{\theta}_3\end{bmatrix}_{t_{i-1}}\Delta T^2 + \begin{bmatrix}o(\Delta T^2) \\ o(\Delta T^2) \\ o(\Delta T^2)\end{bmatrix}\end{aligned} \tag{12}$$

where

$$\boldsymbol{\kappa}_1 = 2\begin{bmatrix} \dot{\theta}_1 & 0 & 0 \\ 0 & \dot{\theta}_2 & 0 \\ 0 & 0 & \dot{\theta}_3 \end{bmatrix}_{t_{i-1}}, \quad \boldsymbol{\kappa}_2 = \begin{bmatrix} \dot{\theta}_2 & \dot{\theta}_1 & 0 \\ 0 & \dot{\theta}_3 & \dot{\theta}_2 \\ \dot{\theta}_3 & 0 & \dot{\theta}_1 \end{bmatrix}_{t_{i-1}}, \quad \boldsymbol{\kappa}_3 = \chi_{4,\theta} - \frac{1}{2}\dot{\theta}(t_{i-1})$$

The truncation errors caused in Eqs (7), (9), (10) and (11) are all $o(\Delta T^2)$. The motion quantities $z(t_{i-1})$, $\dot{z}(t_{i-1})$, $\ddot{z}(t_{i-1})$ at the previous time instant are all known at time instant t_i. Thus, these quantities χ_1, $\chi_{2,\xi}$, χ_3, $\chi_{4,\xi}$, $\overline{c}$ and $\overline{s}$ etc. are all definable for any subsystem for the time interval $(t_i - t_{i-1})$, and hence above formulations are valid.

2.3 State Vectors, Transfer Equations and Transfer Matrices of Elements

According to the dynamics equations of elements and structure of multibody systems, the state vectors of the connection point among any rigid bodies and hinges moving in space are defined as

$$\boldsymbol{z} = [x, y, z, \theta_1, \theta_2, \theta_3, m_x, m_y, m_z, q_x, q_y, q_z, 1]^T \tag{13}$$

where x, y, z, θ_1, θ_2 and θ_3 are the position coordinates of the connection point with respect to the inertial reference system and the corresponding orientation angles rotating in the directions of x, y and z successively defined in section 2.1; m_x, m_y, m_z, q_x, q_y, q_z are the corresponding interior torques and interior forces in the same reference system respectively.

The state vectors of the connection point among any rigid bodies and hinges moving in plane are defined as

$$\boldsymbol{z} = [x, y, \theta_3, m_z, q_x, q_y, 1]^T \tag{14}$$

where, the meaning of the elements in the state vectors are similar to the meaning as described in Eq. (13).

The state vectors of the connection point among any flexible bodies and their hinges moving in space are defined as

$$\boldsymbol{z} = [x, y, z, \theta_1, \theta_2, \theta_3, m_x, m_y, m_z, q_x, q_y, q_z, q_1, \cdots, q_n, 1]^T \tag{15}$$

where, x, y, z, θ_1, θ_2, θ_3, m_x, m_y, m_z, q_x, q_y, q_z have the same meaning as described in Eq. (13), $q_1, \cdots, q_n$ are generalized coordinates described deformation of flexible bodies with mode method, n is the highest order of the model considered.

The state vectors of the connection point among any flexible bodies and their hinges moving in plane are defined as

$$\boldsymbol{z} = [x, y, \theta_3, m_z, q_x, q_y, q_1, \cdots, q_n, 1]^T \tag{16}$$

where, the meaning of the elements in the state vectors are similar to the meaning described in Eq. (13).

The dynamics equations of the jth element that have been linearized using numerical integration procedure can be assembled into a single transfer equation

$$z_{j,j+1}(t_i) = U_j(t_i) z_{j,j-1}(t_i) \tag{17}$$

The meaning of the subscripts of the state vectors z follows the convention in section 2.1. The transfer equation describes the mutual relationship between the state vectors at two ends of the *j*th element and has the similar form in contrast with a general transfer equation used in MS-TMM, DT-TMM or TMM. Here; the matrix $U_j(t_i)$ is the transfer matrix of the *j*th element. It is the functions of the motion quantities ($z(t_k)$, $\dot{z}(t_k)$ and $\ddot{z}(t_k)$, $k = i-1, i-2, \cdots$) which are all known at time instant t_i .

2.4 Transfer Equation and Transfer Matrix and Solutions of Overall System

The overall system transfer equation and transfer matrix U , which relates the state vectors at ends of the system, can be assembled and calculated. That is,

$$z_{n,n+1} = U z_{1,0} \tag{18}$$

$$U = U_n U_{n-1} \cdots U_2 U_1 \tag{19}$$

For a chain system, the order of the overall transfer matrix of the system is equal to the order of the transfer matrix of the element, and it does not increase when the degrees of freedom of system increase. Irrespective of the size of a multibody system, the highest order of the overall transfer matrix U is the same with the order of transfer matrix of single body, that is (13×13) for dynamics of chain multi-rigid-body system moving in space, or ($(13+n)\times(13+n)$) for dynamics of chain multi-rigid-flexible-body system moving in space, or (7×7) for dynamics of chain multi-rigid-body system moving in plane, or ($(7+n)\times(7+n)$) for dynamics of chain multi-rigid-flexible-body system moving in plane. So, the matrices involved in the MS-DT-TMM are always small, which greatly reduces the computational time and the memory storage requirement.

Boundary conditions of the system can then be applied and the unknown quantities in the boundary state vectors can be computed. Now, knowing the boundary state vectors completely, the state vectors and hence the motion quantities at each element at time t_i can be computed by the repeated use of corresponding transfer equations of element similar to Eq. (17). The velocity, angular velocity, acceleration and angular acceleration quantities at time t_i are then obtained using Eq. (1) and Eq. (2) respectively. Then entire procedure can be repeated for time t_{i+1} and so on.

3 Transfer Matrix Library of Elements

By creating a library of transfer matrices for typical elements and by assembling these at the required locations, various configurations of multibody system can be modeled easily. The transfer matrices of elements can be developed directly from dynamics equations of elements or its solutions using numerical integration procedures. The transfer matrices of some elements have been developed in [8-10]. The transfer matrices of other elements, such as flexible body moving in space or in plane, elastic beams moving in space or in plane, elastic hinge moving in space or in plane, damper moving in space or in plane, are developed in this paper.

3.1 Transfer Matrices of Rigid Bodies, Elastic Hinges, Damper, Smooth Hinges

The transfer matrices of rigid bodies moving in space or in plane, transfer matrix of elastic hinge and damper moving in space or in plane, transfer matrix of smooth ball-and-socket hinge whose inboard body and outboard body are rigid bodies, transfer matrix of smooth pin hinge whose inboard body and outboard body are rigid bodies can be found in [8-10].

3.2 Transfer Matrix of A Beam Moving in Plane

The motion of a beam can be decomposed into large motion of overall system and its elastic deformation motion. The dynamics equation of constant section area Euler-Bernoulli beam moving in plane can be deduced [10]

$$\overline{m}\dot{\theta}\int_0^l 2u\dot{u}\mathrm{d}x_2 + \overline{m}\ddot{\theta}\int_0^l (u^2 + x_2^2)dx_2 + \overline{m}\int_0^l x_2\ddot{u}dx_2 + m\ddot{y}_{O_2}x_{O_1C} - m\ddot{x}_{O_2}y_{O_1C} = \sum m_i \quad (20)$$

$$\frac{\partial q_{2,y}(x_2,t)}{\partial x_2} = f_{2,y}(x_2,t) - \overline{m}(\ddot{u} + x_2\ddot{\theta} - u\dot{\theta}^2 + \ddot{y}_{O_2}\cos\theta - \ddot{x}_{O_2}\sin\theta) \quad (21)$$

$$\frac{\partial q_{2,x}(x_2,t)}{\partial x_2} = f_{2,x}(x_2,t) - \overline{m}(-\ddot{\theta}u - 2\dot{\theta}\dot{u} - \dot{\theta}^2 x_2 + \ddot{y}_{O_2}\sin\theta + \ddot{x}_{O_2}\cos\theta) \quad (22)$$

Using mode method, the transverse deformation u of simple supported beam is expressed as

$$u(x_2,t) = \sum_{k=1}^{n} \sin\frac{k\pi x_2}{l} q^k(t) \quad (23)$$

Substitute Eq. (23) to Eqs. (21) and (22), then multiply $\sin\frac{p\pi x_2}{l}$ ($p = 1,2,3$) in the two sides of the equations, integrate along the axial direction of beam, considering the generalized coordinates on two ends of the beam are equal, define state vector of the beam

$$z(x,t) = [x, y, \theta, m, q_x, q_y, q^1, q^2, q^3, 1]^T \quad (24)$$

Then the transfer equation and transfer matrix of the beam can be obtained

$$z(l,t) = \boldsymbol{U}z(0,t) \quad (25)$$

$$\boldsymbol{U}=\begin{bmatrix} 1 & 0 & u_{1,3} & 0 & 0 & 0 & 0 & 0 & 0 & u_{1,10} \\ 0 & 1 & u_{2,3} & 0 & 0 & 0 & 0 & 0 & 0 & u_{2,10} \\ 0 & 0 & 1 & 0 & 0 & 0 & 0 & 0 & 0 & 0 \\ u_{4,1} & u_{4,2} & u_{4,3} & 1 & u_{4,5} & u_{4,6} & u_{4,7} & u_{4,8} & u_{4,9} & u_{4,10} \\ u_{5,1} & 0 & u_{5,3} & 0 & 1 & 0 & u_{5,7} & 0 & u_{5,9} & u_{5,10} \\ 0 & u_{6,2} & u_{6,3} & 0 & 0 & 1 & u_{6,7} & 0 & u_{6,9} & u_{6,10} \\ 0 & 0 & 0 & 0 & 0 & 0 & 1 & 0 & 0 & 0 \\ 0 & 0 & 0 & 0 & 0 & 0 & 0 & 1 & 0 & 0 \\ 0 & 0 & 0 & 0 & 0 & 0 & 0 & 0 & 1 & 0 \\ 0 & 0 & 0 & 0 & 0 & 0 & 0 & 0 & 0 & 1 \end{bmatrix} \tag{26}$$

Due to the limit of the paper space, the meanings of the symbols in the transfer matrix are omitted.

3.3 Transfer Matrix of Smooth Pin Hinge Connected With Beams Moving in Plane

The transfer matrix of any smooth hinge whose outboard hinge is also a smooth hinge or a fictitious hinge can be developed as follows. The transfer matrix of smooth hinge, whose outboard hinges are neither smooth hinge nor fictitious hinge, can be obtained by using the method introduced in [8].

Substitute Eq. (20) and Eq. (23) into Eq. (21), multiplying $\sin\frac{p\pi x_2}{l}$ ($p=1,2,3$) to the two sides of the equation and integrating along x_2, then linearlizing in this equation, we can obtain

$$\begin{cases} \Delta_{1i}=\begin{vmatrix} H_{1i} & & & D_{14} \\ H_{2i} & D_{22} & & D_{24} \\ H_{3i} & & D_{33} & D_{34} \\ H_{4i} & D_{42} & D_{43} & D_{44} \end{vmatrix}, \quad \Delta_{2i}=\begin{vmatrix} D_{11} & H_{1i} & & D_{14} \\ & H_{2i} & & D_{24} \\ & H_{3i} & D_{33} & D_{34} \\ D_{41} & H_{4i} & D_{43} & D_{44} \end{vmatrix} \\ \Delta_{3i}=\begin{vmatrix} D_{11} & & H_{1i} & D_{14} \\ & D_{22} & H_{2i} & D_{24} \\ & & H_{3i} & D_{34} \\ D_{41} & D_{42} & H_{4i} & D_{44} \end{vmatrix}, \quad \Delta_{4i}=\begin{vmatrix} D_{11} & & & H_{1i} \\ & D_{22} & & H_{2i} \\ & & D_{33} & H_{3i} \\ D_{41} & D_{42} & D_{43} & H_{4i} \end{vmatrix} \end{cases} \quad (i=1,2,3,4,5)$$

$$\Delta=\begin{vmatrix} D_{11} & & & D_{14} \\ & D_{22} & & D_{24} \\ & & D_{33} & D_{34} \\ D_{41} & D_{42} & D_{43} & D_{44} \end{vmatrix} \tag{27}$$

where

$$D_{kk}=\frac{l}{2}[EI\frac{k^4\pi^4}{l^4}-\overline{m}\dot{\theta}^2(t_{i-1})]+\frac{l}{2}(\rho I\frac{k^2\pi^2}{l^2}+\overline{m})A \quad (k=1,2,3)\ ,$$

$$D_{k4}=(-1)^{k-1}\overline{m}\frac{l^2}{k\pi}A-\overline{m}l\dot{\theta}(t_{i-1})q^k(t_{i-1})C \quad (k=1,2,3)\ ,$$

$$D_{41}=u_{4,7}\ ,D_{42}=u_{4,8}\ ,D_{43}=u_{4,9}\ ,D_{44}=u_{4,3}\ ,$$

$$H_{k1}=[1-(-1)^k]\overline{m}\frac{l}{k\pi}\overline{s}A,\quad H_{k2}=-[1-(-1)^k]\overline{m}\frac{l}{k\pi}\overline{c}A$$

$$H_{k5}=\int_0^l[f_{2,y}(x_2,t)-\frac{\partial}{\partial x_2}m'(x_2,t)]\sin\frac{k\pi x_2}{l}dx_2-\frac{l}{2}(\overline{m}+\rho I\frac{k^2\pi^2}{l^2})B_{q^k}-\overline{m}l\dot{\theta}^2(t_{i-1})q^k(t_{i-1})$$

$$-(-1)^{k-1}\frac{\overline{m}l^2}{k\pi}B_\theta-[1-(-1)^k]\frac{\overline{m}l}{k\pi}(\overline{c}B_{y_{O2}}-\overline{s}B_{x_{O2}})+\overline{m}l\dot{\theta}(t_{i-1})q^k(t_{i-1})D_\theta,$$

$$H_{41}=-u_{4,1},H_{42}=-u_{4,2},\ H_{43}=-u_{4,5},\ H_{44}=-u_{4,6},\ H_{45}=-u_{4,10}$$

$u_{i,j}$ is the element of transfer matrix of outboard beam of this hinge.

Define state vector of smooth pin hinge whose inboard body and outboard body are beams moving in plane

$$z_I=[x,y,\theta,m,q_x,q_y,q^1,q^2,q^3,1]^T,\ z_O=[x,y,\theta,m,q_x,q_y,q^1,q^2,q^3,1]^T \quad (28)$$

Transfer equation and transfer matrix of the smooth pin hinge can be obtained

$$z_O=\boldsymbol{U}_j z_I \quad (29)$$

$$\boldsymbol{U}_j=\begin{bmatrix} 1 & 0 & 0 & 0 & 0 & 0 & 0 & 0 & 0 & \\ 0 & 1 & 0 & 0 & 0 & 0 & 0 & 0 & 0 & \\ \Delta_{41}/\Delta & \Delta_{42}/\Delta & 0 & 0 & \Delta_{43}/\Delta & \Delta_{44}/\Delta & 0 & 0 & 0 & \Delta_{45}/\Delta \\ 0 & 0 & 0 & 0 & 0 & 0 & 0 & 0 & 0 & \\ 0 & 0 & 0 & 0 & 1 & 0 & 0 & 0 & 0 & \\ 0 & 0 & 0 & 0 & 0 & 1 & 0 & 0 & 0 & \\ \Delta_{11}/\Delta & \Delta_{12}/\Delta & 0 & 0 & \Delta_{13}/\Delta & \Delta_{14}/\Delta & 0 & 0 & 0 & \Delta_{15}/\Delta \\ \Delta_{21}/\Delta & \Delta_{22}/\Delta & 0 & 0 & \Delta_{23}/\Delta & \Delta_{24}/\Delta & 0 & 0 & 0 & \Delta_{25}/\Delta \\ \Delta_{31}/\Delta & \Delta_{32}/\Delta & 0 & 0 & \Delta_{33}/\Delta & \Delta_{34}/\Delta & 0 & 0 & 0 & \Delta_{35}/\Delta \\ 0 & 0 & 0 & 0 & 0 & 0 & 0 & 0 & 0 & 1 \end{bmatrix} \quad (30)$$

State vectors of input end and output end of the smooth pin hinge whose inboard body and outboard body are rigid body and beam moving in plane are defined respectively as

$$z_I=[x,y,\theta,m,q_x,q_y,1]^T,\quad z_O=[x,y,\theta,m,q_x,q_y,q^1,q^2,q^3,1]^T \quad (31)$$

The transfer equation and transfer matrix of the smooth pin can be obtained

$$z_O=\boldsymbol{U}_j z_I \quad (32)$$

$$
\boldsymbol{U}_j = \begin{bmatrix}
1 & 0 & 0 & 0 & 0 & 0 & 0 \\
0 & 1 & 0 & 0 & 0 & 0 & 0 \\
\Delta_{41}/\Delta & \Delta_{42}/\Delta & 0 & 0 & \Delta_{43}/\Delta & \Delta_{44}/\Delta & \Delta_{45}/\Delta \\
0 & 0 & 0 & 0 & 0 & 0 & 0 \\
0 & 0 & 0 & 0 & 1 & 0 & 0 \\
0 & 0 & 0 & 0 & 0 & 1 & 0 \\
\Delta_{11}/\Delta & \Delta_{12}/\Delta & 0 & 0 & \Delta_{13}/\Delta & \Delta_{14}/\Delta & \Delta_{15}/\Delta \\
\Delta_{21}/\Delta & \Delta_{22}/\Delta & 0 & 0 & \Delta_{23}/\Delta & \Delta_{24}/\Delta & \Delta_{25}/\Delta \\
\Delta_{31}/\Delta & \Delta_{32}/\Delta & 0 & 0 & \Delta_{33}/\Delta & \Delta_{34}/\Delta & \Delta_{35}/\Delta \\
0 & 0 & 0 & 0 & 0 & 0 & 1
\end{bmatrix} \tag{33}
$$

State vectors of input end and output end of the smooth pin hinge whose inboard body is beam and outboard body is rigid body are defined respectively as

$$
\boldsymbol{z}_I = [x, y, \theta, m, q_x, q_y, q^1, q^2, q^3, 1]^T, \quad \boldsymbol{z}_O = [x, y, \theta, m, q_x, q_y, 1]^T \tag{34}
$$

The transfer equation and transfer matrix of the smooth pin hinge can be obtained.

$$
\boldsymbol{z}_O = \boldsymbol{U}_j \boldsymbol{z}_I \tag{35}
$$

$$
\boldsymbol{U}_j = \begin{bmatrix}
1 & 0 & 0 & 0 & 0 & 0 & 0 & 0 & 0 & 0 \\
0 & 1 & 0 & 0 & 0 & 0 & 0 & 0 & 0 & 0 \\
-\frac{u_{4,1}}{u_{4,3}} & -\frac{u_{4,2}}{u_{4,3}} & 0 & 0 & -\frac{u_{4,5}}{u_{4,3}} & -\frac{u_{4,6}}{u_{4,3}} & 0 & 0 & 0 & -\frac{u_{4,7}}{u_{4,3}} \\
0 & 0 & 0 & 0 & 0 & 0 & 0 & 0 & 0 & 0 \\
0 & 0 & 0 & 0 & 1 & 0 & 0 & 0 & 0 & 0 \\
0 & 0 & 0 & 0 & 0 & 1 & 0 & 0 & 0 & 0 \\
0 & 0 & 0 & 0 & 0 & 0 & 0 & 0 & 0 & 0
\end{bmatrix} \tag{36}
$$

The meanings of the elements of the transfer matrix are the same as reference [8].

3.4 Transfer Matrix of Beam Moving in Space

Considering transverse vibration of a Euler - Bernoulli beam moving in space in large range. One end of the beam is fixed on the original point of body fixed coordinate system $o_2x_2y_2z_2$, and the axis o_2x_2 along with the axis of undeformation beam. Define state vector of the beam

$$
z = [x, y, z, \theta_1, \theta_2, \theta_3, m_x, m_y, m_z, q_x, q_y, q_z, q_{y_2}^1, q_{y_2}^2, \cdots, q_{y_2}^{n_y}, q_{z_2}^1, q_{z_2}^2, \cdots, q_{z_2}^{n_z}, 1]^T \tag{37}
$$

According to the dynamics equation of mass center, traversal vibration equation, rotation equations of the beam moving in space in the inertia coordinate system [10, 11], let

$$
v(x_2,t) = \sum_{k=1}^{n_2} Y^k(x_2) q_{y_2}^k(t), \quad w(x_2,t) = \sum_{k=1}^{n_3} Z^k(x_2) q_{z_2}^k(t) \tag{38}
$$

By linearization for the solution of dynamics equations of the beam in time step, then the transfer equation and transfer matrix of the beam moving in space can been obtained

$$z_O = U z_I \tag{39}$$

$$U = \begin{bmatrix} I_3 & U_{12} & O_{3\times3} & O_{3\times3} & U_{15} & U_{16} & U_{17} \\ O_{3\times3} & I_3 & O_{3\times3} & O_{3\times3} & O_{3\times n_2} & O_{3\times n_3} & O_{3\times1} \\ U_{31} & U_{32} & I_3 & U_{34} & U_{35} & U_{36} & U_{37} \\ U_{41} & U_{42} & O_{3\times3} & I_3 & U_{45} & U_{46} & U_{47} \\ O_{n_2\times3} & O_{n_2\times3} & O_{n_2\times3} & O_{n_2\times3} & I_{n_2} & O_{n_2\times n_3} & O_{n_2\times1} \\ O_{n_3\times3} & O_{n_3\times3} & O_{n_3\times3} & O_{n_3\times3} & O_{n_3\times n_2} & I_{n_3} & O_{n_3\times1} \\ O_{1\times3} & O_{1\times3} & O_{1\times3} & O_{1\times3} & O_{1\times n_2} & O_{1\times n_3} & 1 \end{bmatrix} \tag{40}$$

Due to the limit of the paper space, the meanings of the symbols in the transfer matrix are omitted.

3.5 Transfer Matrix of Smooth Hinge Connection with Beam Moving in Space

For a smooth ball-and-socket hinge, when the interior torque of its outboard hinge is zero, the relationship between the orientation angles of the two body fixed coordinate systems connected with the smooth ball-and-socket hinge can be obtained

$$[\theta_1, \theta_2, \theta_3]_O = [\bar{U}_{3,2}^{-1}\bar{U}_{3,1} \quad O_{3\times3} \quad O_{3\times3} \quad \bar{U}_{3,2}^{-1}\bar{U}_{3,4} \quad \bar{U}_{3,2}^{-1}\bar{U}_{3,7}] z_I \tag{41}$$

Where, submatrices $\bar{U}_{i,j}$ are the elements of the transfer matrix of the beam.

The transfer equation and the transfer matrix of the smooth ball-and-socket hinge whose inboard body and outboard body are rigid body and beam can be obtained

$$z_O = U z_I$$

$$U = \begin{bmatrix} I_3 & O_{3\times3} & O_{3\times3} & O_{3\times3} & O_{3\times1} \\ \bar{U}_{3,2}^{-1}\bar{U}_{3,1} & O_{3\times3} & O_{3\times3} & \bar{U}_{3,2}^{-1}\bar{U}_{3,4} & \bar{U}_{3,2}^{-1}\bar{U}_{3,7} \\ O_{3\times3} & O_{3\times3} & I_3 & O_{3\times3} & O_{3\times1} \\ O_{3\times3} & O_{3\times3} & O_{3\times3} & I_3 & O_{3\times1} \\ U_{51} + U_{52}\bar{U}_{3,2}^{-1}\bar{U}_{3,1} & O_{n\times3} & O_{n\times3} & U_{52}\bar{U}_{3,2}^{-1}\bar{U}_{3,4} & U_{55} + U_{52}\bar{U}_{3,2}^{-1}\bar{U}_{3,7} \\ U_{61} + U_{62}\bar{U}_{3,2}^{-1}\bar{U}_{3,1} & O_{n\times3} & O_{n\times3} & U_{62}\bar{U}_{3,2}^{-1}\bar{U}_{3,4} & U_{65} + U_{62}\bar{U}_{3,2}^{-1}\bar{U}_{3,7} \\ O_{1\times3} & O_{1\times3} & O_{1\times3} & O_{1\times3} & 1 \end{bmatrix} \tag{42}$$

Where, z_O is state vector of beam transverse vibration in space, z_I is state vector of rigid body moving in space.

4 Discrete Time Transfer Matrix Method of Rigid-flexible Multibody System

Using the transfer matrices of the flexible bodies, rigid-flexible multibody system dynamics can be studied easily using MS-DT-TMM. The steps to study rigid-flexible multibody system dynamics and to study rigid multibody system dynamics [10] both are similar. Due to the limit of the paper space, a numerical example of a rigid-flexible multibody system studied using MS-DT-TMM is omitted.

5 MS-DT-TMM-FEM

Mixed method of MS-DT-TMM and FEM (MS-DT-TMM-FEM) is developed[10, 11]. A multibody system is divided to various subsystems. Some subsystems which including rigid bodies and simple flexible bodies can be simulated by DT-TMM-MS, and other subsystems, such as flexible bodies, can be simulated by FEM. Then global dynamics equations of the subsystems including flexible bodies can be deduced by FEM, and overall transfer matrix equation of the subsystem including rigid bodies and simple flexible bodies can be obtained by DT-TMM-MS. The connected points among the subsystems can be regarded as boundaries of these subsystems. The position coordinates and the orientation angles of the ends of the connected points, the interior forces and the interior torque acted on the ends of these points are equal to each other. Solving these transfer equations and global dynamics equations of the subsystems together, the dynamics analysis of global system can be implemented. Due to the limit of the paper space, a numerical example of a rigid-flexible multibody system studied using MS-DT-TMM-FEM is omitted.

6 MS-DT-TMM-MSDM

Mixed method of MS-DT-TMM and MSDM (MS-DT-TMM-MS-MSDM) is developed[10]. A multibody system can be divided into various subsystems, some subsystems can be simulated by DT-TMM-MS, and other subsystems can be simulated by ordinary MSDM. Then the global dynamics equations of some subsystems can be deduced by MSDM, and the overall transfer matrix equation of other subsystem can be obtained by MS-DT-TMM. The connected points among the subsystems can be regarded as boundaries of these subsystems. The position coordinates and the orientation angles of the ends of the connected points, the interior forces and the interior torque acted on the ends of these points are equal to each other. Solving these transfer equations and global dynamics equations of the subsystems together, the dynamics analysis of global system can be implemented. Due to the limit of the paper space, a numerical example of a rigid-flexible multibody system studied using MS-DT-TMM-MSDM is omitted.

7 MS-R-DT-TMM

By introducing Riccati transform in MS-DT-TMM, develop Riccati MS-DT-TMM (MS-R-DT-TMM)[11]. Generally speaking, in the state vectors of system boundary z_O, z_I, half of the elements is known and the rest is unknown. They are represented by z^1 and z^2 respectively. The state vector can be represented as

$$z = [z^{1T}, z^{2T}, 1]^T \tag{43}$$

So overall transfer equations of system an be rewritten as

$$\begin{bmatrix} \boldsymbol{z}_i^1 \\ \boldsymbol{z}_i^2 \end{bmatrix} = \begin{bmatrix} \bar{\boldsymbol{U}}_{11}^i & \bar{\boldsymbol{U}}_{12}^i \\ \bar{\boldsymbol{U}}_{21}^i & \bar{\boldsymbol{U}}_{22}^i \end{bmatrix} \begin{bmatrix} \boldsymbol{z}_{i-1}^1 \\ \boldsymbol{z}_{i-1}^2 \end{bmatrix} + \begin{bmatrix} \bar{\boldsymbol{U}}_i^1 \\ \bar{\boldsymbol{U}}_i^2 \end{bmatrix} \tag{44}$$

where $\bar{\boldsymbol{U}}_i^1, \bar{\boldsymbol{U}}_i^2$ are corresponding with the seventh column of subsystem transfer matrix $\boldsymbol{U}_i$. According as the partition of z^1 and z^2 , the submatrices $\bar{\boldsymbol{U}}_{jk}^i, j,k = 1,2$ can be obtained directly. Using Riccati transform [13], that is

$$\boldsymbol{z}_{i-1}^1 = \boldsymbol{S}_{i-1}\boldsymbol{z}_{i-1}^2 + \boldsymbol{P}_{i-1} \tag{45}$$

To substitute Eq. (45) into Eq. (44), we can obtain

$$\boldsymbol{z}_i^2 = \left(\bar{\boldsymbol{U}}_{21}^i \boldsymbol{S}_{i-1} + \bar{\boldsymbol{U}}_{22}^i\right)\boldsymbol{z}_{i-1}^2 + \bar{\boldsymbol{U}}_{21}^i \boldsymbol{P}_{i-1} + \bar{\boldsymbol{U}}_i^2 \tag{46}$$

Then to transpose some items in Eq. (46), we can obtain

$$\boldsymbol{z}_{i-1}^2 = \left(\bar{\boldsymbol{U}}_{21}^i \boldsymbol{S}_{i-1} + \bar{\boldsymbol{U}}_{22}^i\right)^{-1} \boldsymbol{z}_i^2 - \left(\bar{\boldsymbol{U}}_{21}^i \boldsymbol{S}_{i-1} + \bar{\boldsymbol{U}}_{22}^i\right)^{-1} \left(\bar{\boldsymbol{U}}_{21}^i \boldsymbol{P}_{i-1} + \bar{\boldsymbol{U}}_i^2\right) \tag{47}$$

Let [13, 16]

$$\boldsymbol{T}_i = (\bar{\boldsymbol{U}}_{21}^i \boldsymbol{S}_{i-1} + \bar{\boldsymbol{U}}_{22}^i)^{-1} \tag{48}$$

$$\boldsymbol{Q}_i = -(\bar{\boldsymbol{U}}_{21}^i \boldsymbol{P}_{i-1} + \bar{\boldsymbol{U}}_i^2) \tag{49}$$

The Eq. (47) can be rewritten as

$$\boldsymbol{z}_{i-1}^2 = \boldsymbol{T}_i \boldsymbol{z}_i^2 + \boldsymbol{T}_i \boldsymbol{Q}_i \tag{50}$$

To substitute Eqs (45) and (50) into Eq. (46), we can obtain

$$\boldsymbol{z}_i^1 = \bar{\boldsymbol{U}}_{11}^i \left[\boldsymbol{S}_{i-1}\left(\boldsymbol{T}_i \boldsymbol{z}_i^2 + \boldsymbol{Q}_i\right) + \boldsymbol{P}_{i-1}\right] + \bar{\boldsymbol{U}}_{12}^i \left(\boldsymbol{T}_i \boldsymbol{z}_i^2 + \boldsymbol{Q}_i\right) + \bar{\boldsymbol{U}}_i^1 \tag{51}$$

Then to substitute Eq. (49) into Eq. (51) again, we can obtain

$$\boldsymbol{z}_i^1 = \left(\bar{\boldsymbol{U}}_{11}^i \boldsymbol{S}_{i-1} + \bar{\boldsymbol{U}}_{12}^i\right)\boldsymbol{T}_i \boldsymbol{z}_i^2 + \bar{\boldsymbol{U}}_{11}^i \boldsymbol{P}_{i-1} + \bar{\boldsymbol{U}}_i^1 - \left(\bar{\boldsymbol{U}}_{11}^i \boldsymbol{S}_{i-1} + \bar{\boldsymbol{U}}_{12}^i\right)\boldsymbol{T}_i(\bar{\boldsymbol{U}}_{21}^i \boldsymbol{P}_{i-1} + \bar{\boldsymbol{U}}_i^2) \tag{52}$$

By compare with Eq. (45) and Eq. (52), the transfer formula can be found

$$\boldsymbol{S}_i = (\bar{\boldsymbol{U}}_{11}^i \boldsymbol{S}_{i-1} + \bar{\boldsymbol{U}}_{12}^i)\boldsymbol{T}_i \tag{53}$$

$$\boldsymbol{P}_i = \bar{\boldsymbol{U}}_{11}^i \boldsymbol{P}_{i-1} + \bar{\boldsymbol{U}}_i^1 + \boldsymbol{S}_i \boldsymbol{Q}_i \tag{54}$$

Due to the limit of the paper space, a numerical example of a huge multibody system including one hundred thousand rigid bodies connected with spring hinges studied using MS-R-DT-TMM is omitted.

8 CMS-DT-TMM

Take the controlled multibody system shown in Fig. 12 as example of discrete time transfer matrix method of controlled multibody system (CMS-DT-TMM).

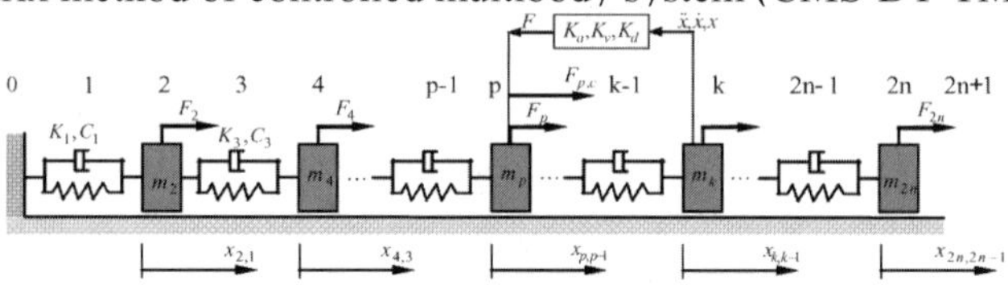

Fig. 12 Controlled multiby system

Transfer equations and transfer matrices of lumped mass and spring damper hinge can be developed respectively, see reference [10, 11]. The controlled force can be expressed as

$$f_{p,c} = -(K_a A + K_v C + K_d) x_{k,k+1} - K_a B_{x_{k,k+1}} - K_v D_{x_{k,k+1}} \tag{55}$$

Define state vector

$$z = \begin{bmatrix} x & q & 1 \end{bmatrix}^T \tag{56}$$

For the lumped mass m_p with controlled force, there is

$$z_{p-1,p} = U_p z_{p,p+1} + U_{p,c} z_{k,k+1} \tag{57}$$

where,

$$U_p = \begin{bmatrix} 1 & 0 & 0 \\ -m_p A & 1 & m_p B_{x_{p,p+1}} + f_p \\ 0 & 0 & 1 \end{bmatrix},$$

$$U_{p,c} = \begin{bmatrix} 0 & 0 & 0 \\ -(K_a A + K_v C + K_d) & 0 & -K_a B_{x_{k,k+1}} - K_v D_{x_{k,k+1}} \\ 0 & 0 & 0 \end{bmatrix} \tag{58}$$

Overall transfer equation and overall transfer matrix of the controlled system can be obtained

$$z_{0,1} = U_{all} z_{2n,0} \tag{59}$$

$$\begin{aligned} U_{all} = (U_{1\leftarrow 2} U_{3\leftarrow 4} \cdots U_{p-3\leftarrow p-2} U_{p-1} U_p U_{p+1\leftarrow p+2} \cdots U_{k\leftarrow k+1} + U_{p,c}) \\ U_{k+1\leftarrow k+2} \cdots U_{2(n-1)-1\leftarrow 2(n-1)} U_{2n-1\leftarrow 2n} \end{aligned} \tag{60}$$

Due to the limit of the paper space, a numerical example of a multibody system with control and without control studied using CMS-DT-TMM is omitted.

References

1. Schiehlen, W., 'Multibody system dynamics: roots and perspectives', *Multibody System Dynamics* 1, 1997, 149-188
2. Shabana, A.A., 'Flexible multibody dynamics: review of past and recent

developments', *Multibody System Dynamics* 1, 1997, 189-222

3. Rui, X.T. and Lu, Y.Q., 'Transfer matrix method of vibration of multibody system', *Chinese Journal of Astronautics* 16(3), 1995, 41-47
4. Rui, X.T., Sui, W.H. and Shao, Y.Z., 'Transfer matrix of rigid body and its application in multibody dynamics', *Chinese Journal of Astronautics* 14(4), 1993, 82-87
5. Lu, Y.Q. and Rui, X.T., 'Eigenvalue problem, orthogonal property and response of multibody system', *ICAPV 2000, Proceedings of International Conference on Advanced Problems in Vibration Theory and Applications*, Zhang, J.H. and Zhang, X.N., Science Press, Beijing, 2000
6. Rui Xiaoting, Yun Laifeng, Tang Jingjing, Lu Yuqi, He Bin, 'Transfer matrix method for 2-dimension system', *Proceedings of the international conference on mechanical engineering and mechanics, Nanjing in China*, Qian Lingfang, 2005, 93-99
7. Kumar, A.S. and Sankar, T.S., 'A new transfer matrix method for response analysis of large dynamic systems', *Computers & Structures* 23, 1986, 545-552
8. Xiaoting Rui, Yuqi Lu, Ling Pan, and Wenguang Lu, 'Discrete time transfer matrix method for mutibody system dynamics', Euromech Colloquium 404 on Advances in Computational Multibody Dynamics, Lisboa, Portugal, 1999: 93-108.
9. Xiaoting Rui, Bin He, Yuqi Lu, Wenguang Lu and Guoping Wang, 'Discrete time transfer matrix method for mutibody system dynamics', *Multibody System Dynamics* 14(3, 4), 2005, 317–344
10. Xiaoting Rui, Yun Laifeng, Yuqi Lu and He Bin, *Transfer matrix method of multibody system dynamics and its application*, publishing house of southeast university, Nanjing in china, 2006
11. He Bin, '*Study on* Transfer *Matrix Method for Multibody System Dynamics and Its Application in Launch Dynamics*'(Ph. D dissertation), Nanjing university of science and technology, 2006
12. Li Chunming, 'Study *on improving accuracy of discrete time transfer matrices method for multibody system dynamics*'(Technology report), Nanjing university of science and technology, 2003
13. He Bin, Rui Xiaoting, Lu Yuqi, 'Riccati discrete time transfer matrix method for huge chain multi-rigid-body system dynamics', *Proceedings of the international conference on mechanical engineering and mechanics, Nanjing in China*, Qian Lingfang, 2005, 726-731
14. Kane, T.R., Likine, P.W. and Levinson, D.A., *Spacecraft Dynamics*, McGraw-Hill, New York, 1983
15. Valasek, M., 'On the Efficient Implementation of Multibody System Formulations', Technical Report institutsbericht, IB-17, University Stuttgart, 1990
16. G C Horner and W D Pilkey, 'The Riccati transfer matrix method', *Journal of Mechanical Design*, 1(2), 1978, 297-302
17. A.K. Banerjee and S. Nagarajan, 'Efficient simulation of large overall motion of beams undergoing large deflection', *Multibody System Dynamics*, 1(1), 1997, 113-126
18. I Sharf, 'Geometric stiffening in multibody dynamics of cantilever beam attached to moving base', *Journal of Guidance, Control, and Dynamics*, 18(4), 1995, 882-890
19. Dokainish, M.A. and Subbaraj, K., 'A study of direct time-integration methods in computational structural dynamics-I. Explicit methods', *Computers & Structures* 32(6), 1989, 1371-1386.
20. Subbaraj, K. and Dokainish, M.A., 'A study of direct time-integration methods in computational structural dynamics-II. Implicit methods', *Computers & Structures* 32(6), 1989, 1387-1401.

Impact Studies of Gears in Combustion Engines

Pascal Ziegler[1], Peter Eberhard[1], and Bernhard Schweizer[2]

[1] Institute of Engineering and Computational Mechanics
University of Stuttgart
Pfaffenwaldring 9
70569 Stuttgart, Germany
[ziegler, eberhard]@itm.uni-stuttgart.de

[2] Fachbereich Maschinen- und Energietechnik
HTWK Leipzig
Koburger Straße
04416 Markkleeberg, Germany
schweizer@me.htwk-leipzig.de

1 Introduction

Gears are commonly used design elements, typically used to convert torque. However, gears are also used in mechanisms or gear drives to transmit motion. Typical applications of gear drives are gear trains, used to drive the camshaft by the crankshaft in large-scale diesel engines. There, normally the transmitted rated torque is relatively small compared to the dynamical loads. They often introduce vibrations of the entire drive train, caused by gas forces or auxiliary devices. These dynamic loads cause the flanks of teeth to lift-off. The re-establishment of contact is mostly in the form of impacts and may occur on both sides of the teeth. Because of the noise induced by these impacts, this phenomenon is called gear hammering. In gear trains, the gears are often designed with thin gear bodies to reduce inertia effects and the total weight of the engine. As a result not only noise, but also endurance problems may arise due to the high peaks in the contact forces. Thus, the precise knowledge of the contact forces is necessary for the design process. However, contact forces between rotating gears are extremely difficult and expensive to measure. Therefore, the simulation of contact forces inheres great importance. Nowadays, the contact simulations are mainly done with commercial multibody packages, assuming rigid gears connected by elastic elements. These elastic elements somehow describe the contact stiffness as well as the elasticity of the gear bodies. However, due to thin gear bodies the elasticity of the gear may introduce dynamical effects that are important even during the short time of

Peter Eberhard (ed.), Multiscale Problems in Multibody System Contacts, 243–256.

impact, but may not be sufficiently described by multibody models with rigid bodies.

For a better understanding for the physical phenomena during contact and of the influence of different boundary conditions on the contact forces, impact simulations are carried out for two gear pairs. The first one consists of the crankshaft wheel and the idler wheel from a gear train of a mid-scale diesel engine. It is chosen because of the gear body geometry of the idler gear, which is very thin and additionally weakened by radial holes. Thus, distinct gear body effects are expected. The second pair consists of two identical gears with cylindrical gear bodies and of considerably smaller diameter. For both gear pairs impact simulations are performed for different boundary conditions using the method of nonlinear finite elements. Further, the influence of different impact velocities, friction and misalignment, as well as stresses in the dedendum are investigated. Some of the results are compared to results obtained with a multibody model similar to those used in commercial program packages. Due to the very high simulation times for the nonlinear finite element model, the investigations are only carried out for one impact and contact is only considered between one single pair of teeth.

2 System and model description

The first investigated gear pairing, denoted by gear pair 1, consists of an idler gear and a crankshaft gear. In the following the gears are denoted by the subscripts i and c, respectively. As shown in Figure 1 (left), the idler gear has a very thin gear body with axial holes. Both gears have $z_i = z_c = 97$ teeth, with modulus $m_i = m_c = 6$mm, $\alpha = 20°$ and no addendum modifications. The pitch circles follow as $d_i = d_c = 582$mm, and the face widths are $b_i = 50$mm and $b_c = 57$mm, the mass moments of inertia are $I_i = 2.27\text{kgm}^2$ and $I_i = 4.08\text{kgm}^2$, see [13]. The centre distance is $a_1 = 582$mm and the overlap ratio is $\epsilon_1 = 1.85$.

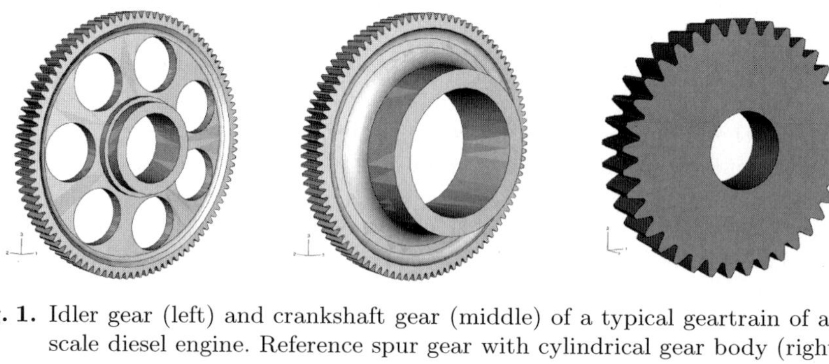

Fig. 1. Idler gear (left) and crankshaft gear (middle) of a typical geartrain of a mid-scale diesel engine. Reference spur gear with cylindrical gear body (right).

Gear pair 2 consists of two identical gears, denoted by the subscript r. As shown in Figure 1 (right), the gears have cylindrical gear bodies, with $z_r = 37$ teeth of modulus $m_r = 4.9$mm, a profile modification coefficient of $x_r = 0.56178$, a pitch circle of $r_r = 181.3$mm and a face width of $b_r = 36$mm. The mass moment of inertia follows as $I_i = 0.033$kgm^2. This pairing is mainly chosen as a reference system due to its symmetry. As centre distance $a_2 = 186.2$mm is investigated, resulting in a working angle of $\alpha_w = 23.8°$, and an overlap ratio $\epsilon_2 = 1.53$.

For all investigations a fully elastic material with Youngs modulus $E = 210000$MPa, density $\rho = 7850$kg/m^3 and Poisson ration $\nu = 0.3$ is assumed. Material damping effects are neglected.

2.1 Finite element model

For the impact simulations a fully nonlinear, transient finite element model is used, [3], [5]. To obtain precise results an appropriate spatial discretization is very important. This includes the precise description of the contact geometry, namely the flank profile. Therefore, the actual production process is mimicked, by describing the rolling of the milling and grinding tools on the production pitch circle, [11]. This has the advantage that even for special purpose gear hobs, commonly used in industrial practise, the exact profile with all addendum modifications, as well as the exact dedendum diameter and the exact root fillet radius are obtained. For the gear bodies an overall element length of 7mm is chosen, since impact problems often show wave propagation phenomena up to frequencies over 50kHz. However, since the contact area is in the order of 1mm, it obviously demands a much smaller element size. Thus, an element size of 0.1mm is chosen for the contact area. As elements, solely hexahedral elements with reduced integration and hourglass control are used.

Clearly, this fine discretization is only needed in the vicinity of the contact area and a rapid transition from the contact area to the coarser discretization of the gear bodies is required to avoid unnecessarily many nodes, especially for gear pair 1. After unsatisfying results for gear pair 1 to obtain a reasonable mesh using automatic meshing algorithms, the following meshing strategy was used. The contacting teeth are amply cut from the remaining gear body and both parts are meshed independently. During simulation both parts are assembled back together by constraining the motion of the nodes of the finer mesh to those of the nearest node of the coarse mesh. As described in [2], this allows for a very efficient solution while the results are valid, if the stress gradient and the transition of element size are small across the intersection. Due to the smaller dimensions, for the gear pair 2 a homogeneous mesh could be achieved. This results in a finite element model of about 80000 nodes and 65000 elements for each gear of gear pairing 1, and about 60000 nodes and 50000 elements for each gear of gear pairing 2

For the contact a two-pass algorithm with node-to-segment elements is used with a penalty formulation as well as the kinematic contact provided by

ABAQUS, see [1]. An efficient application of boundary conditions is achieved by constraining the nodal degrees of freedom of the nodes on the fit bores to reference nodes defined in the centres of the wheels and by applying the rigid body motion to these reference points.

2.2 Multibody model

The multibody model used here is similar to those used in many commercial programs and impact investigations like [8], [9], or [12]. It consists of two rigid bodies with one rotational degree of freedom each, connected with a spring element with backlash and a damper, see Figure 2.

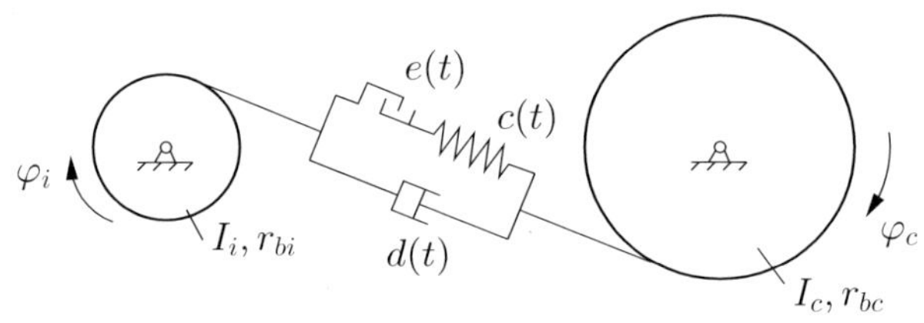

Fig. 2. Multibody model for gear pairing.

The stiffness c accounts for the contact stiffness, the bending of the teeth and the elasticity of the gear body, see [10]. The damper d accounts for material damping or damping effects of a possible oil film. The equations of motion read as

$$I_i\ddot{\varphi}_i + r_{bi}d(r_{bi}\dot{\varphi}_i + r_{bc}\dot{\varphi}_c) + r_{bi}c(r_{bi}\varphi_i + r_{bc}\varphi_c - e) = 0, \tag{1}$$

$$I_c\ddot{\varphi}_c + r_{bc}d(r_{bi}\dot{\varphi}_i + r_{bc}\dot{\varphi}_c) + r_{bc}c(r_{bi}\varphi_i + r_{bc}\varphi_c - e) = 0 \tag{2}$$

and can be reduced to

$$m_e\ddot{x} + d\dot{x} + c\,(x - e) = 0, \quad m_e = (r_{bi}^2/I_i + r_{bc}^2/I_c)^{-1} \tag{3}$$

by introducing the dynamic transition error $x = \varphi_i r_{bi} + \varphi_c r_{bc}$, see [4]. To account for a single unilateral contact, the backlash function e is chosen as

$$e = \begin{cases} 0 \text{ for } x \geq 0, \\ x \text{ for } x < 0. \end{cases} \tag{4}$$

For good results an appropriate choice of the stiffness c is crucial. Here, a physically motivated choice of the stiffness is preferred. Thus, the tooth stiffness is used. It is determined according to method A and B defined in the industry standard [7], and is assumed to be constant during the impact. In method A, the tooth stiffness is determined by comprehensive analyses like measurements, investigations based on the theory of elasticity or finite element analyses. This is done by using the finite element model described above for static analyses. Method B provides a diagram based on studies of the elastic behaviour of spur gears. The tooth stiffness can then easily be determined with several geometric parameters of the gears. Deviant gear body geometries are considered by correction factors.

3 Simulation results

First, static analyses using the finite element model are carried out to determine the so-called single stiffness and the single tooth stiffness. These results are further compared to those determined using method A, [7]. Further, impact simulations are carried out with a nonlinear transient finite element model. Boundary and initial conditions are varied to estimate their influence on the contact forces and the stresses in the dedendum fillet. Then, the influence of friction, misalignment and tooth modifications are investigated. Some results are compared to simulation results using the multibody model.

3.1 Static results

In method A of the industry standard [7], the tooth stiffness c is defined as the load rated to the face width, along the line of action to produce a deformation of 1μm in line with the load. This deformation is equal to the base circle length of arc, corresponding to the load-induced rotational angle of the pair when the mating gear is held fast. Thus, to determine the tooth stiffness, both gears are setup according to their centre distance and rotated such that they touch in the contact point. Further, one gear is held fast by fully constraining its reference point and a rotational angle γ is prescribed to the mating gear. Then, the resultant contact force F_k is related to the face width and the base circle arc length, corresponding to γ. This is done for different contact points $CP1, \ldots, CP5$, distributed over the heights of the flanks, as shown in Figure 3 (left).

In Figure 4 the respective contact forces for gear pair 1 and 2 are shown. Due to the symmetry of gear pair 2 only the contact points CP1, ..., CP3 are investigated. Obviously, the relationships are almost linear which endorses the choice of a constant stiffness c in the multibody model. Since here only the contact between one tooth pair is considered, the resultant stiffness $c = F_k/\gamma$ is called single tooth stiffness, see [10]. However, this is equal to the tooth stiffness defined in [7], if only one tooth pair is engaged. Figure 3 (right) shows the single tooth stiffness for both gear pairs. The curve connecting the calculated values is a least square fitted polynomial of order five. Very remarkable is the fact that the level of the single tooth stiffness for gear pair 1 is significantly lower as for gear pair 2. Further investigations reveal that due to the thin body in combination with the radial holes the outer rim is radially

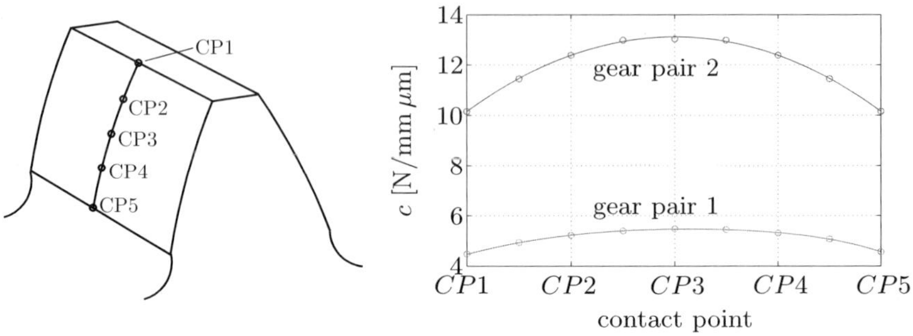

Fig. 3. Contact points distributed over the height of the flank (left) and single tooth stiffness c for gear pair 1 and 2 (right).

deformed. This leads to the much lower tooth stiffness and furthermore to a polygonisation of the idler gear body, see [14].

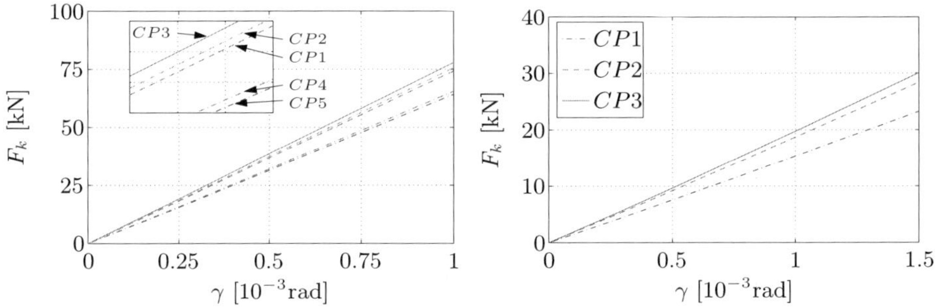

Fig. 4. Contact forces for different contact points over rotational angle γ for gear pair 1 (left) and gear pair 2 (right).

For the overlap ratios of both gear pairs, contact only occurs between one tooth pair for contacts between contact points CP2, ..., CP4. Therefore, the single stiffness c' is the maximum of the single tooth stiffness. From Figure 3 the single stiffnesses for both pairs follow as

$$\begin{aligned} c'_{1,A} &= 5.47 \frac{\mathrm{N}}{\mu\mathrm{m\,mm}} \quad , \\ c'_{2,A} &= 13.11 \frac{\mathrm{N}}{\mu\mathrm{m\,mm}} \quad . \end{aligned} \tag{5}$$

Method B in [7] provides an equation to determine the single stiffness c' of spur gears, including correction factors to account for differences between theoretical and measured values, the gear body geometry and profiles different from the basic rack profile. For both gear pairs the single stiffnesses according to method B follow as

$$c'_{1,B} = 15.66 \frac{\mathrm{N}}{\mu\mathrm{m\,mm}} \quad ,$$
$$c'_{2,B} = 13.76 \frac{\mathrm{N}}{\mu\mathrm{m\,mm}} \quad . \tag{6}$$

For gear pair 2 both methods yield nearly equal values, for gear pair 1 both solutions differ significantly. This is clearly a result of the very compliable idler gear body. Again, this is due to the particular design of the idler gear body that can not be sufficiently considered by the industry norm [7], since it is based on results from cylindrical gear bodies.

3.2 Dynamic results

For the impact simulations the gears are set up similarly as for the static simulations. Instead of a rotational angle γ an initial rotational velocity ω_0 is prescribed to the idler gear and one cylindrical gear, respectively. Both gears are pivoted in their centre.

Finite element model

From existing simulations of an entire engine model using a commercial multi-body program package which were already carried out by the manufacturer of the engine, a typical impact velocity of $\omega_{01} = 100°/\mathrm{s}$ is known for gear pair 1. Therefore, it is used as initial rotational velocity in the following simulations. In Figure 5 (left), the normal contact forces F_k, considering contact points $CP1, \ldots, CP5$ are shown for one impact.

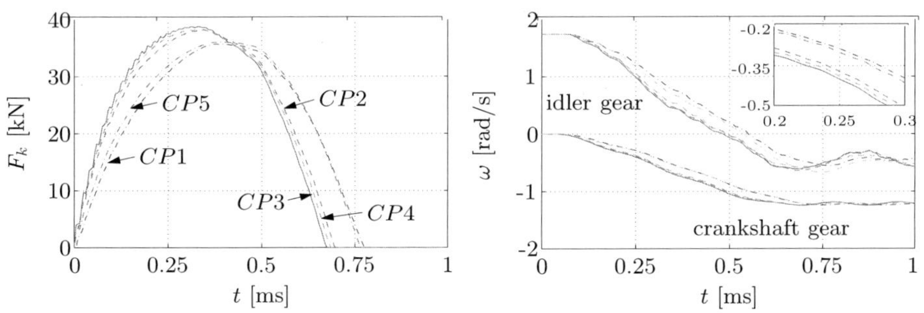

Fig. 5. Contact forces (left) and rotational velocities (right) for the finite element analyses of gear pair 1 considering different contact points.

For the contact points $CP1$ and $CP5$ a much softer contact can be observed, resulting in a contact time of about $\tau = 0.77\mathrm{ms}$ and a maximal normal contact force of $\hat{F}_k = 36\text{kN}$, whereas the other contact points show a contact time of about $\tau = 0.68\text{ms}$ and a normal contact force of $\hat{F}_k = 38\text{kN}$. A very remarkable phenomenon can be observed in the rotational velocities, shown in Figure 5 (right). Especially the idler gear shows a vibration after the impact. It turns out that this vibration is characterised by an out of phase torsional vibration of the outer rim with the teeth against the inner rim of the fit bore. Obviously, this is an effect due to the elasticity of the gear body.

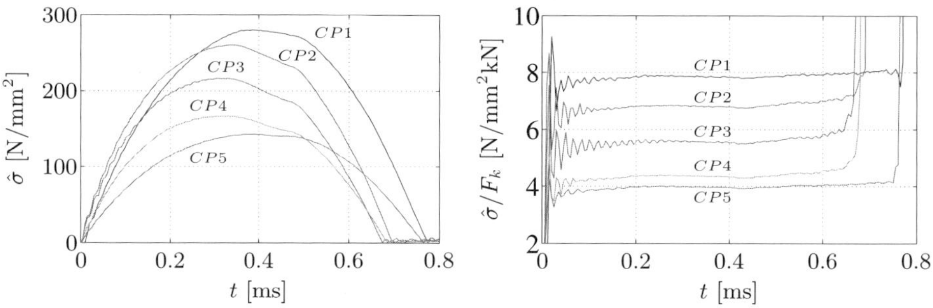

Fig. 6. Maximal von Mises stress in the dedendum fillet of the idler gear of gear pair 1 (left) and von Mises stress rated to the contact force (right).

One very important quantity for the design of gears is the stress in the dedendum fillet. Therefore, the point with the highest stress in the dedendum fillet of the idler gear is determined and the von Mises stress is determined in this point, see Figure 6 (left). Even though, the maximum stress increases with the heights of the contact point above the dedendum, their progression resembles the force curves. In Figure 6 (right) the dedendum stresses are plotted in relation to the respective contact forces. Almost during the entire contact a nearly constant ratio can be observed which agrees very well with the ratio determined according to [7]. The deviant values at the beginning and end of the contact are due to the division by very small contact forces, leading to a fraction going to infinity.

In Figures 7 and 8 the same results are shown for gear pair 2. Since it is completely symmetric, only the contact points $CP1, \ldots, CP3$ have to be investigated. For comparable results, an initial rotational velocity of $\omega_{02} = 300°/\text{s}$ is chosen. This results in a maximal contact pressure of the same magnitude as for gear pair 1 with ω_{01}. Of course, for the smaller gears the contact forces as well as the contact times are much smaller. For $CP1$, $\hat{F}_k = 14\text{kN}$ and $\tau = 0.23\text{ms}$, for $CP3$ $\hat{F}_k = 16\text{kN}$ and $\tau = 0.21\text{ms}$ are obtained. As a result of the cylindrical and therefore stiffer gear bodies, a significant torsional out of phase vibration is not observed. Again, the rated maximal dedendum stresses show a nearly constant ratio during contact, see Figure 8.

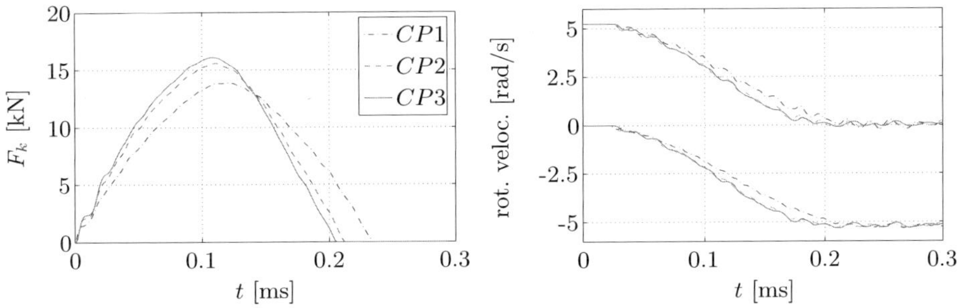

Fig. 7. Contact force (left) and rotational velocities (right) of the finite element analyses of gear pair 2 considering different contact points.

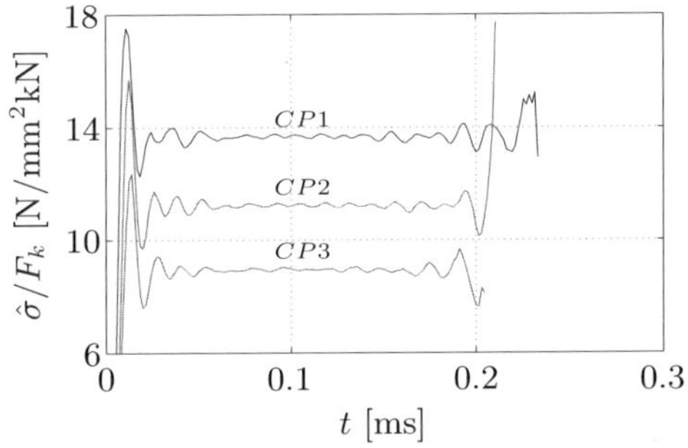

Fig. 8. Maximal von Mises stress in the dedendum fillet of gear pair 2 (left) and von Mises stress rated to the contact force (right).

To estimate the influence of the impact velocity on the contact forces, simulations are carried out with half and twice the initial rotational velocity. The results for both gear pairs, considering $CP3$ are shown in Figure 9. Here, the contact forces are related to the respective initial rotational velocities. It shows that at least for the investigated gear pairs, the normal contact force scales linearly with the impact velocity.

In Figure 10 results for impact simulations including friction are shown for both gear pairs. Here, Coulomb friction is assumed between the contacting teeth, and friction coefficients of $\mu_1 = 0.05$ and $\mu_2 = 0.1$ are considered.

Friction in the contact results in a slight decrease of about about 2% of the contact force. This is mostly due to the fact that during impact the contact point is not moving on the flank and no significant relative motion between the contact partners occurs. Therefore, friction can be neglected, when contacts occur mostly in the form of impacts. Of course, this is not the case for usual mating gears where flanks do not lift off.

In the following the influence of different boundary conditions is investigated. Therefore, the simulations are carried out for a fixed crankshaft wheel.

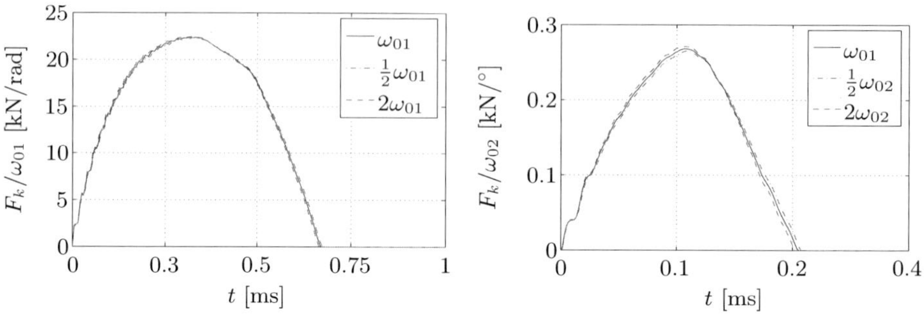

Fig. 9. Contact forces for gear pair 1 (left) and gear pair 2 (right), related to the initial rotational velocity.

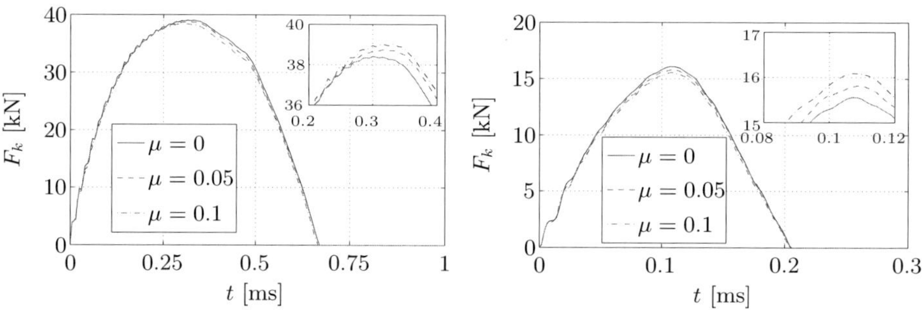

Fig. 10. Contact forces for gear pair 1 (left) and gear pair 2 (right), considering different friction coefficients.

Furthermore, the crankshaft gear is replaced by a fixed rigid plate oriented tangentially to the flank profile of the idler gear. The results are shown in Figure 11.

The fixation of the crankshaft wheel causes a significantly higher contact force of $\hat{F}_k = 52$kN and a longer contact time of $\tau = 0.8$ms. The contact with the rigid plate causes an even higher contact force of $\hat{F}_k = 67$kN, whereas the contact time is nearly the same as with a pivoted crankshaft wheel. Very remarkable is an indentation in the force progression near $t = 0.25$ms and a following increase of the contact force. This overshot is particularly noticeable for contact with the rigid plate in $CP5$. Again, this phenomenon is due to a radial deformation of the outer idler gear rim, similar to the polygonisation effect mentioned above. In the compression phase the outer rim is radially deformed, the contact force lowers and energy is stored in the structure. During decompression, this energy is released, causing the increase of the contact force observed at $t \approx 0.4$ms.

Another topic investigated is the influence of misalignment caused by production tolerances or deformation of axles under load. Since this misalignment

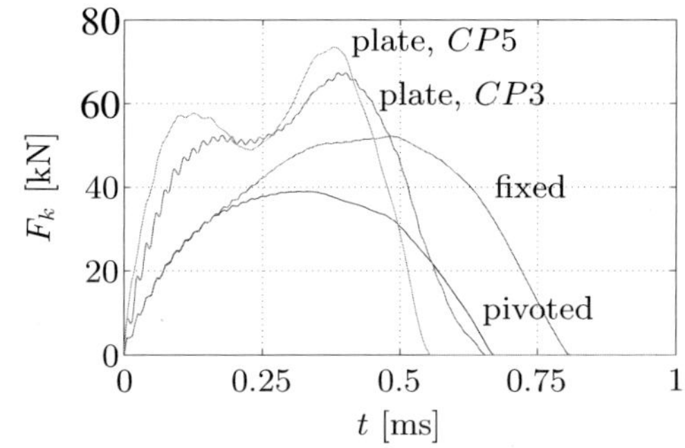

Fig. 11. Contact forces for the idler gear of gear pair 1 for contact with a fixed crankshaft gear considering CP3 and contact with a rigid plate considering CP3 and CP5.

results in a skew between the flanks, longitudinal form modifications and longitudinal profile errors are included in the investigation of misalignment. As typical values for gear pair 1, an offset of the centre distance of $\Delta a = 0.1$mm and an axial misalignment of $\alpha = 0.1°$ is chosen. The resulting contact forces are shown in Figure 12 (left).

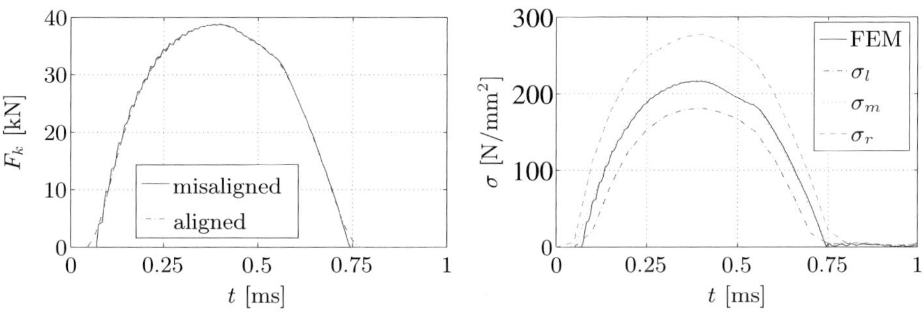

Fig. 12. Contact force of gear pair 1 for a typical misalignment (left) and maximal von Mises stresses for three characteristic points in the dedendum fillet (right).

The skew causes a softer beginning and ending of the contact, but the maximal contact force is essentially the same. However, the stresses in the dedendum fillet differ significantly. In Figure 12 (right), the stresses are evaluated in three points of the dedendum fillet, distributed equally over the face width. As a result of the misalignment, at first, the contact only establishes for only one side of the flank. This results in about 30% higher von Mises stresses on this side of the dedendum compared to the well aligned gears.

Multibody model

Since in the finite element analyses material damping is neglected and an oil film is not considered, the damping ratio d is set to zero for comparable results. For the tooth stiffness the values according to Equations (5) and (6) are used. For a single impact the solution of Equation (3) can easily be derived as the first sinusoidal half cycle

$$F_k = c\frac{\dot{x}_0}{\nu_0}\sin\sqrt{\frac{c}{m_e}}t. \tag{7}$$

Figure 13 shows the contact forces and rotational velocities of gear pair 1 considering $CP3$.

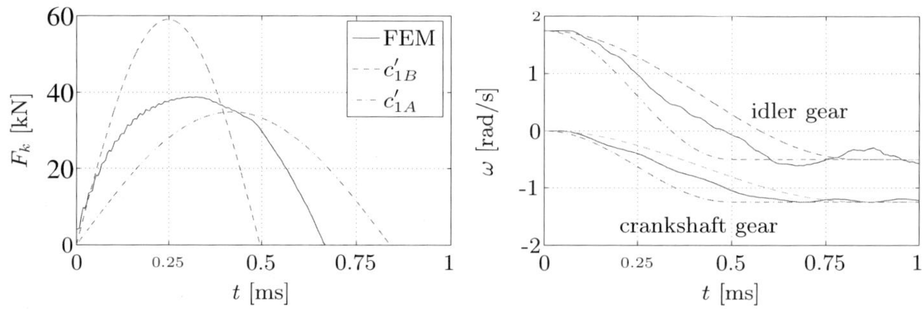

Fig. 13. Contact forces (left) and rotational velocities (right) for gear pair 1 considering $CP3$, calculated using the multibody model.

Clearly, the practical choice of the stiffness c'_{1B} yields an unrealistically high contact force and the contact duration is much too short. Even though the use of the more sophisticated choice c'_{1A} yields a fairly good estimate of the maximum contact force, the contact duration is too long. The comparison of the rotational velocities shows considerable differences as well. At least, the mean value of the rotational velocity after contact is met very well. Of course, the rigid body approach is not capable to account for the torsional vibrations explained before.

The corresponding results for gear pair 2 show a rather good agreement between the finite element analyses and the multibody results. Obviously, no significant gear body effects are present and the model assumption of rigid bodies yields good results.

4 Conclusions

The simulation results clearly show that for certain gear body designs the compliance of the gear body may not be neglected. In the static analyses of the

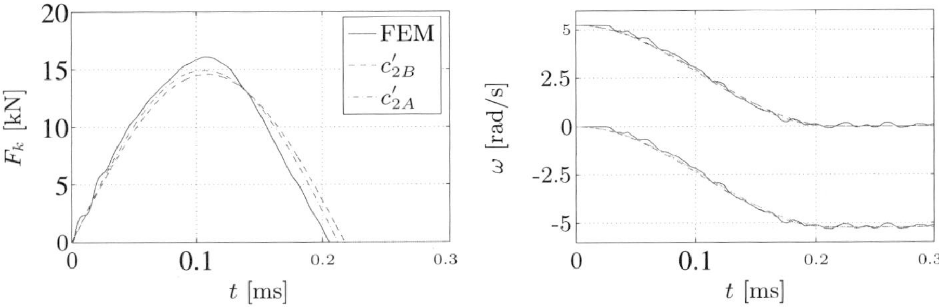

Fig. 14. Contact forces (left) and rotational velocities (right) for gear pair 2 considering $CP3$, calculated using the multibody model.

tooth stiffness this shows in significantly lower stiffnesses than for gears with cylindrical bodies. In the dynamic analyses gear body effects can be observed as torsional vibrations of the gear body even after impact or as superimposed vibrations in the force progression, as shown for very hard impacts with the rigid plate. Even if they are hardly recognisable, as for the results for impacts with typical impact velocities, they can not be neglected, as shown in the comparison with the multibody results. Of course, such gear body effects can not be considered by classical rigid body models. However, with enhanced multibody models, introducing additional degrees of freedom, a much better agreement can be achieved, see [6].

For gear pair 2 with cylindrical gear bodies, a rather good agreement between a finite element and a multibody model was achieved. Here, gear body effects are of minor influence and the compliance of the gear body is negligible.

Further investigations on the influence of the impact velocity, friction and misalignment show a linear dependence between the contact force and the impact velocity. Neither friction nor misalignment show a significant influence on the contact forces during impacts. However, regarding the stresses in the dedendum fillet, misalignment causes an increase of the stresses of about 30% for a typical misalignment.

Acknowledgements

The authors want to thank MTU Friedrichshafen and ZF Friedrichshafen, especially Dr.-Ing. U. Schmid (MTU), Dr.-Ing. J. Weber (MTU) and Dipl.-Math. J. Tasche (ZF) for their support, many useful discussions and for sharing their insight with us.

The project 'Rädertriebsimulation' is supported by the Forschungsvereinigung Verbrennungskraftmaschinen (FVV) and the Forschungsvereinigung Antriebstechnik (FVA) since January 2005. This support is highly appreciated and we also want to thank all the industrial members of the 'Arbeitskreis Rädertriebsimulation' for their contributions and ideas.

References

1. ABAQUS, (2004) Theory Manual. ABAQUS Inc., Providence.
2. ABAQUS, (2004): Verification Manual, ABAQUS Inc., Providence.
3. Bathe K.-J., (1996): Finite Element Procedures, Prentice Hall, New Jersey.
4. Blankenship, G.W., Kahraman A., (1995): Steady State Forced Response of a Mechanical Oscillator with Combined Parametric Excitation and Clearance Type Non-Linearity. Journal of Sound and Vibration, Vol. 185(5), pp. 743–765.
5. Cook, R.D., Malkus D.S., Plesha M.E., (1989): Concepts and Applications of Finite Element Analysis, John Wiley & Sons, New York.
6. Ebrahimi, S., Eberhard, P., (2006): Rigid-Elastic Modelling of Gear Wheels in Multibody Systems, Accepted for publication by Multibody System Dynamics. Springer, Berlin.
7. ISO 6336-1, (2000): Calculation of load capacity of spur and helical gears - Part 1, International Organisation for Standardisation.
8. Kahraman A., Blankenship, G.W., (1997): Experiments on Nonlinear Dynamic Behavior of an Oscillator with Clearance an Periodically Time-Varying Parameters. Journal of Applied Mechanics, Vol. 64, pp. 217–226.
9. Kücükay, F., (1987): Dynamik der Zahnradgetriebe (in German). Springer-Verlag, Berlin.
10. Niemann, G., Winter, H., (1989): Maschinenelemente Band 2 (in German). Springer-Verlag, Berlin.
11. Padieth, R., (1978): Exakte Ermittlung der Zahnform (in German), Antriebstechnik 17 (10), pp. 434–436.
12. Prestl, W., (1991): Zahnhämmern in Rädertrieben von Dieselmotoren (in German). Fortschrittsberichte Reihe 11 (145). VDI-Verlag, Düsseldorf.
13. Ziegler, P. (2005): Statische und dynamische Zahnkontaktuntersuchungen (in German), DIPL-104, Institute B of Mechanics, University of Stuttgart.
14. Ziegler, P., Eberhard P., Schweizer B. (2006): Simulations of Impacts in Geartrains Using Different Approaches, Submitted to Archive of Applied Mechanics. Springer, Berlin.

Comparison of Building Collapse Simulation Results From Finite Element and Rigid Body Models

Steffen Mattern[1], Gunther Blankenhorn[1], Michael Breidt[2], van Vinh Nguyen[2], Sebastian Höhler[3], Karl Schweizerhof[1], Dietrich Hartmann[2], and Friedhelm Stangenberg[3]

[1] Institut für Mechanik, Universität Karlsruhe (TH), 76131 Karlsruhe, Germany `mechanik@ifm.uni-karlsruhe.de`
[2] Lehrstuhl für Ingenieurinformatik im Bauwesen, Ruhr Universität Bochum, 44780 Bochum, Germany `hartus@inf.bi.rub.de`
[3] Lehrstuhl und Institut für Stahlbeton- und Spannbetonbau, Ruhr-Universität Bochum, 44780 Bochum, Germany `friedhelm.stangenberg@rub.de`

1 Introduction

In case of planning a building demolition, the information about geometry, quality of building materials, the design of the load carrying system and documentation of the structural calculation is often incomplete and imprecise. Thus for the analysis of a collapse event, engineers are forced to consider the uncertainty of primary parameters influencing e. g. the resistance of structural elements of a building. This kind of uncertainty can be described using suitable data models such as fuzziness and fuzzy randomness [6]. Within such an 'uncertain' structural analysis the deterministic fundamental solution is applied repeatedly. A comprehensive overview over algorithms of fuzzy analysis and fuzzy stochastic analysis is given in [5]. First applications of uncertainty collapse analyses can be found in [7, 8]. However, considering several uncertain parameters in an analysis the problem dimension and the necessary effort can be quite high. To receive a good prediction for a complex building collapse, several hundred or even more deterministic solutions are needed. This requires an efficient and fast scheme to perform the analysis for highly nonlinear problems, concerning geometry, material and changing boundary conditions such as contact.

In particular, efficient multi rigid body models for the simulation are therefore created, that are supported and validated by finite element methods. The investigation presented in the following shows results and comparisons of building collapse simulation from finite element as well as rigid body models elaborated in a joint research project. The partners in this project are stem-

Peter Eberhard (ed.), Multiscale Problems in Multibody System Contacts, 257–267.

ming from different institutes (Computational/Structural Mechanics, Reinforced Concrete and Engineering Informatics). A common goal of the joint project is the realistic and efficient deterministic simulation of real-world structures that builds a foundation for continuative studies like the aforementioned fuzzy analysis.

2 Analysis concept

As mentioned in section 1 an efficient analysis is needed to predict the holistic demolition process especially the collapse sequence of a building. Commonly used numerical analysis tools are finite element programs, which allow a very flexible discretization, approximating arbitrary geometries and materials to predict the behavior of real structures under estimated loads and boundary conditions. However, even if highly efficient algorithms and implementations are applied, efficiency problems occur for complex structures concerning the calculation time, e.g. a six storied reinforced concrete building with a simple cubic geometry, discretized with 85000 solid elements, needs several hours for the solution on an eight processor (Intel® Itanium® 2) node of a Linux system. One possibility to reduce the calculation time considerably is to use multibody models instead of finite elements. Such models are highly efficient with respect to the calculation time.

The main problem while using rigid body models is the proper modeling if various contact situations happen during a collapse event or if local zones of accumulated damage appear, which cannot be represented and detected by a rigid body. A solution concept to overcome this problem and to develop an efficient scheme which is used in this investigation includes both, finite element analysis as well as rigid body models. The finite element analysis supports the modeling process of the rigid body model; it also allows to predict the behavior of the local zones of accumulated damage and the region of moderate deformation which can be estimated as rigid. This is also taken to gain experience defining suitable rigid body models. Once the rigid parts are identified, the hinges between the various rigid bodies must be linked to a characteristic resistance curve which represents the resistance against relative rotations or displacements of the individual bodies. To get a good approximation of the behavior of the real structure in zones of local damage, finite element submodels are separately created with a fine discretization and rather sophisticated material models. Via this approach, characteristic resistance curves are produced for some standard load cases which are expected in the collapse event.

Summarizing the concept, in a first step a finite element analysis is used to support the setup of a special multi rigid body model for efficient simulation of collapses. Thereby, one of the main goals is to develop suitable multibody systems such that parts with moderate deformations can be treated as rigid, defining the different rigid body parts of the multibody system. Then the zones of local damages are identified and modeled using tailored multibody subsystems that are built up with multibody elements using nonlinear force elements according to the characteristic resistance curves. These curves are calculated a-priori by a separate detailed finite element submodel of the structure. The following sections show fundamentals of (i) the global finite element analysis of the entire structure, (ii) the finite element analysis of the local zones of accumulated damage applying specialized material models for dynamically reinforced concrete and (iii) the developed multi body based simulation system.

3 Example description / Fundamentals of applied finite element and multibody models

3.1 Model description / Reference System

In order to compare results within the development of a procedure for the generation of a multi body system (MBS), a selected reference model is used. For simplicity, all investigations presented in this paper are based on a fictitious, three-storied framework structure of reinforced concrete as shown in Figure 1(a).

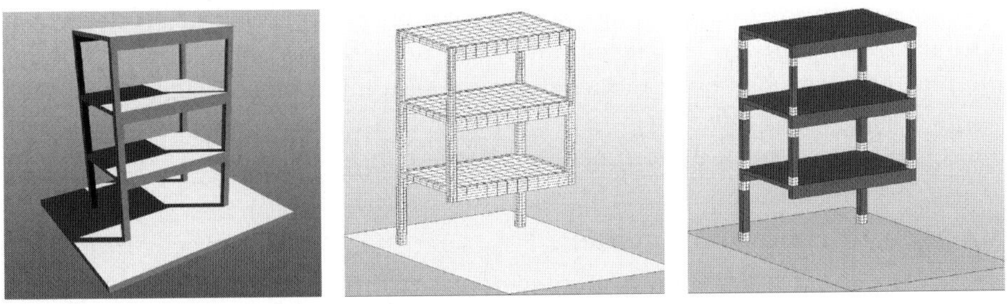

(a) Reference geometry (b) FE-discretization with front columns removed (c) Developed 'hybrid rigid body model'

Fig. 1. Three story framework – geometry, pure FE-model and hybrid rigid body model – removal of front columns for blasting simulation

The destruction of this structure is initialized by removing the two front side columns, which leads to a collapse under dead load. This rather moderately complex reference model was chosen to develop the simulation process in all details, taking advantage of reduced simulation requirements and representing the expected exchange of data between the different simulation concepts. The described simulation process can be later taken in an identical fashion for far more complex building simulations.

3.2 Finite Element model

The geometry in Figure 1(a) has been discretized with 2330 eight-node hexahedral finite elements (Figure 1(b)) and computed with explicit time integration, using a central difference scheme [4, 3]. The rather coarse mesh leads to a fairly good approximation, mainly because of the one-point under-integrated solid elements, which do not show locking. However, for these elements a stabilization against unphysical kinematics, the so-called hourglass modes is necessary, for which the assumed strain co-rotational stiffness form by Belytschko/ Bindemann [3, 4, 1] was chosen. The blasting process was simulated — as mentioned above — by removing the two front side columns at the beginning of the computation. Under dead load, this leads to a certain collapse kinematics with the appearance of hinges, i. e. zones of local accumulated damage, which are modeled with appropriate multibody subsystems e.g. joints combined with nonlinear springs in an MBS-Simulation. In order to detect these local zones in a rough manner, element failure was introduced, here by introducing a critical plastic strain $\varepsilon_{pl,crit}$. Every element, which fulfills the condition

$$\varepsilon_{pl} \geq \varepsilon_{pl,crit} \tag{1}$$

at any time of the simulation, is removed (eroded) from the computation. This means, in regions with high plastic strains, many elements erode, which automatically leads to the development of the described local zones. The reinforced concrete has been modeled with a simplified, homogeneous material — *piecewise linear plasticity* — where the parameters concerning plasticity and erosion have been determined by simple experiments.

3.3 Hybrid rigid body model

Based on the Finite Element (FE) simulations from section 3.2, several parts of the model which show small deformations, compared to the local zones of accumulated damage during the whole simulation, are modeled now as rigid bodies to reduce the numerical effort. As criterion for rigidity of a body, the strain rate in the flexible parts of the FE structure is chosen, following the proposal of [2], where

$$\dot{\varepsilon} \leq \dot{\varepsilon}_{crit} \tag{2}$$

defines a structural component as rigid. Parts which do not exceed the value $\dot{\varepsilon}_{crit}$ could be treated as rigid for the full simulation time, the rest of the structure is still modeled with finite elements as described in section 3.2. The 'hybrid rigid body model' as a result of the finite element simulation based on the initial fully FE model given in Figure 1(b) is shown in Figure 1(c), where the black parts are rigid bodies and white parts are local finite element meshes.

3.4 Multi rigid body model

Based on the finite element analyses, rigid parts with initially little deformation and parts with larger deformation could be localized. For the multi body analyses a suitable subsystem with similar kinematic and dynamic behavior such as the 'costly' FE model has to be defined to simulate the failure process. As found in the FE simulation, the main failure areas are at the bottom and at the top of the columns (Fig. 2a). For that it is appropriate to model a column as one rigid body with hinges and spring elements on bottom and top (Fig. 2b+ 2c). To correctly achieve the failure characteristics of reinforced concrete, the spring elements are defined by pre-calculated nonlinear characteristic resistance curves.

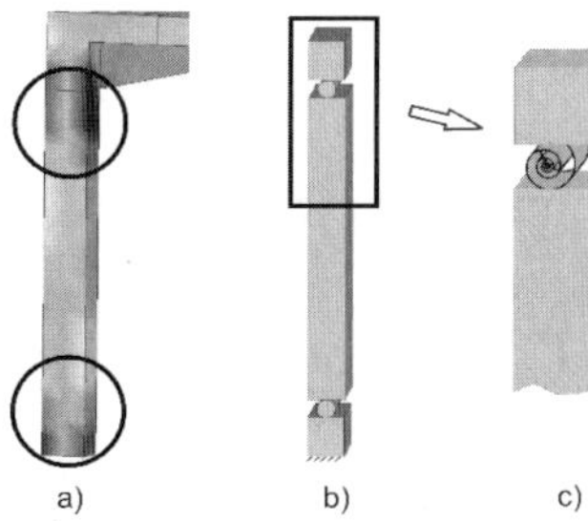

Fig. 2. Column modeled as rigid body with hinges and nonlinear spring elements

3.4.1 Determining the characteristic resistance curves

To determine the nonlinear characteristic resistance curves for the spring elements a detailed finite element analysis is applied. Compared to the finite element analyses of the global structure, the size of the elements is rather small, leading to up to 5000 hexahedral elements for one column. In the local zones the stress resultants in a cross section are computed in every step with a stress integration over the area. Also the curvature/rotation according to the different stress resultants are stored. The implemented material model, described in [10] is a 'close-to practice' elastic-plastic damage theory model for reinforced concrete. It is applied to solid and solid-shell elements, used in the analyses.
To describe the behavior of concrete under compression a yield/damage potential of Drucker-Prager-type is taken:

$$\Phi_c(\sigma, \alpha) = \frac{1}{\frac{1}{\sqrt{3}} - \mu} \left[\mu\, I_1 + \sqrt{J_2}\right] - \alpha_c(q_c), \tag{3}$$

where I_1 is the first invariant of the stress tensor σ and J_2 the second deviator invariant. The term left of the brackets guarantees that during plastic loading

α_c always corresponds to the negative uniaxial compression stress. For tension, a damage potential of Rankine type is used for all three main stress axes

$$\Phi_{t(i)}(\sigma, \alpha_t) = \xi_i - f_t \leq 0, \quad i = 1, 2, 3. \tag{4}$$

The reinforcement is modeled by truss elements with elasto-plastic behavior. To simplify matters only tension softening is considered. The model is very efficient for uncertainty analyses. The computed characteristic resistance curves are stored in the model database for the multi body system program which will be described in the next section.

3.4.2 Assembling the multi rigid body system

For the multibody simulation, appropriate subsystems are assembled to a special multibody model for the discussed reference system. This is carried out by a simulation platform using information and data generated by the different aforementioned finite element analyses.

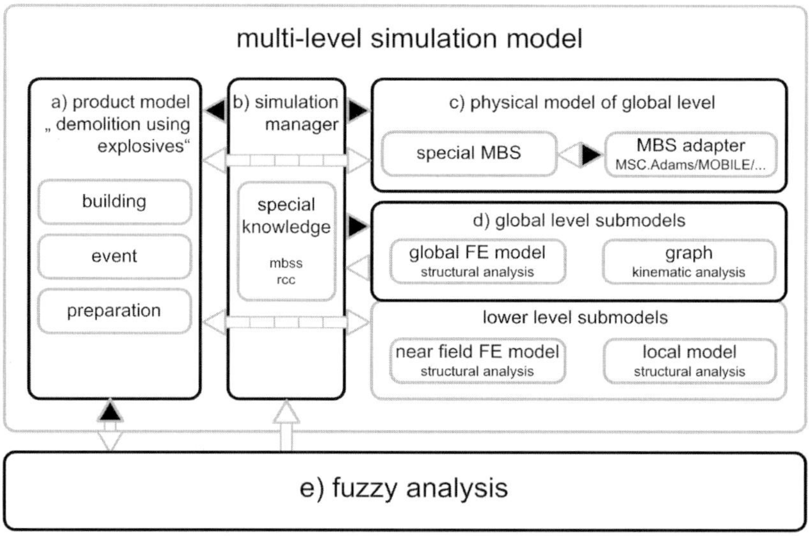

Fig. 3. Schematic presentation of the simulation model

In Figure 3, schematically the concept of the simulation model of the platform is depicted. The simulation model is based on different submodels. The submodel a) in Figure 3, represents the product model for a so-called 'demolition using explosives'. This submodel serves as a database and contains all relevant data needed for the global level simulation, such as the position, the geometry and material data of the parts of the building, the details of the preparatory work (modifications of the static structure of the building before ignition of the explosives) and the potential events (locations and ignition times of explosive loads). Using these data along with the results of the different submodels of the global level (d) as well as the lower levels, the submodel 'simulation manager' (b) creates a model description of the special multibody system (c). This modeling process is carried out by using special knowledge gained from the various finite element analyses. The creation and solution of the system equations is accomplished by a multibody system (MBS) software that is applied by the special MBS submodel via a specific MBS adapter (c). Currently, the program system MSC.Adams [9] is applied. The described simulation platform provides interfaces, here, exemplarily shown for the fuzzy analysis [5].

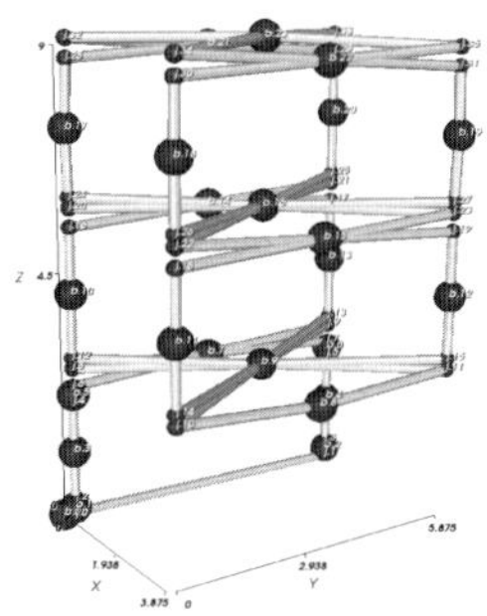

Fig. 4. Schematic presentation of the special multibody model showing the kinematic skeleton and the mass properties

In Figure 4, the resulting multibody model of the reference system is demonstrated. According to the results of the finite element analyses, all columns are connected to the neighboring structural elements by revolute joints with distinct rotational springs with nonlinear characteristics defined by the characteristic resistance curves. Figure 5 shows exemplarily the characteristic resistance curve for the bottom rotational spring.

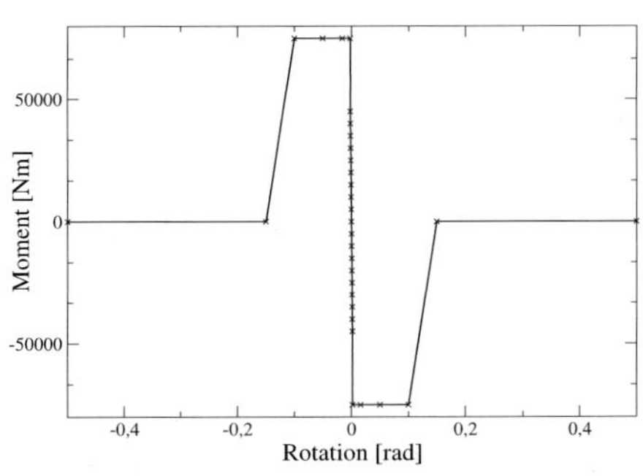

Fig. 5. Characteristic resistance curve

4 Numerical Analysis and results

4.1 Finite Element model

In Figure 6, two states of the finite element simulation with a critical strain rate of $\dot{\varepsilon}_{crit} = 0.25$ are shown. The black parts of the structure show strain rates, larger than the critical value, which indicates local zones of accumulated damage. The bright parts show rigid body like behavior, because $\dot{\varepsilon} \leq \dot{\varepsilon}_{crit}$. This investigation of the strain rate over the whole collapse, which is about 2 s, leads to a combined model of rigid bodies and finite elements as shown in Figure 1(c), there labeled as 'hybrid rigid-body model'.

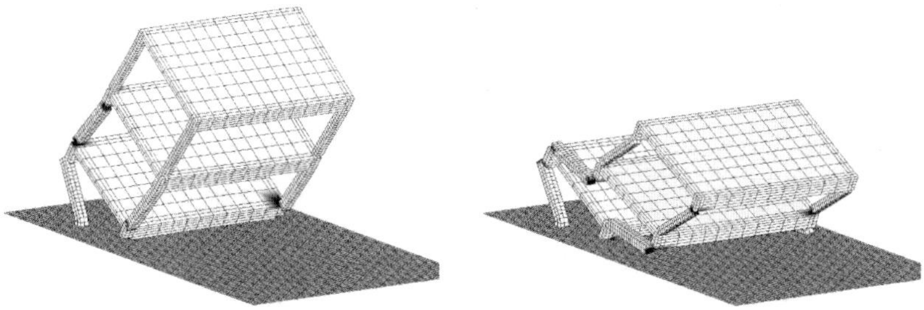

Fig. 6. Visualization of the strainrate at two states of the collapse simulation with full FE-model

4.2 Hybrid rigid body model

Both, FE- and hybrid rigid body model, seem to correlate very well especially at the beginning of the collapse, as can be seen in Figure 7. Differences become more visible at the end of the collapse; here the subdivision into smaller rigid bodies would be necessary to obtain better correlation. In further investigations, the influence of contact on the kinematics of the collapse has been investigated. Including obstacles leads to different kinematics, different strain rates and consequently to a different hybrid rigid body model. This is implied in Figure 8, where a rigid wall as new obstacle has been introduced. This example shows, that the hybrid rigid body model has to be checked for contact. Each time when contact appears during the simulation, the formation of rigid bodies and local finite elements changes and so the hybrid rigid body model has to be adapted. Such modifications are also required in the case of failures or fractures of components. The latter concept has later to be transferred to the multi rigid body analyses with an MBS analysis system.

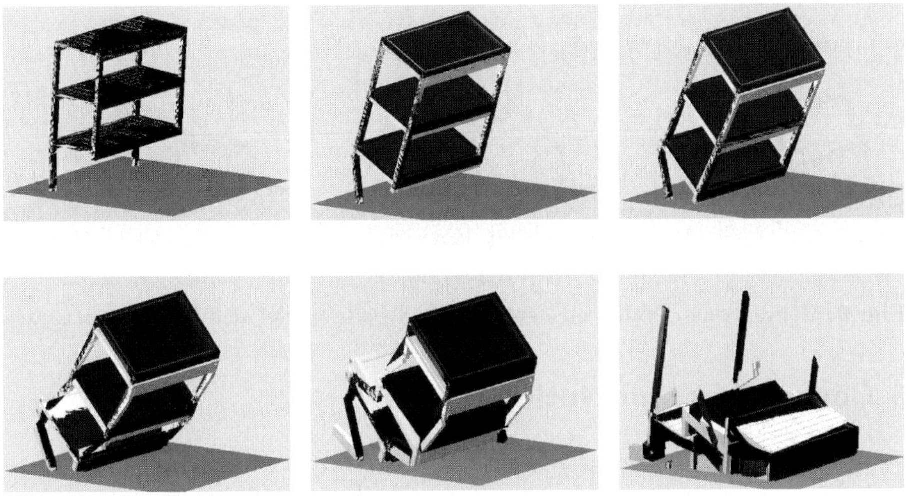

Fig. 7. Pure finite element model (black) and hybrid rigid-body model (white) compared at different time states

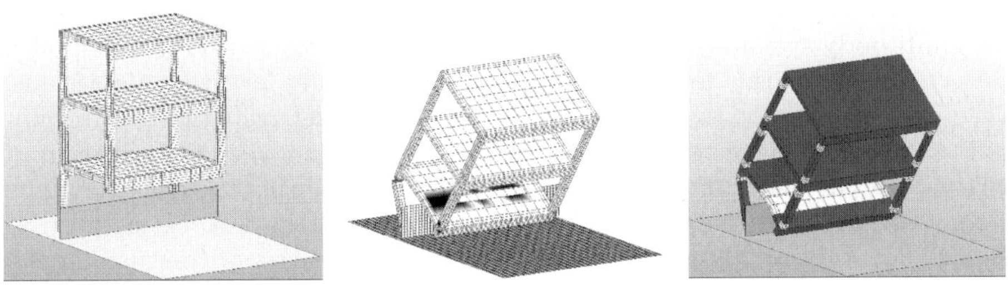

Fig. 8. FE-simulation, strain rate analysis and hybrid rigid-body model with rigid wall as obstacle

4.3 Rigid body model

In Figure 9, the results of the finite element and the rigid body model are compared. Up to the first contact with the ground, the correlation is excellent. Due to different contact models for the finite element and the rigid body simulation, the collapse starts to differ. Another point to be mentioned is that the connections of the rigid bodies are only cylindrical joints so far. The use of translational joints and a more detailed multi body system may reduce the differences.

However, at this point of work in the research program it is attempted to gain experience and find rules to create appropriate rigid body models for a realistic simulation of collapses.

Fig. 9. Comparison of finite element and rigid body model at different time states

5 Conclusions and Outlook

Within this contribution, the general possibilities of modeling building collapse by means of multi body systems taking advantage of the experience from fully FE simulations is discussed. It has been shown, that rather simple multi body systems, which require far less computational effort than corresponding finite element models lead to very similar results, if the multibody system contains the proper subsystems with hinges and nonlinear spring elements at the correct locations. This approach allows to develop a simulation process for multi body systems, hedged by finite element solutions, taking finally also uncertainties by fuzzy analysis into account. Especially for fuzzy algorithms, very efficient simulations are needed because extremely repetitive deterministic simulations have to be carried out. Important for the discretization of the MBS is the proper detection of evolving contact during the collapse. Each time a contact event happens, subsystems with hinges have to be included at specific locations of the MBS. This also holds in the case of fracture events. With the experience gained through the global FE simulation and the nonlinear characteristic resistance curves from the detailed FE models, it is possible to create a database of subsystems, from which optimal MBS for the simulation of the collapse process can be generated, almost automatically, or with little effort by the analyst who uses the developed program.

6 Acknowledgment

The financial support of the German Research Foundation (DFG) (Project DFG-FOR 500 - 'Computer aided destruction of complex structures using controlled explosives') is greatfully acknowledged.

References

1. T. Belytschko and L. P. Bindemann. Assumed strain stabilization of the eight node hexahedral element. *Computer Methods in Applied Mechanical Engineering*, 105:225–260, 1993.
2. B. Göttlicher. *Effiziente Finite-Element-Modellierung gekoppelter starrer und flexibler Strukturbereiche bei transienten Einwirkungen.* Universität Karlsruhe(TH), Institut für Mechanik, 2002.
3. J.O. Hallquist. *LS-DYNA Theoretical Manual.* Livermore Software Technology Corporation, 1991-1998.
4. J.O. Hallquist. *LS-DYNA Keyword User's Manual.* Livermore Software Technology Corporation, 1992-2005.
5. B. Möller and M. Beer. *Fuzzy Randomness - Uncertainty in Civil Engineering and Computational Mechanics.* Springer, Berlin, 2004.
6. B. Möller, W. Graf, and M. Beer. Safety assessment of structures in view of fuzzy randomness. *Computers and Structures*, 81:1567–1582, 2003.
7. B. Möller, A. Hoffmann, and M. Liebscher. Modeling of blasting processes in view of fuzzy randomness. In *9th ASCE Specialty Conference on Probabilistic Mechanics and Structural Reliability (PMC2004)*, 2004.
8. B. Möller and M. Liebscher. Fuzzy multi body systems and fuzzy probabilistic multi body systems and their application for the numerical simulation of controlled demolitions of structures. In K. J. Bathe, editor, *Compilation of Abstracts of the Third M.I.T. Conference Computational Fluid and Solid Mechanics*, page 265, Boston, 2005. M.I.T.
9. MSC.Adams. www.mscsoftware.com, 2006.
10. R. Poelling. *Eine praxisnahe, schädigungsorientierte Materialbeschreibung von Stahlbeton für Strukturanalysen.* Ruhr-Universität Bochum, Lehrstuhl fuer Statik und Dynamik, 2002.

Computational Analysis and Experimental Investigation of Impacts in Multibody Systems

Robert Seifried and Werner Schiehlen

Institute of Engineering and Computational Mechanics, University of Stuttgart
Pfaffenwaldring 9, 70569 Stuttgart, Germany
[seifried, schiehlen]@itm.uni-stuttgart.de

1 Introduction

For many engineering applications the multibody system approach proves to be most efficient for the analysis of the overall motion. Collisions between moving bodies might result in impacts which interrupt the overall motion. During impact kinetic energy is lost by wave propagation in the bodies and/or by plastic deformation of the contact region, see Goldsmith [5], Johnson [10] and Stronge [20]. For multibody system simulation on the macroscale, these kinetic energy losses during impact are represented by the coefficient of restitution, see e.g. Glocker [4] and Pfeiffer and Glocker [12]. However, the coefficient of restitution cannot be computed within the multibody system approach and must be estimated from costly experiments or experience. Alternatively the coefficient of restitution may be evaluated by additional numerical simulations on a fast time scale, resulting in a multiscale simulation approach as presented in References [9, 14, 15, 18].

This paper deals mainly with the computational analysis and experimental investigation of impact processes for multibody systems on a fast time scale in order to determine the kinetic energy loss during impact and to evaluate the corresponding coefficient of restitution. The computational impact analysis on the fast time scale requires accurate and efficient models which include the phenomena of wave propagation and plasticity. The experimental verification on the fast time scale requires advanced measurement techniques capable for catching high frequency phenomena resulting from impact. Computational and experimental results are presented for the longitudinal impact of steel spheres on different aluminum rods made of elastic as well as elastic-plastic material. Also the trustiness of the multiscale simulation approach is demonstrated for the impact of an elastic steel pendulum on an aluminum body with elastic-plastic material behavior.

Peter Eberhard (ed.), Multiscale Problems in Multibody System Contacts, 269–280.

2 Computational Analysis

The computational impact analysis on the fast time scale can be carried out by using a complete Finite Element (FE) model of the impacting bodies, see e.g. Wu et al. [21]. However, for the FE-model a small overall element length is required to comprise the wave propagation with an additional refinement necessary for the modeling of the contact region, see Reference [16]. Therefore, FE-models for impact analysis are very time consuming.

On the other hand impact processes can be divided into two parts, the contact itself and the resulting wave propagation. The contact is a nonlinear problem which is limited to a small boundary region, while the wave propagation is a linear problem encompassing the entire bodies, see References [17, 18]. Based on this decomposition, combined models are developed in which the elastodynamic behavior of the bodies is represented by a modally reduced linear FE-model of the impacting bodies while the deformation of the contact region is presented by a local contact model. This approach is presented schematically in Figure 1. The two submodels are coupled by exchanging force and displacement values. In each integration time step the modal model is used firstly to compute the displacement vector $\boldsymbol{u}$ of the boundaries of the contact region. These values are then passed to the contact model where they are used to compute the contact force vector $\boldsymbol{F}$. The contact force is then feed back to the modal model where it is used as excitation for the next time step.

For the local contact model three different contact models can be used as presented in References [15, 18]. The first contact model is a FE-model of the contact region which is computed concurrently with the modal model. This approach includes all nonlinear effects of the complete FE-model, such as strain rate material behavior of the contact region. The second model is confined to quasi-static contact and rate-independent material behavior. The behavior of the contact region is pre-computed by static FE-analysis and presented in force-displacement diagrams. For the impact simulation with the

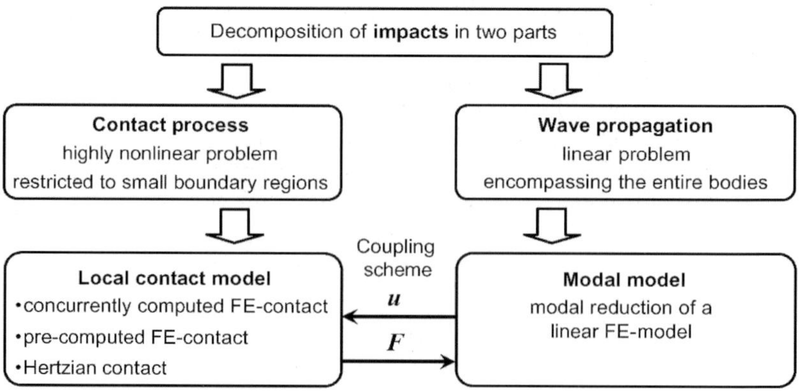

Fig. 1. Schematic representation of the simulation models

modal model these pre-computed diagrams are then used to determine the impact force. The third contact model is based on the elastostatic Hertzian contact law. A comparison shows the consistency of the modal models with local contact models and the complete FE-model. However, compared to complete FE-models, the modal models with local contact models are much more time efficient, see Reference [15]. Furthermore, the modally reduced models with FE-contact can modified to simulate repeated impacts and multiple impacts, including the deformation history of the contact region due to plasticity, see References [15, 17, 18] for more details.

3 Experimental Investigation

Even if the computational analysis has been performed with different approaches, experimental investigations provide the verification required for responsible engineering practice. Therefore, extensive experimental investigations on a fast time scale are carried out. The experimental setup is an advancement of the setup originally developed by Hu et al. [8] and is shown in Figure 2 for the longitudinal impact of a steel sphere on an aluminum rod. The impacting bodies are suspended by kevlar wires in a frame. For initiating the experiment, the steel sphere is released by a magnet from a predefined height, and it impacts on the resting aluminum body.

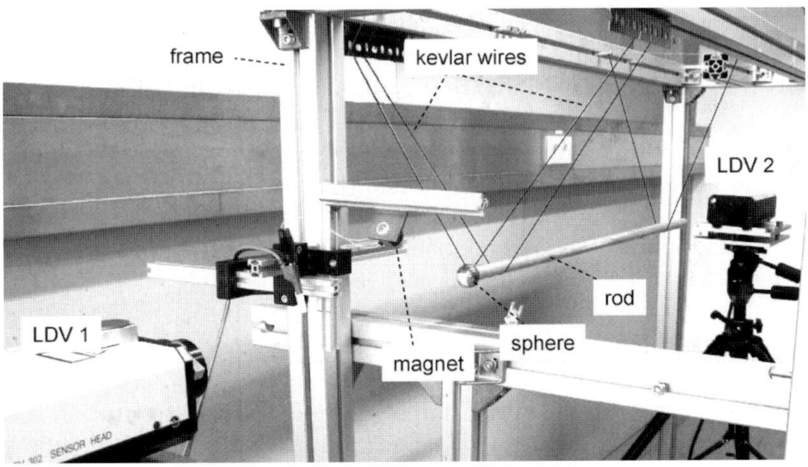

Fig. 2. Experimental setup

3.1 Measurements using Laser-Doppler-Vibrometers

The velocities and displacements of the impacting bodies are measured using two Laser-Doppler-Vibrometers (LDVs) of type OFV-3000/OFV-302 made by Polytec GmbH [13]. The LDVs are placed on opposite sides of the impact systems along the impact line. So it is possible to measure directly the rebound velocity of the steel sphere and the response of the aluminum body. The LDVs utilize an interferometric technique to measure vibrational signals, see Lewin et al. [11]. Interference is a phenomenon observed when two beams of light from a common origin are made to coincide. In the LDV measurement the laser beam is split into a measuring beam and a reference beam. The measuring beam is focused on the surface of interest where it is reflected. Then, the reflected beam and the reference beam are made to coincide. After re-combination the interference signal produces a bright-dark pattern which is measured by a photodetector. Thereby one bright-dark-cycle corresponds to the displacement of the measurement point by half a wave length of the laser light. The displacement measurement is based on counting theses bright-dark-cycles in combination with an interpolation technique. For the velcoity measurement the modulation frequency of the interference signal is determined, which corresponds to the Doppler-frequency and is proportional to the velocity of the measurement point.

For the impact of a steel sphere (radius $15mm$) on an elastic aluminum rod (radius $10mm$, length $1000mm$) the measured displacements and velocities are presented in Figure 3 for a series of 30 experiments. The very good reproducibility of the measurements is clearly shown. It turns out that for the LDVs used the low frequency phenomena, such as the initial rigid body velocity before impact, are captured most accurately by the displacement signals, while the high frequency phenomena, like the impact process and wave propagation, are measured most accurately by the velocity signals, see Reference [18]. The standard deviation of the measured initial velocity of the sphere is 1.1%, with a maximal deviation of 2%. The velocity of the sphere after impact shows a standard deviation of 1.4% and a maximal deviation of 3.3%. The measurements also show that the high frequency wave propagation phenomena in the rod are captured with a high accuracy. This proofs the high reliability of the measurement equipment and the high quality of the experimental setup with the bodies suspended as pendulum.

3.2 Measuring the Coefficient of Restitution

A common method for the evaluation of the coefficient of restitution is based on the optical measurement of the rigid body motion on a slow time scale. For the measurement of the rigid body motion multiple exposed photographs might be used, see e.g. Foerster et al. [3], or high speed video cameras, see e.g. Stoianovici and Hurmuzlu [19]. By measuring the relative velocities before and after impact on the slow time scale the kinematic coefficient of restitution is determined.

In this paper the kinetic coefficient of restitution is determined form the measurements on the fast time scale. The kinetic coefficient of restitution is defined as the ratio of impulse during the compression and restitution phases

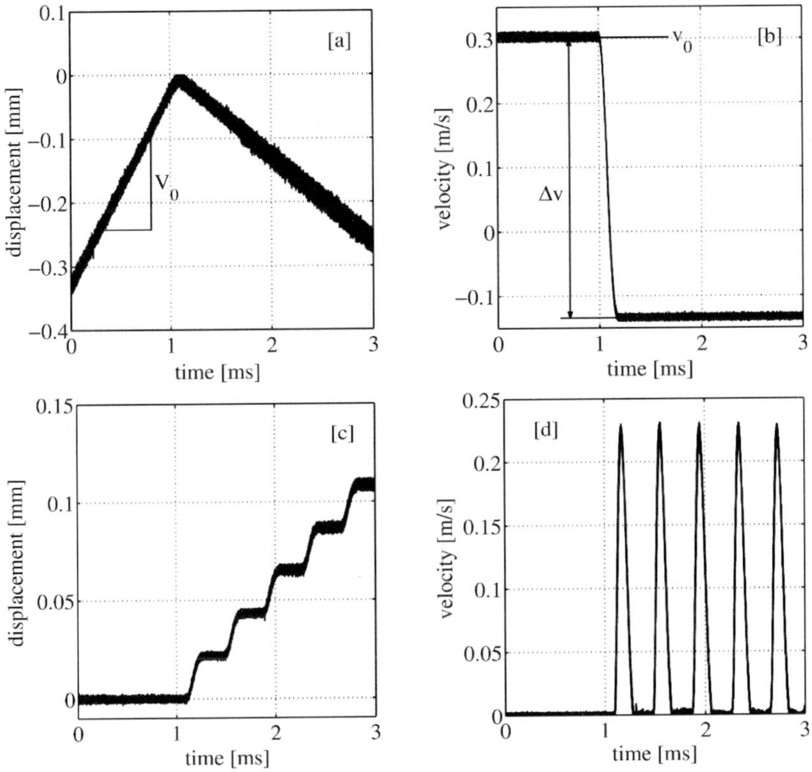

Fig. 3. Displacment and velocity of sphere (top) and rod (bottom) for 30 measurements

of the impact, see Stronge [20]. Since the aluminum bodies are initially at rest, the kinetic coefficient of restitution can be evaluated experimentally purely from the measurements of the sphere‘s velocity. From the change of velocity Δv due to impact, see Figure 3b, and the mass m_s of the sphere the impulse $\Delta p = m_s \Delta v$ produced by impact is determined. Based on the assumption for rigid body impacts the impact duration is infinitesimally small and all forces except the impact force are negligibly small. Following Reference [14] and establishing the linear momentum balance in the central impact line of two colliding bodies during the compression and restitution phases the coefficient of restitution reads as

$$e = \frac{(m_a + m_s) m_s \Delta v}{m_a \; m_s v_0} - 1. \tag{1}$$

where m_a and v_0 are the mass of the aluminum body and the initial velocity of the sphere, respectively.

3.3 Measuring the Impact Force for Impacts on Rods

The impact force is very useful to investigate the micromechanical processes during impact. For the measurement of impact forces of arbitrary shaped bodies piezo-electric force sensors might be used, see Falcon et al. [2]. However the force sensors have to be applied on the contact region and therefore will influence the impact process itself. For impacts on long rods the impact force can be determined from the measurement of the wave propagation in the rod without influencing the impact process, see e.g. Cunningham and Goldsmith [1]. During impact kinetic energy is transformed into strain energy which propagates as a wave away from the contact region. If the investigated rods are long enough the contact ends before the reflected wave returns to the struck end. After the wave passes a point of the rod this point remains at rest until the wave passes through it again, see Figure 3d. Therefore, the impact force can be determined from a strain measurement ε at some point of the rod or from the velocity measurement v_{fe} at the free end of the rod, respectively. Using the strain measurement the impact force is determined from

$$F_{\varepsilon} = AE\varepsilon, \tag{2}$$

where A and E denote the rod's cross section and Young's modulus, respectively. Based on the wave equation for a rod, see e.g. Graff [6], the impact force can be computed from the velocity of the free end by LDV2 as

$$F_v = \frac{AEv_{fe}}{2c} \quad with \quad c = \sqrt{\frac{E}{\rho}}, \tag{3}$$

where c is the wave speed in the rod and ρ denotes the rod's density.

For the impact of a steel sphere (radius $15mm$, initial velocity 0.3m/s) on an aluminum rod (radius $10mm$, length $1000mm$) the left plot of Figure 4 shows the measured rod velocity. The right plot shows the impact force determined from this velocity measurement in comparison with the impact force determined from a strain measurement. The technique for the strain measurement using strain gauges is described in Hu et al. [8]. Due to measurement noise in the strain signal the plot shows the still irregular mean values of six measurements. Both methods agree very well. However, despite of the averaging over six measurements, the impact force computed from the strain signals shows a noticeable measurement noise. A statistical analysis of the measurement noise shows that the accuracy of the velocity signal measured by the LDVs is about ten times higher than the accuracy of the strain measurements using strain gauges, see Hu and Eberhard [7]. It turns also out that the signal to noise ratio for the measured strain signals is about 20 dB while the signal to noise ratio for the measured velocity signals is up to 40 dB. Therefore, in the following the impact force is determined using the velocity measurement.

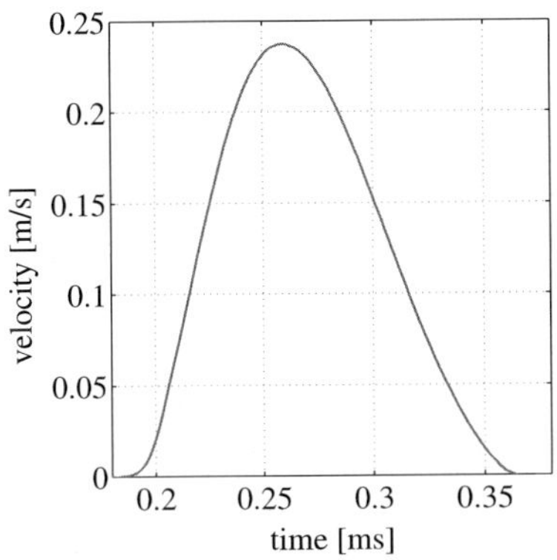

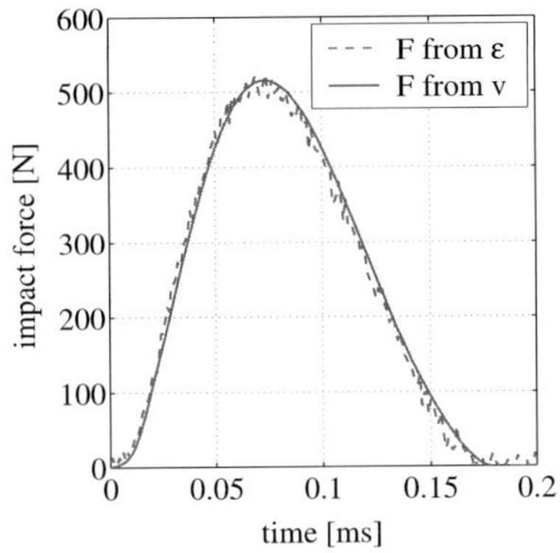

Fig. 4. Measured velocity of the rod and the resulting impact force in comparison with impact force determined from strain measurement

4 Comparison of Computational and Experimental Results

In this section computational and experimental results for three different impact systems are presented. Firstly the longitudinal impact of two steel spheres with different masses on an aluminum rod with high yield stress, representing elastic material behavior, is presented. Then results for repeated impact of a steel sphere on an aluminum rod with low yield stress, representing elastic-plastic material behavior, are shown. Finally the multiscale simulation approach is demonstrated for a simple multibody system.

4.1 Elastic Impact on Rods

The impact of two different spheres on an elastic aluminum rod (radius $10mm$, length $1000mm$) is investigated. The first sphere has radius $15mm$ and a mass of $m_s = 110g$, the second sphere has radius $20mm$ and a mass $m_s = 261g$. Figure 5 shows the simulated and measured sphere velocity and impact force. The very good agreement of simulations and measurements is obvious. The heavier sphere yield a longer impact duration and higher impact force and a significantly lower rebound velocity after impact. In both cases over 50% of the initial kinetic energy is transformed into wave propagation. This yields for the heavier sphere to a coefficient of restitution of $e = 0.58$ in the experiments and $e = 0.61$ in the simulations. The difference observed is mainly due to measurement noise of the velocity difference Δv. However, for engineering application this difference is of minor importance. The lighter sphere yields to coefficients of $e = 0.62$ in the measurements and $e = 0.64$ in the simulations. This shows the good consistency of simulations and experiments.

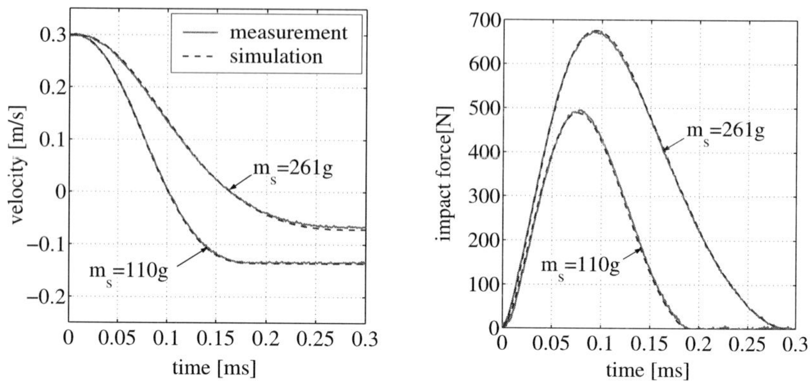

Fig. 5. Velocity of the sphere and impact force for impacts on the rod with elastic material behavior

4.2 Elastic-plastic Impact on Rods

Repeated impacts of a steel sphere (radius $15mm$) on an aluminum rod (radius $10mm$, length $1000mm$) with elastic-plastic material behavior and yield stress $205MPa$ are performed. In a series of repeated impacts the contact region gets plastically deformed by the first few impacts until a stationary configuration is reached. In this stationary phase no additional plastic deformation occurs and all kinetic energy loss is purely due to wave propgation, see References [17, 18].

For the investigated impact system a stationary configuration is reached after approximately 5-6 impacts. For the experiments five measurement series of 50 repeated impacts are performed and evaluated by computing the mean value. For the comparison of the velocity of the sphere and the impact force in the stationary phase the $10th, 20th, 30th, 40th$ and $50th$ impact are averaged and denoted in the following as impact $\bar{30}$. Figure 6 shows the velocity of the sphere and the impact force for impact 1 and $\bar{30}$. The simulations and measurements show very good agreement. The first impact differs significantly from the tenth impact which represents the stationary phase. The increase of the rebound velocity and impact force with the number of impacts is obvious in simulations and experiments and yield an increase of the coefficient of restitution. For the first impact a coefficient of restitution of $e = 0.50$ is consistently computed and measured. For the stationary phase of plastification the computed coefficient is $e = 0.54$ and agrees very well with the measured value of $e = 0.55$.

4.3 Multiscale Investigation of an Elastic Pendulum

The approach of the multiscale simulation for multibody systems is verified experimentally for a simple impact system presented in Figure 7. This system consists of an elastic steel pendulum with square cross section which is

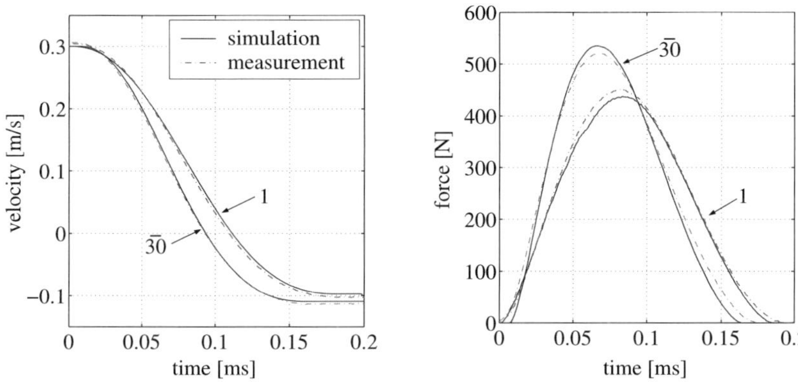

Fig. 6. Velocity of the sphere and impact force for repeated impacts on the rod with elastic-plastic material behavior

connected via a ball bearing with the frame and a compact aluminum body suspended by kevlar wires in the same frame. The steel pendulum can be considered elastic, the aluminum body has again a yield stress of $205MPa$ and shows elastic-plastic material behavior.

The pendulum is released from an initial angle of $\alpha = 10.3^o$ and impacts after $0.259s$ on the resting aluminum body. Then, in the multiscale simulation on the slow time scale, the time integration is interrupted and a simulation on the fast time scale is performed using the modal model with pre-computed

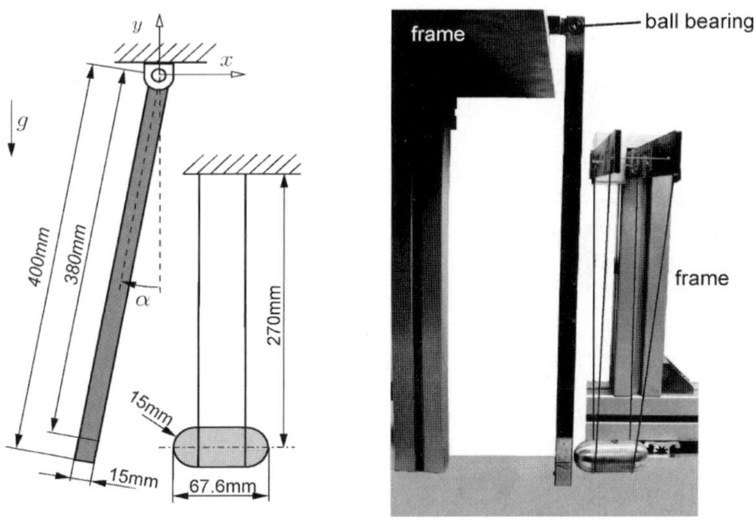

Fig. 7. Impact of an elastic steel pendulum on an aluminum body: schematic representation (left) and experimental setup (right)

FE contact. Thereby the rigid body velocity of the pendulum is used as initial condition for the fast time scale.

Figure 8 shows for the fast time scale the simulated and measured velocity of both bodies in the line of impact. The impact duration is about $0.12ms$ and excites strong vibrations in the pendulum, while the compact aluminum body shows no elastodynamic phenomena. The simulated and measured impact behavior shows an overall good agreement. However in detail there are some small variations of the vibrations of the pendulum after impact. A coefficient of restitution of $e = 0.185$ is computed and feed back to the slow time scale and the multibody simulation continues. The computed coefficient of restitution is in good agreement with the measured coefficient of $e = 0.162$.

For the slow time scale the simulated and measured velocity of the pendulum and the aluminum body are presented in Figure 9. Also on the slow time

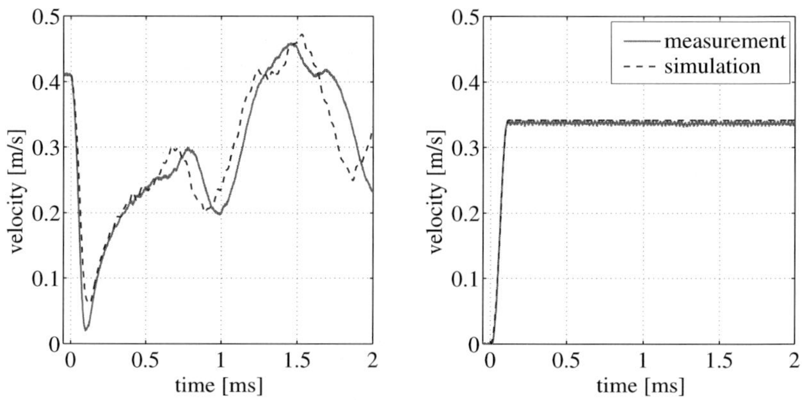

Fig. 8. Pendulum velocity (left) and velocity of aluminum body (right) on the fast time scale

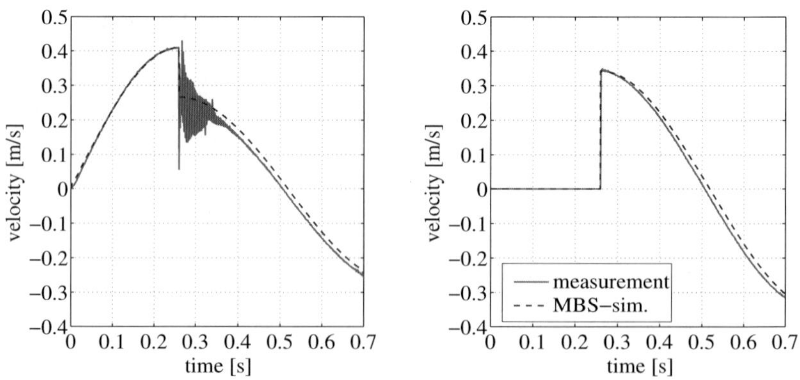

Fig. 9. Pendulum velocity (left) and velocity of aluminum body (right) on the slow time scale

scale the good agreement of computational and experimental results is obvious in the velocity plots. The occurrence of the impact is clearly indicated by the velocity jump. In the measurements the pendulum velocity shows structural vibrations initiated by the impact. These vibrations vanish after a short time period due to material damping. These vibrations cannot be represented by the rigid body model.

5 Summary

Impacts in multibody systems can be analyzed efficiently if the coefficient of restitution is known. A multiscale simulation is an efficient approach to evaluate the coefficient of restitution numerically by additional simulation on a fast time scale. This approach requires detailed numerical models on the fast time scale including the phenomena of wave propagation within the bodies due to impact and elastic-plastic deformation of the contact region. These two phenomena are presented by different numerical models, which are based on modal approach and nonlinear finite elements, respectively.

For the experimental impact investigation on the fast time scale a test bench is presented using Laser-Doppler-Vibrometer for displacement and velocity measurements, catching the high frequency phenomena resulting from impact. The performed measurements show a very good reproducibility. From the measurements the computed coefficient of restitution is verified and, in addition, for rod impacts the impact force is determined.

Comparisons of computational and experimental results for impacts on simple shaped bodies involving elastic as well as elastic-plastic material behavior prove the high accuracy of the numerical models. Investigations of impacts of steel spheres of different size on an elastic aluminum rod show that a substantial amount of kinetic energy might be transformed into wave propagation. For repeated impacts involving elastic-plastic material behavior, the influence of the deformation history of the contact region on the impact process is consistently observed in measurements and simulations. In this case the contact region gets deformed plastically during the first impacts until a stationary configuration is reached without any additional plastic deformation. Correspondingly this yields an increase of the coefficient of restitution with the number of impacts until a stationary value is reached. Finally, the impact of an elastic pendulum on a compact aluminum body with elastic-plastic material behavior is investigated experimentally and numerically. For this simple multibody system the high reliability of the multiscale simulation approach is proven on both time scales.

References

1. Cunningham DM, Goldsmith W (1958) Short-time impulses produced by longitudinal impact. Proceedings of the Society of Experimental Stress Analysis

16:153-162
2. Falcon E, Laroche C, Fauve S, Coste C (1998) Behavior of one inelastic ball bouncing repeatedly off the ground. The European Physical Journal B 3:45-57
3. Foerster SF, Louge MY, Chang H, Allia K (1994) Measurements of the collision properties of small spheres. Physics of Fluids 3:1108-1115
4. Glocker C (2001) On frictionless impact models in rigid-body systems. Philosophical Transactions of the Royal Society of London A359: 2385-2404
5. Goldsmith W (1960) Impact: The theory and physical behaviour of colliding solids. Edward Arnold Ltd, London
6. Graff KF (1991) Wave motion in elastic solids. Dover Publications, Mineola
7. Hu B, Eberhard P (1999) Experimental and theoretical investigation of a rigid body striking an elastic rod. Institutsbericht IB-32, Institute for Computational and Engineering Mechanics, University of Stuttgart
8. Hu B, Eberhard P, Schiehlen W (2003) Comparison of analytical and experimental results for longitudinal impacts on elastic rods. Journal of Vibration and Control 9:157-174
9. Hu B, Schiehlen W (2003) Multi-time scale simulation for impact systems: From wave propagation to rigid-body motion. Archive of Applied Mechanics 72:885-898
10. Johnson KL (1985) Contact Mechanics. Cambridge University Press, Cambridge
11. Lewin A, Mohr F, Selbach H (1990) Heterodyn-Interferometer zur Vibrationsanalyse (in German). Technisches Messen 57:9 335–345
12. Pfeiffer F, Glocker C (1996) Multibody dynamics with unilateral contacts. Wiley, New York
13. Polytec GmbH (1994), Vibrometer's manual for Polytec Vibrometer Series OFV-3000/OFV-302, OFV501 and OFV502. Manual No. VIB-MAN-9308-e04/01, Polytec
14. Schiehlen W, Seifried R (2004) Three approaches for elastodynamic contact in multibody systems. Multibody System Dynamics 12:1-16
15. Schiehlen W, Seifried R, Eberhard P (2006) Elastoplastic phenomena in multibody impact dynamics. Computer Methods in Applied Mechanics and Engineering, in press [doi:10.1016/j.cma.2005.08.011]
16. Seifried R, Hu B, Eberhard P (2003) Numerical and experimental investigation of radial impacts on a half-circular plate. Multibody System Dynamics 9:265-281
17. Seifried R, Schiehlen W, Eberhard P (2005) Numerical and experimental evaluation of the coefficient of restitution for repeated impacts. International Journal of Impact Engineering 32:508-524
18. Seifried R (2005) Numerische und experimentelle Stoßanalyse für Mehrkörpersysteme (in German). PhD-Thesis, Schriften aus dem Institut für Technische und Numerische Mechanik der Universität Stuttgart, Vol. 2, Shaker, Aachen
19. Stoianovici D, Hurmuzlu Y (1996) A critical study of the applicability of rigid-body collision theory. Journal of Applied Mechanics 63:307-316
20. Stronge WJ (2000) Impact Mechanics, Cambridge University Press, Cambridge
21. Wu C-Y, Li L-Y, Thornton C (2003) Rebound behavior of spheres for plastic impacts. International Journal of Impact Engineering 28:929-946

Eccentric Flows Registered by DPIV in a Model of Silo

Irena Sielamowicz[1] and Tomasz Kowalewski[2]

[1] Bialystok Technical University, Civil Engineering Department, Wiejska 45 E, 15-351 Bialystok, Poland sieliren@pb.bialystok.pl
[2] Institute of Fundamental Technological Research, Polish Academy of Sciences, Swietokrzyska 21, 00-049 Warsaw, Poland tkowale@ippt.gov.pl

Summary. The paper presents results of eccentric flows of amaranth and flax-seed in a flat-bottomed Plexiglas model of silo. The images of the flowing grains were registered by a camera of high resolution. The Digital Particle Image Velocimetry technique was used to analyse the seed flow. For this purpose long sequences of digital images of the flowing grains were processed to obtain full velocity vector fields as well as selected velocity profiles for the analysed discharge processes. Deformations occurred in the flowing material are also evaluated. The experiments were made in the model of three different roughness of the walls, the smooth and lined with sand paper of two different roughness. To investigate effects of eccentricity, both during filling and discharge, the model was filled centrally at the symmetry axis, close to the left edge of the silo, and near its right edge. The filling pipe was also located centrally or eccentrically above the model. The differences between different modes of flow are described and analysed in detail.

1 Introduction

Structures like silos and tanks for bulk storage of granular materials, powders and liquids are commonly used in industry. The walls of such structures are expected to withstand the pressure exerted by the granular material stored inside. Silos are filled from above, the material is stored and at a proper time discharged. During storing, filling and discharge the wall pressures of the stored material must be precisely determined. The values of wall pressures vary in different stages of acting load coming from the stored grains. But this parameter is a basic one in designing the thickness of the silo wall. When calculating wall stresses, one must also consider the flow patterns which depend on the way of discharge. Even tiny eccentricity of the filling or discharge may lead to quite unexpected behavior of the structure. A very important issue in such an analysis is the fact if we recognize symmetric or asymmetric loads to the wall, symmetric or asymmetric flow patterns and the form of the stagnant zones boundaries. During asymmetrical processes, flow patterns and wall stresses may be quite different. It is therefore crucial to recognize flow patterns developed in the material during eccentric filling or discharge, and to determine the wall stresses occurring under such state of loads.

Peter Eberhard (ed.), Multiscale Problems in Multibody System Contacts, 281–290.

2 Literature Review on Eccentric Filling and Discharge

The International standards, as ASAE Standards [1] or ENV 1991-4 [2] relate only to axial symmetric states of stresses and even avoid defining discharge pressures and flow patterns because of continuing uncertainties. And the use of eccentric discharge is even discouraged. Some codes and guides include eccentric discharge but in a very different way - ACI 313 [3], AS 3774 [4], Rotter [5], Jenike [6]. The European Standard ENV 1991-4, 1995 [2] includes a remark on the eccentricity of the outlet, but no calculation method is proposed. So far few approaches have been developed for the design of bins under eccentric discharge, Rotter, [7], Ayuga et al.,[8]. That issue was found to be a major cause of hopper failures, given by Carson [9]. Eccentricity of the flow to the silo axis causes the pressure patterns to become much more complicated than in the centric case. There are a few theories which have been advanced for the design of silos under eccentric discharge, Jenike [10], Wood [11], Rotter [7]. In the field of silo investigations three main issues are usually analysed: pressures under eccentric discharge, flow patterns in eccentric discharge and the stagnant zone boundaries. Works on eccentric discharge have been published for many years and a few researchers have dealt with this complicated problem. Anon [12] presented eccentric discharge silo loads and wall loads as a function of discharge rate. Thompson et al. [13] measured wall loads in a corrugated model grain bin when unloaded eccentrically. Pokrant and Britton [14] investigated the effect of eccentricity draw off and flow rate in a model grain bin. Haydl [15] analysed eccentric discharge and the calculation of bending moments in circular silos. Safarian and Harris [16] dealt with post-tensioned circular silos for modern industry and presented irregularities of pressure intensity caused by flow problems, eccentric discharge, or multiple discharge openings. Rotter et al. [17] discussed experiments with buckling failures problems in which the wall stresses are directly induced by stored solids. De Clercq [18] studied flow patterns in a silo with concentric and eccentric outlet, and also a steel silo with two types of outlets and its susceptibility to buckling. He found that a thin-walled silo with concentric outlet tends to be well behaved, and one with eccentric outlet susceptible to buckling and collapse. It was also stated that none of existing theories adequately addresses the issue of buckling of an eccentrically-emptied silos. Blight [19] investigated the behaviour of two steel silos under eccentric emptying. Borcz and Hamdy [20] presented the experimental results of pressure measurements in the wall in a full-scale silo under eccentric discharge. Shalouf and Kobielak [21] investigated eccentric discharge in silos and reduction of the dynamic flow pressures in grain silo by using discharge tubes. Ayuga et al. [8] investigated discharge and the eccentricity of the hopper influence on the silo wall pressures. Molenda et al. [22] presented investigation of bin loads by both central and eccentric filling and discharge of grains in a model of silo. Chou et al. [23] made some experiments in two-dimensional flat-bottomed model and recognized flow patterns and stresses on the wall during centric and eccentric discharge. The flowing material was recorded using a digital camcorder and the normal and shear stresses were measured using pressure gauges. Chou et al. [24] also observed the pressure pulsation phenomena. Chou and Hsu [25] measured experimentally the heights of the stagnant zones for two kinds of granular materials after hopper eccentric and centric discharge. In the analysis, Chou et al. [26] using the kinematic model, proposed by Nedderman and Tüzün [27], constructed two boundary-value problems. Guaita et al.[28] applied FEM modelling in the analysis of influence of hopper eccentricity on wall pressures. In the proposed model the distribution of plastic areas according to eccentricity was analysed. Song and Teng [29] analysed the results in FEM into a buckling strength of steel silo subject to code-specified pressures for eccentric discharge with the wall loads predicted by four codes where the pressure asymmetry was determined by local pressure increases or reductions and described by the authors as patch loads. Most of these investigations were made on laboratory models.

3 Flow Pattern Investigations Review

Simulations of silo filling and discharge leads to a better understanding of the bulk solids behaviour. Many experiments are aimed at investigating types of flow and shapes of the flowing regions. Behaviour of flowing materials was observed by numerous investigators and some new classifications of funnel flow were introduced, namely semi-mass flow and internal or pipe flow given by Watson and Rotter in [30]. It was also found that the flowing zone predicted by Jenike's radial velocity field (RVF) appeared far narrower than that observed in reality [31]. In all experiments researchers considered some flow factors, such as: hopper geometry, height of packed materials, size of particles, material density, material-wall interface friction of flow patterns [30, 32, 33, 34, 35]. The problem of predicting the shape of stagnant zone boundaries was investigated by Nedderman [36]. In the case of eccentric filling and discharge the problem is more complicated. Carson [9] described errors which can cause silos and bins to fail with a probability much higher than that observed for any other industrial equipment. The latest paper on the flow pattern measurement but in a full scale silo was presented by Chen et al. [37]. One may find a long list of references there concerning investigation of eccentric discharge. Ooi et. al. [38] mentioned different techniques used to investigate the flow of granular solids in small model silos. "These include visual observations through transparent walls, insertion of markers to measure residence times, freezing of the flow with paraffin and dissection, photographic techniques, radio pills and positron emission". Modern video techniques were also used by Khelil [39]. Ooi et al. [38] states that "Because flow pattern depends on the solid's properties, which vary in a non-linear manner with the stress state, the flow pattern tends to vary considerably between the different stress levels occurring at full scale and those in a model."

4 Experimental Setup

This paper presents results obtained in a flat-bottomed silo model. Central and eccentric filling and discharge were investigated. The images of the flowing material were recorded by a high resolution camera using the new experimental technique Digital Particle Image Velocimetry. This technique is an optical diagnostic tool to investigate velocities in one and multi-phase flows. The flow velocities can be obtained instantaneously for the whole area and the accuracy of the measurements are only limited by the optical resolution of the camera. This method allows the measurement of the instantaneous in-plane velocity field within a planar section of the flow field, the spatial gradients, dissipation of turbulent energy, spatial correlations, and the like. In [40] one can find a detailed description of the history, possibilities and development of PIV technique and the reasons for introducing an alternative approach – Digital Particle Image Velocimetry (DPIV). PIV technique was used by Waters and Drescher in [41] and Lueptow et al. [42]. Recently the PIV technique in measurements of velocity fileds was used among others by Medina [43,44], Ostendorf and Schwedes [45], Steingart and Evans [46], Sielamowicz et al. [47]. Here, the following flow characteristics were obtained using DPIV technique: the range of stagnant boundaries in the flowing material, velocity vector fields, velocity magnitude contours, displacement fields, velocity profiles, and traces of individual particles. The experimental setup used for the flow analysis consists of a Plexiglas box, a set of illumination lamps, and a high speed CCD camera (PCO1200HS) with the objective 50mm lens. Sequences of 12-bit images with resolution of 1280 pixels x 1024 pixels were acquired by Pentium 4 based personal computer using IEEE1394 interface. The system allows to acquire up to 1000 images with at time interval of 1.5ms (670fps). For the purpose of the velocity evaluation long sequences of 100-400 images were taken at variable time intervals covering the whole discharge time. The quasi two-dimensional model of silo used in the experimental study has a height of 80 cm, a depth of 10 cm, and a width of 26 cm. The granular flow was observed in the vicinity of the transparent front wall of the model. The box with a rectangular bottom opening was placed on a stand with a collecting vessel below. The material was supplied from the top through a feeding pipe. For this purpose a 30 l container was suspended above the model and connected to the pipe. To investigate the effects of eccentricity the model was filled centrally at the axis of symmetry, close to the left edge of the silo, and near its right edge. The discharge opening was also located centrally or eccentrically. To analyze the effects of wall roughness, the side walls of the model were lined with sand paper and the experiments were compared with analogous runs made for the model with smooth lateral walls. The degree of roughness was varied by the use of two different sand papers. To obtain medium rough walls ABRATEX X P60A 194 sand paper was used. In the model with very rough lateral walls, the paper ABRATEX X P16A 432 was used. Wall friction for amaranth and flax seed were φ_w 25° and 26°, respectively. Angle of internal friction φ_e is 28° and 25°, respectively. Young modulus was 12.96 MPa and 6.11 MPa, respectively.

5 Results and Discussion

This paper presents collection of experiments in a model of silo. The flow was investigated for amaranth seed and for flax seed in different configurations, e.g.

the silo was filled from the left, in the symmetry axis of the model and from the right. Discharge was investigated also in three positions of the outlet, from the left, from the right and in the symmetry axis. Only some of the results such as velocity profiles and velocity distributions are presented here to show the evolution of the flow in all discussed cases. The influence of both the position of the outlet and the roughness of the wall on the flow pattern were observed and registered.

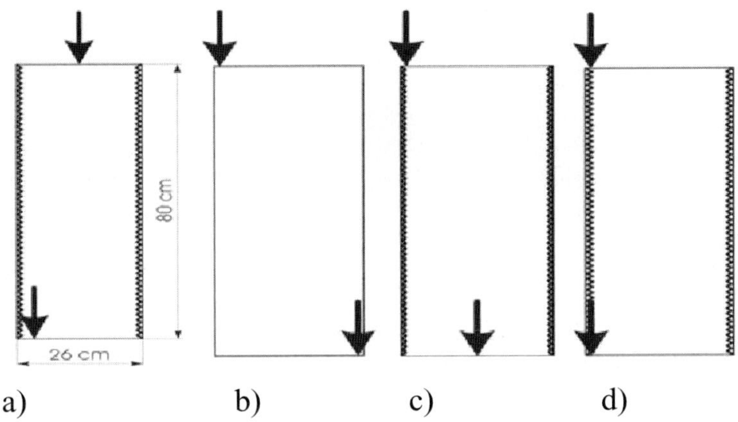

Fig.1: Schematic images of filling and discharge modes for described here flows: a) amaranth seed, the model with rough walls, b) amaranth seed in the model with smooth walls, c) flax seed in the model with medium rough walls, d) flax seed in the model with rough walls

5.1 Velocities in the Flowing Material

Fig. 2 illustrates velocity contours for the eccentric flow with the model filled with amaranth seed in the symmetry axis and discharged eccentrically. The walls of the model were rough. The discharge outlet was situated 0.8 cm from the left lateral wall in the model. The set of pictures in Fig.2 presents the flow after 3.75, 30, and 52.5 sec. The total time of the flow was 68 sec. The total mass of the grains was 18.6 kg and the moisture was found to be 10.2%.

After centrally filling, the grains formed the natural angle of repose. The total height of the packed material was 78 cm at the symmetry axis. As it can be seen the free surface natural angle collapses along the symmetry axis. Soon after the flow initiation one may observe a nearly uniform channel flow for the entire height of the silo. The region of the highest velocity is called "plug flow".

The rough model walls cause the material near the lateral wall to adhere to it, creating zone of a higher density. The rough wall behaves like a "brake" holding the lateral parts of the material. The boundary between the flowing material and the stagnant zones is clearly recognized. However, in the initial phase of the flow, in the upper part the velocity vectors are less inclined than in the lower part of the flowing material. The flow evidently accelerates, the velocity vectors in the initial phase of the flow are shorter than in the advanced phase of the flow.

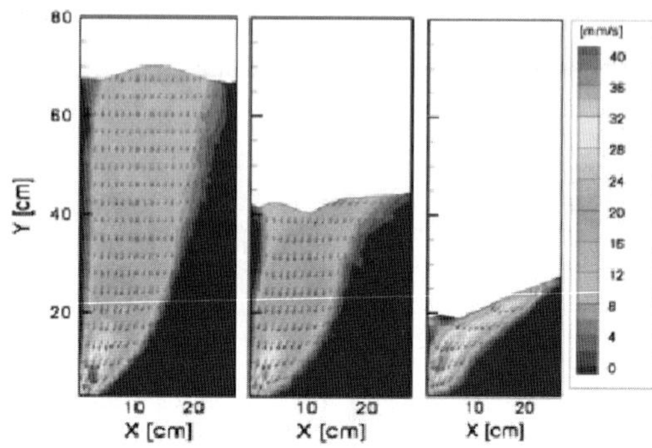

a) 3.75 sec b) 30sec c) 52.5 sec

Fig. 2: Velocity distributions and vector fields for central filling and eccentric discharge of amaranth seed from the left in the model through the outlet situated 0.8 cm from the left edge in the model with rough walls, after a) 3.75 sec, b) 30 sec, c) 52,5 sec after the beginning of the flow.

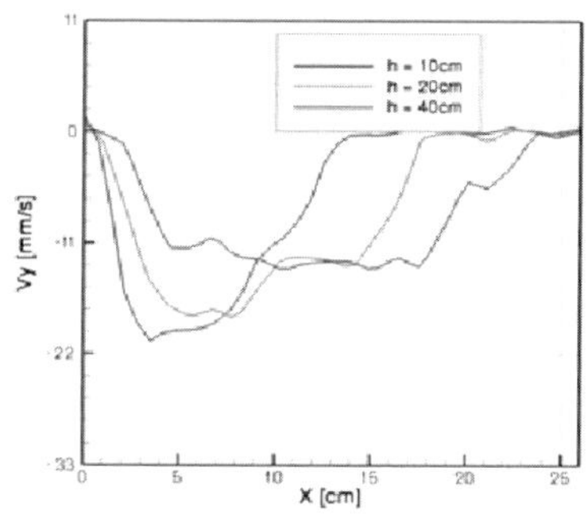

Fig.3: Velocity profiles for the flow of amaranth in the central filling and discharge from the left in the model with rough walls, at 15th sec of the flow.

Full field velocity measurements were used to extract selected velocity profiles. Figure 3 presents velocity profiles obtained for the flowing amaranth in the model with rough walls. The profiles of the vertical velocity components across the cavity were extracted at three different heights 10 cm, 20 cm and 40 cm (indicated in legend) and at the time step 15th sec after the beginning of the experiments.

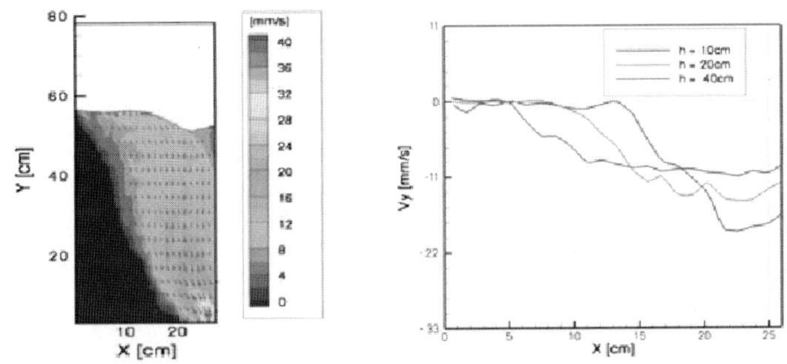

Fig. 4: Velocity distributions and vector fields for discussed cases of the flow of amaranth in the model with the smooth walls, after 15 sec of the flow.

5.2 Deformations in the Flowing Material in the Model with Smooth Walls

Having dense velocity vector field obtained in DPIV one may evaluate velocity gradients and calculate strains in the flowing material. The strain tensor is calculated in the terms of partial derivatives dU / dX where U denotes a velocity component and X coordinate. Measuring the displacement field $u_i(x, y)$, the strain component can be calculated, namely the Green strain tensor

$$E_{ij} = \frac{1}{2}\left(\frac{\partial u_i}{\partial x_j} + \frac{\partial u_j}{\partial x_i} + \frac{\partial u_k}{\partial x_i}\frac{\partial u_k}{\partial x_i}\right).$$

Figure 5 shows evaluated deformation profiles of the flax flow recorded after the first second of the flow. The flow took place in the model with smooth wall and the feeding was performed eccentrically at the left side.

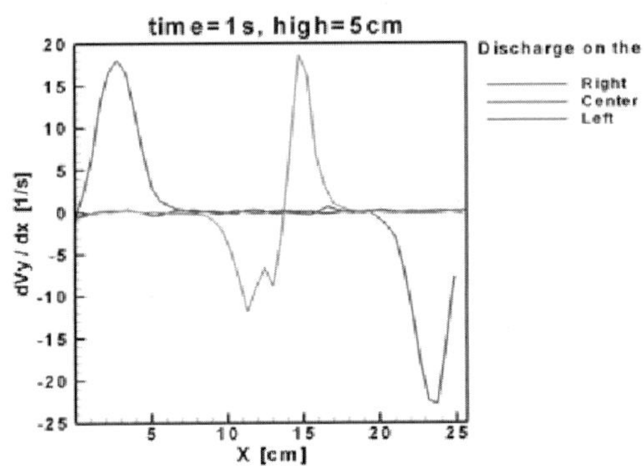

a)

Fig. 5 Deformation profiles evaluated for the flowing flax seed in the case of eccentric filling for three discharge locations (see legend), and observed for the initial flow time (1s) at the level 5cm above the bottom.

5.3 The flow in the Model with Medium Rough Walls (Flax-Seed)

To present the effects of the wall roughness another experimental runs were made for the model with medium rough walls (sand paper P60A194) and in the model with rough walls (sand paper P16A 432). The total time of the flow in the case of the model with medium rough walls was found to be 88 sec, and in the model with rough walls – 125 sec. In Fig. 6 the region of the highest velocity – "plug flow" can be well identified.

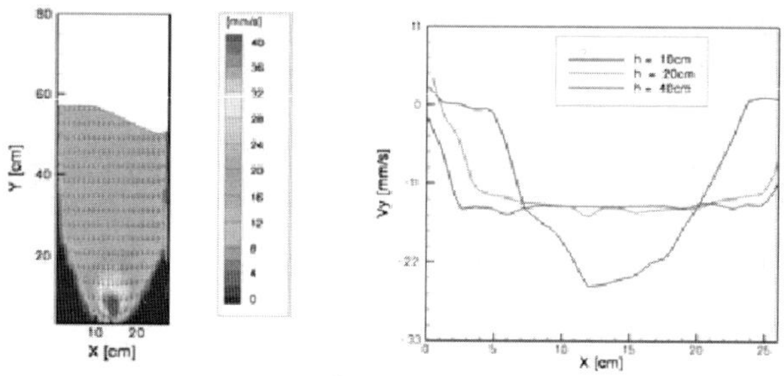

Fig. 6: Velocity distributions and vector fields for filling from the left and central discharge of flax seed in the model of medium-rough walls after 15th sec of the flow.

Velocity profiles (Fig. 6) confirm the mass flow in the model with medium rough walls. The blue and the green line present the constant velocities, about 14 mm/s. Near the left wall, the stagnant zone forms but it is a little wider than on the right wall. The red line represents profile of the velocity at the height 10cm, here one may notice a clear funnel flow mode. In the advanced phase of the flow still at the heights 20 and 40 cm the mass flow is observed and below at the height 10 cm the funnel flow with some disturbances is registered.

6 Conclusions

This paper presents a series of experiments made in the plane model which was filled and discharged eccentrically and in some cases centrally. The walls of the model were lined with sand paper of different roughness. This fact changed the mode of the flow and the behavior of the flowing material. Some interesting observations of the flow are shown. To register the flows the Digital Particle Image Velocimetry technique was applied. This technique appears to be a useful diagnostic tool to recognize the flow and describe different factors having influence to the character of the flow.

7 Acknowledgements

This paper was prepared with financial support by the grant 4 T07E 054 30 and using the funds 2006-2008 from the Polish Government, is acknowledged. The special thanks are addressed to Prof. Zenon Mroz of the Institute of Fundamental

Technological Research of Polish Academy of Sciences for the valuable advices and suggestions during the experiments. The authors also express their special thanks to Mr. Andrzej Cybulski and Slawomir Blonski of the Institute of Fundamental Technological Research of Polish Academy of Sciences for their help in conducting experiments.

References

1. ASAE EP433 (1997) Loads exerted by free-flowing grains on bins, ASAE Engineering Practice EP433, ASAE Standards
2. ENV 1991 ENV 1991-Part 4 Eurocode 1 (1995) Basis of design and actions on structures. Part 4, Actions on silos and tanks, Brussels, Belgium
3. ACI 313 ACI 313 (1989) Alternate Design Procedure, Discussion Document before ACI Committee 313 on Concrete Bins, Silos and Bunkers for Storing Granular Materials, ACI, Detroit
4. AS 3774 AS 3774 (1986) Loads on Bulk Solids Containers, Australian Standards Association of Australia, Sydney
5. Rotter JM (1998) Guide for the Economic Design of Metal Silos, E&FN Spon, London
6. Jenike AW, Johanson JR (1969) Transactions ASME, **91**, Ser. B 2, 339-344
7. Rotter JM (1986) Proc., 2nd Inter. Conference on Bulk Materials Storage Handling and Transportation, Ins. of Eng., Wollongong, Australia, July, 264-271
8. Ayuga F, Guaita M, Aguado PJ, Couto A (2001) J. Eng. Mechanics, **127**, 10, 1067-1074
9. Carson JW (2000) Silo failures: Third Israeli Conference for Conveying and Handling of Particulate Solids, Dead Sea Israel, May
10. Jenike AW (1967) Journal of the Structural Division ASCE 93 (ST1) 27-35
11. Wood JGM (1983) Proc.,2nd Int. Conf. on Design of Silos for Strength and Flow, Stratford-upon-Avon, 132-144
12. Anon (1985) EFCE Publication Series (European Federation of Chemical Engineering), 49
13. Thompson SA, Usry JL, Legg JA (1986) Transactions of the ASAE, 29, 2
14. Pokrant DK, Britton MG (1986) Paper-American Society of Agricultural Engineers, 86, 40-76
15. Haydl HM (1987) Proceedings of the Institution of Civil Engineers, 83, 2
16. Safarian SS, Harris EC (1987) Bulk Solids Handling, 7, 2
17. Rotter JM , Jumikis PT, Fleming SP, Porter SJ (1988) Research Report-University of Sydney, R570
18. de Clercq H (1990) Civil Engineers in South Africa, 32, 3
19. Blight GE (1991) Bulk Solids Handling, 11, 2, 451-457
20. Borcz A, el Rahim, Hamdy Abd (1991) Bulk Solids Handling, 11, 2, 469-476
21. Shalouf F, Kobielak S (2001) Powder Handling Processing, 13, 1

22. Molenda M, Horabik J, Thompson SA, Ross LR (2002) Transactions of the ASAE,45, 3 781-785
23. Chou CS, Hsu JY, Lau JD (2003) Journal of Chinese Institute of Engineers, Transactions of the Chinese Institute of Engineers, A, 26, 4, 397-408
24. Chou CS, Chuang YC, Smid J, Hsiau SS, Kuo JT (2002) Advanced Powder Technology, 13, 1, 1-13
25. Chou CS, Hsu JY (2003) Advanced Powder Technology, 14, 3, 313-331
26. Chou CS, Hsu JY, Lau YD (2002) Physica A Statystical Mechanics and its Applications, 308, 1-4, 46-58
27. Nedderman RM, Tüzün U (1979) Powder Technology, **22**, 243-253
28. Guaita JS, Couto A, Ayuga F (2003) Biosystem Engineering, 85,1, 101-109
29. Song CY, Teng JG (2003) Engineering Structures, 25, 1397-1417
30. Watson GR, Rotter JM (1996) Chemical Engineering Science, 51, 3967-3978
31. Cleaver JAS, Nedderman RM (1993) Chem Eng Sci, 48, 21, 3693-3703
32. Giunta JS (1969) Journal of Engineering for Industry, Transactions of the ASME, 91, 406-413
33. Nguyen TV, Brennen C, Sabersky RH (1979) Journal of Applied Mechanics, Transactions of the ASME, 46, 529-535
34. Takahashi H, Yanay H (1973) Powder Technology, **7**, 205-214
35. Waters AJ, Drescher A (2000) Powder Technology, 113, 168-175
36. Nedderman RM (1995) Chemical Engineering Science, **50**, 6, 959-965
37. Chen JF, Rotter JM, Ooi JY, Zhong Z (2005) Chemical Engineering Science, 60, 3029-3041
38. Ooi JY, Chen JF, Rotter JM (1998) Powder Technology, 99, 272-284
39. Khelil A, Weisse D, Roth JC (1998) Analysis of the flow of granular materials with gravitationally emptying a silo-a study of the experimental and theoretical fields, in: Biarez, Grouve (Eds.), Powders and Grains, Balkema, Rotterdam, 461-468
40. Quenot GM, Pakleza J, Kowalewski T A (1998) Experiments in Fluids, **25**, 177-189
41. Waters AJ, Drescher A (2000) Powder Technology, 113, 168-175
42. Lueptow RM, Akonur A, Shinbrot T (2000) Experiments in Fluids, 28, 2, 183-186
43. Medina A, Cordova JA, Luna E, Trevino C (1998) Physics Letters A 250, 111-116
44. Medina A, Cordova JA, Luna E, Trevino C (1998) Physics Letters A 273, 109-116
45. Ostendorf M, Schwedes J (2005) Powder Technology, 158, 69-75
46. Steingart DA, Evans JW (2005) Chemical Engineering Science, 60, 1043-1051
47. Sielamowicz I, Blonski S, Kowalewski TA (2005) Chemical Engineering Science, 60, 2, 589-598

Regularization of 2D Frictional Contacts for Rigid Body Dynamics

Wolfgang Stamm[1] and Alexander Fidlin[2]

[1] LuK GmbH & Co. oHG, Bühl, Germany wolfgang.stamm@luk.de
[2] LuK GmbH & Co. oHG, Bühl, Germany alexander.fidlin@luk.de

Summary. Classic rigid body mechanics does not provide frictional forces acting in a 2D contact interface between two bodies during sticking. This is due to the statical undeterminacy related to this problem. Many technical systems, e.g. disk clutches, have such surface-to-surface contacts and it is sometimes desirable to treat them as rigid body systems despite the 2D contact. Alternatively, it is possible to model the systems using elastic instead of rigid bodies, but this might lead to certain drawbacks. A new regularization model of such 2D contacts between rigid bodies is proposed. It is based on the similarity to a material model for elasto-plasticity in continuum mechanics. Only dry friction is taken into account.

1 Introduction

Many mechanical systems incorporate 2D contact interfaces, i.e. bodies contacting each other on a plane surface. In some cases, the distribution of frictional forces along these contact interfaces affect the systems' dynamics or the phenomena of interest. Technical examples are dry friction clutches [5], diaphragm springs and hydro-mechanical torque sensors used in automotive transmissions [2]. Often elastic deformations can be neglected. The surface-to-surface contacts are not treated in classic rigid body theory because of the statical undeterminacy related with multiple contacts during sticking. This undeterminacy is mostly irrelevant; the onset of motion can be predicted uniquely by investigating systematically the inequalities of the contact problem [3, 9]. However, it was shown that under some geometric conditions, the undeterminacy cannot be resolved using only information from rigid body theory [7]. In such cases, no decision on the onset of motion can be made—even when infinite forces are acting on the system.

Modeling the bodies as elastic is common engineering practice to solve the contact problem. Unfortunately this requires a high number of degrees of freedom resulting from the discretization of volumes. The goal of this work is to incorporate less degrees of freedom when the volumes' elasticity can be neglected but to obtain the frictional forces with high accuracy.

Peter Eberhard (ed.), Multiscale Problems in Multibody System Contacts, 291–300.

2 An Elasto-Plastic Regularization Model

The basic idea of the approach presented here is to split the system into two subsystems. The first subsystem deals with the rigid body motions as a macroscale system and the second one describes the contact interface model. The latter system spans an infinitesimal thin layer between the rigid bodies. It is similar to material models usual in continuum mechanics and therefore common discretization methods (meshes) can be applied. The following advantages can be achieved due to this approach. Firstly, the framework of rigid body mechanics remains unaffected for the macro system. Secondly, the material properties of the introduced contact layer can be chosen independently from the properties of the body itself, whereas the tangential contact stiffness in the conventional finite-element approach is ruled by the bodies' bulk elastic properties. Finally, the number of degrees of freedom remains relatively small due to discretization of the 2D contact layer instead of the 3D contacting bodies.

The contact interface is subdivided into a finite number of contact points. One can adjust the resolution of frictional forces acting in the contact area by choosing a reasonable set of contact points. Then for each contact point, a friction law is formulated. A regularized friction law is used due to the reasons given in Sect. 1 and the fact that COULOMB friction does not predict the distribution of frictional forces during sticking. However, a standard regularization with springs and dampers as e.g. in [10] fails for 2D multiple contact interfaces. To see this, consider a contact point in sticking state, regularized with a spring (cf. Fig. 1). The spring displacement corresponds to a maximal transmittable frictional force F_f^{max}. Now a small perturbation with some arbitrary relative velocity $\boldsymbol{v}_{rel}$ is applied. The perturbation causes the contact point to tear off, i.e. the frictional force represented by the spring reaction force (sticking) exceeds F_f^{max} and has to be replaced by a force opposing relative velocity (sliding). This is a discontinuous jump of the force direction immediately before ($\boldsymbol{F}_f^-$) and immediately after ($\boldsymbol{F}_f^+$) tear-off (cf. Fig. 1). A jump of the *absolute value* of a frictional force is permitted according to some Stribeck curve. But the discontinuities of friction force *orientations* may lead to non-accelerating tear-off situations. Furthermore the dependency of a friction force orientation on a small perturbation seems unrealistic from a macroscopic point of view—although one might accept it for a micro contact.

Here we want to outline the functioning of the proposed regularization method without concentrating on the choice of contact points. As pointed out above, the goal is to use smooth transitions of the frictional force orientations between sticking and sliding states in both directions. This is achieved by regularizing the sticking state with an elasto-plastic element. Fig. 2 specifies the structure of the equations for ideal plasticity with the VON-MISES yield criterion, also called PRANDTL-REUSS equations [4, 8] in continuum mechanics. The equations will be similar to those obtained in [1] for the 1D case, but now the derivation is straightforward from the material model applied to a discretized contact area; no further information is needed.

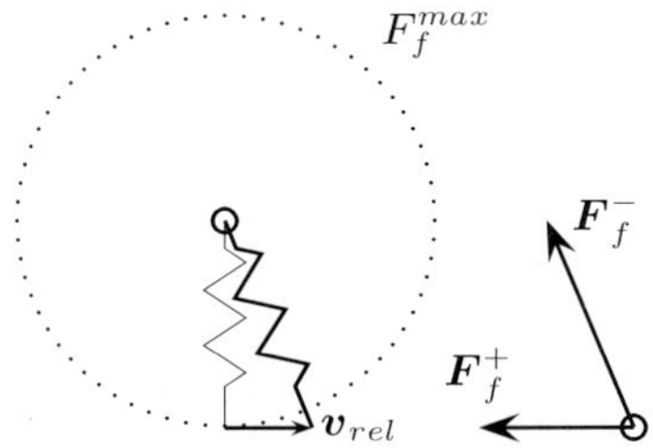

Fig. 1. Regularization with a spring, force direction jump

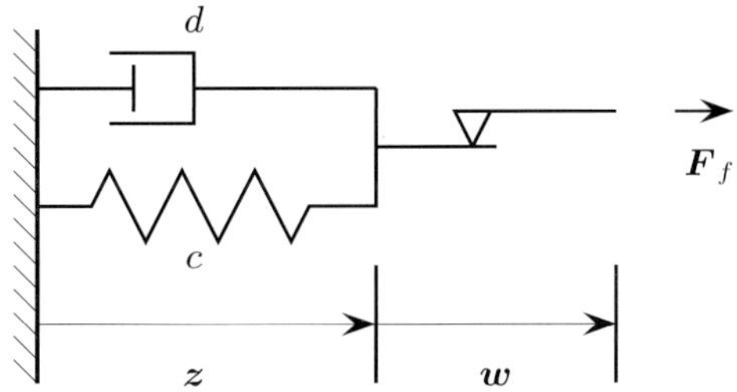

Fig. 2. The PRANDTL-REUSS material model, the friction regularization model

In our case Fig. 2 identifies real spring, damper and plastic elements, but their deformation should be analogous to the material model. The displacement of a contact point $\boldsymbol{x}$ during sticking is divided into elastic displacement $\boldsymbol{z}$ and plastic displacement $\boldsymbol{w}$ (cf. Fig. 2):

$$\boldsymbol{x} = \boldsymbol{z} + \boldsymbol{w}, \quad \dot{\boldsymbol{x}} = \dot{\boldsymbol{z}} + \dot{\boldsymbol{w}} \tag{1}$$

Due to the series connection, the forces in elastic and plastic elements are equal:

$$\boldsymbol{F}_f = \boldsymbol{F}_{el} = \boldsymbol{F}_{pl} \tag{2}$$

The force in the elastic element is the reaction force of spring and dashpot:

$$\boldsymbol{F}_{el} = -c\boldsymbol{z} - d\dot{\boldsymbol{z}} \tag{3}$$

and by (2)

$$\boldsymbol{F}_{pl} = -c\boldsymbol{z} - d\dot{\boldsymbol{z}} \tag{4}$$

The plastic deformation is defined as:

$$-\lambda \boldsymbol{F}_{pl} = \dot{\boldsymbol{w}} \tag{5}$$

where λ controls the occurrence of plastic deformation. The yet undetermined multiplier λ is specified by the VON-MISES yield criterion in the plasticity model [4]. Here a similar 'yield criterion' is introduced as

$$\lambda = \begin{cases} 0 & , \|\boldsymbol{F}_{pl}\| < F_{St}(\|\dot{\boldsymbol{x}}\|) \\ \|\dot{\boldsymbol{w}}\| / F_{St}(\|\dot{\boldsymbol{x}}\|) & , \|\dot{\boldsymbol{w}}\| > \varepsilon \end{cases} \tag{6}$$

with the scalar-valued function $F_{St}(\|\dot{\boldsymbol{x}}\|)$ as a Stribeck curve. Hence any local relative displacement $\boldsymbol{x}$ caused by a force lower than the local steady-state sliding friction force F_{St} should be elastic (reversible)[3]. The force F_{St} leads to plastic (irreversible) displacements.

In continuum mechanics, it is common to justify the choice of λ with the CLAUSIUS-DUHEM inequality as a local expression of the Second Law of Thermodynamics [4, 6]. This leads to the inequality

$$\sigma_{ij}\dot{\varepsilon}_{ij}^{pl} \geq 0 \tag{7}$$

where σ_{ij} are the stress tensor components, $\dot{\varepsilon}_{ij}^{pl}$ the corresponding plastic strain rates and EINSTEIN's convention of summing is used. For a single contact point we get analogously

$$(-\boldsymbol{F}_{pl}) \cdot \dot{\boldsymbol{w}} \geq 0 \tag{8}$$

Inequality (8) is fulfilled for any $\lambda \geq 0$ in (5). This restriction to λ is not surprising, because it is quite intuitive that frictional forces should oppose relative motion in order to render a dissipative system[4]. Inserting (6) in (5) yields:

$$\dot{\boldsymbol{w}} = \begin{cases} \boldsymbol{0} & , \|\boldsymbol{F}_{pl}\| < F_{St}(\|\dot{\boldsymbol{x}}\|) \\ -\|\dot{\boldsymbol{w}}\|\, \boldsymbol{F}_{pl}/F_{St}(\|\dot{\boldsymbol{x}}\|) & , \|\dot{\boldsymbol{w}}\| > \varepsilon \end{cases} \tag{9}$$

which describes the evolution of plastic deformation in the form of an implicit differential equation. In the second line of (9) one can see that the plastic deformation is chosen in such a way that the regularized frictional force approaches the COULOMB frictional force for macroscopic sliding ($\|\dot{\boldsymbol{z}}\| \ll \|\dot{\boldsymbol{w}}\|$):

$$F_{St}(\|\dot{\boldsymbol{x}}\|)\frac{\dot{\boldsymbol{w}}}{\|\dot{\boldsymbol{w}}\|} = -\boldsymbol{F}_{pl} \approx F_{St}(\|\dot{\boldsymbol{x}}\|)\frac{\dot{\boldsymbol{z}}+\dot{\boldsymbol{w}}}{\|\dot{\boldsymbol{z}}+\dot{\boldsymbol{w}}\|} = F_{St}(\|\dot{\boldsymbol{x}}\|)\frac{\dot{\boldsymbol{x}}}{\|\dot{\boldsymbol{x}}\|} \tag{10}$$

In a plasticity model, the magnitude of the multiplier λ is usually determined by a consistency condition of the yield surface. By the definition in (6), the magnitude of λ is introduced in a rather heuristic manner, aiming at COULOMB friction for steady-state sliding as shown in (10). Further investigations of analogies between friction modeling and plasticity, such as yield surfaces, associated or non-associated flow rules etc. are beyond the scope of this paper.

It is possible to give an explicit form of the second line of (9). With (1) one can rewrite (10):

$$F_{St}(\|\dot{\boldsymbol{x}}\|)\frac{\dot{\boldsymbol{w}}}{\|\dot{\boldsymbol{w}}\|} = c(\boldsymbol{x}-\boldsymbol{w}) + d(\dot{\boldsymbol{x}}-\dot{\boldsymbol{w}}) = \boldsymbol{k} - d\dot{\boldsymbol{w}} \tag{11}$$

with the abbreviation

$$\boldsymbol{k} = c(\boldsymbol{x}-\boldsymbol{w}) + d\dot{\boldsymbol{x}} = c\boldsymbol{z} + d\dot{\boldsymbol{x}} \tag{12}$$

[3] called *true stiction* in [1]

[4] an accelerative frictional force as a model artifact is called a *Stribeck slingshot effect* in [1]. This effect is avoided here by $\lambda \geq 0$.

By multiplying scalarly (11) with $\boldsymbol{k}$ and $\dot{\boldsymbol{w}}$ we obtain:

$$F_{St}(\|\dot{\boldsymbol{x}}\|)\,\|\dot{\boldsymbol{w}}\| = \boldsymbol{k}\cdot\dot{\boldsymbol{w}} - d\,\|\dot{\boldsymbol{w}}\|^2 \tag{13a}$$

$$F_{St}(\|\dot{\boldsymbol{x}}\|)\boldsymbol{k}\cdot\dot{\boldsymbol{w}}/\,\|\dot{\boldsymbol{w}}\| = \|\boldsymbol{k}\|^2 - d\boldsymbol{k}\cdot\dot{\boldsymbol{w}} \tag{13b}$$

From (13) we get an expression for $\boldsymbol{k}\cdot\dot{\boldsymbol{w}}$:

$$\boldsymbol{k}\cdot\dot{\boldsymbol{w}} = \frac{\|\boldsymbol{k}\|^2\,\|\dot{\boldsymbol{w}}\|}{F_{St}(\|\dot{\boldsymbol{x}}\|) + d\,\|\dot{\boldsymbol{w}}\|} \tag{14}$$

Resubstituting (14) in (13) gives:

$$F_{St}(\|\dot{\boldsymbol{x}}\|)\,\|\dot{\boldsymbol{w}}\| = \frac{\|\boldsymbol{k}\|^2\,\|\dot{\boldsymbol{w}}\|}{F_{St}(\|\dot{\boldsymbol{x}}\|) + d\,\|\dot{\boldsymbol{w}}\|} - d\,\|\dot{\boldsymbol{w}}\|^2 \tag{15}$$

Equation (15) is a cubic polynomial for $\|\dot{\boldsymbol{w}}\|$. However, we are not interested in the solution $\|\dot{\boldsymbol{w}}\| = \dot{\boldsymbol{w}} = 0$ because this was excluded by the condition of the second case in (9). Therefore we can divide (15) by $\|\dot{\boldsymbol{w}}\| \neq 0$ and get a quadratic equation for $\|\dot{\boldsymbol{w}}\|$:

$$\|\dot{\boldsymbol{w}}\|^2 + \frac{2F_{St}(\|\dot{\boldsymbol{x}}\|)}{d}\,\|\dot{\boldsymbol{w}}\| + \frac{F_{St}^2(\|\dot{\boldsymbol{x}}\|) - \|\boldsymbol{k}\|^2}{d^2} = 0 \tag{16}$$

with the only one positive solution:

$$\|\dot{\boldsymbol{w}}\|_1 = \frac{\|\boldsymbol{k}\| - F_{St}(\|\dot{\boldsymbol{x}}\|)}{d} \tag{17}$$

Next, the existence of the solution $\|\dot{\boldsymbol{w}}\|_1$ is examined. The right hand side of (17) is non-negative for

$$\|\boldsymbol{k}\| = \|c\boldsymbol{z} + d(\dot{\boldsymbol{z}} + \dot{\boldsymbol{w}})\| \geq F_{St}(\|\dot{\boldsymbol{x}}\|) \tag{18}$$

From (10) and (4) we know

$$\|-\boldsymbol{F}_{pl}\| = \|c\boldsymbol{z} + d\dot{\boldsymbol{z}}\| = F_{St}(\|\dot{\boldsymbol{x}}\|) \tag{19}$$

Combining (18) with (19) the following inequality can be obtained:

$$2d\dot{\boldsymbol{w}}\cdot(c\boldsymbol{z} + d\dot{\boldsymbol{z}}) + d^2\dot{\boldsymbol{w}}\cdot\dot{\boldsymbol{w}} \geq 0$$

This inequality is fulfilled, because $d^2\dot{\boldsymbol{w}}\cdot\dot{\boldsymbol{w}}$ is always non-negative, and

$$\dot{\boldsymbol{w}}\cdot(c\boldsymbol{z} + d\dot{\boldsymbol{z}}) = \dot{\boldsymbol{w}}\cdot(-\boldsymbol{F}_{pl}) \geq 0$$

according to the choice of $\lambda \geq 0$ in (6). Therefore (17) is a valid solution of the quadratic polynomial (16). From (11) we know that

$$\dot{\boldsymbol{w}} = \frac{\boldsymbol{k}\,\|\dot{\boldsymbol{w}}\|}{F_{St}(\|\dot{\boldsymbol{x}}\|) + d\,\|\dot{\boldsymbol{w}}\|} \tag{20}$$

Thus substituting (17) in (20) we get the desired explicit form of $\dot{\boldsymbol{w}}$:

$$\dot{\boldsymbol{w}} = \begin{cases} \mathbf{0} & , \|\boldsymbol{F}_{pl}\| < F_{St}(\|\dot{\boldsymbol{x}}\|) \\ \left(\frac{c}{d}(\boldsymbol{x}-\boldsymbol{w})+\dot{\boldsymbol{x}}\right)\left(1-\frac{F_{St}(\|\dot{\boldsymbol{x}}\|)}{\|c(\boldsymbol{x}-\boldsymbol{w})+d\dot{\boldsymbol{x}}\|}\right) & , \left[\|\boldsymbol{k}\| - F_{St}(\|\dot{\boldsymbol{x}}\|)\right]/d > \varepsilon \end{cases} \tag{21}$$

Finally we want to use $\dot{\boldsymbol{z}}$ instead of $\dot{\boldsymbol{w}}$:

$$\dot{\boldsymbol{z}} = \dot{\boldsymbol{x}} - \dot{\boldsymbol{w}} = \dot{\boldsymbol{x}} - \alpha(\boldsymbol{z}, \dot{\boldsymbol{x}})\left(\frac{c}{d}\boldsymbol{z} + \dot{\boldsymbol{x}}\right)\left(1 - \frac{F_{St}(\|\dot{\boldsymbol{x}}\|)}{\|c\boldsymbol{z} + d\dot{\boldsymbol{x}}\|}\right) \tag{22}$$

With the discontinuity function $\alpha(\boldsymbol{z}, \dot{\boldsymbol{x}})$:

$$\alpha(\boldsymbol{z}, \dot{\boldsymbol{x}}) = \begin{cases} 0 & , \|\boldsymbol{k}\| < F_{St}(\|\dot{\boldsymbol{x}}\|) \\ 1 & , \left[\|\boldsymbol{k}\| - F_{St}(\|\dot{\boldsymbol{x}}\|)\right]/d > \varepsilon \end{cases} \tag{23}$$

(22) and (23) build an explicit differential equation for $\boldsymbol{z}$. Note that $\boldsymbol{x}, \dot{\boldsymbol{x}}$ can be considered as given for the contact analysis, because they are state variables of the macroscopic system.

Multiple Contacts

Equation (22) holds for each member of a a given set of contact points. Let N be the number of contact points. Then the equations of motion of a rigid body $\dot{\boldsymbol{q}} = \boldsymbol{f}(\boldsymbol{q}, t)$ are completed by N explicit equations for the inner variables $\boldsymbol{z}_i$ such that

$$\begin{aligned} \dot{\boldsymbol{q}} &= \boldsymbol{f}(\boldsymbol{q}, \boldsymbol{z}_i, t) \\ \dot{\boldsymbol{z}}_i &= \boldsymbol{g}_i(\boldsymbol{q}, \boldsymbol{z}_i, \dot{\boldsymbol{z}}_i, t) \quad , i = 1, \ldots, N \end{aligned} \tag{24}$$

3 Numerical Solutions

Results of two different numerical methods are presented. Both methods are applied to the simple example of a disk moving on a plane. The first numerical solution is a direct implementation of (24). The second one tries to reduce computational costs by temporal removal of the inner-state equations for $\boldsymbol{z}_i$ from the system during macroscopic sliding. The number of differential equations is then time-variant.

Figure 3 shows the example system. A plane ring of 2 cm diameter is moving on a plane. The motion is considered completely two-dimensional, i.e. neither tilting of the body nor collisions between body and plane are taken into account. The body's contact interface was discretized using 12 or 48 surface elements (cf. Fig. 4). The pressure distribution in the contact interface was assumed constant; the normal contact force of each contact point was calculated from the mass m_i associated with each surface element.

To simulate some external normal loading material density was increased until the total mass reached 0.6 kg. At the ring's center of mass a force $\boldsymbol{F}_a$ and a moment $\boldsymbol{M}_a$ were applied. No Stribeck curve was used; instead the sliding friction force was assumed to be constant:

$$F_{St,i}(\|\dot{\boldsymbol{x}}\|) = \mu N_i \tag{25}$$

where N_i denotes the absolute value of the normal force in contact point i. Total simulation duration was 1.5 s.

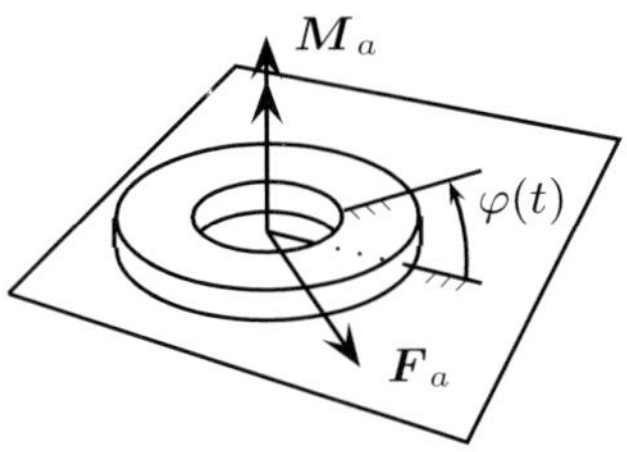

Fig. 3. Rigid ring moving on a plane with 2D frictional contact interface

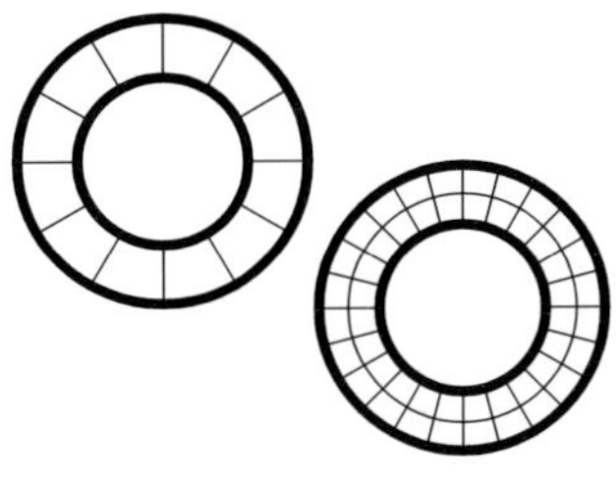

Fig. 4. Discretization of a plane ring into 12 elements and 48 elements

The applied force and torque profiles are shown in Fig. 5 and 6. Both were turned on three times for 0.25 s at the time points $t_0 = 0$ s, $t_1 = 0.4$ s and $t_2 = 0.85$ s. No load was applied to the system in-between; the disk was allowed to slide to a rest position. While the applied force had a constant absolute value during the loading phases, its direction was changed according to the second plot in Fig. 5. The torque was used to turn the disk counterclockwise, clockwise and counterclockwise again (cf. Fig. 6).

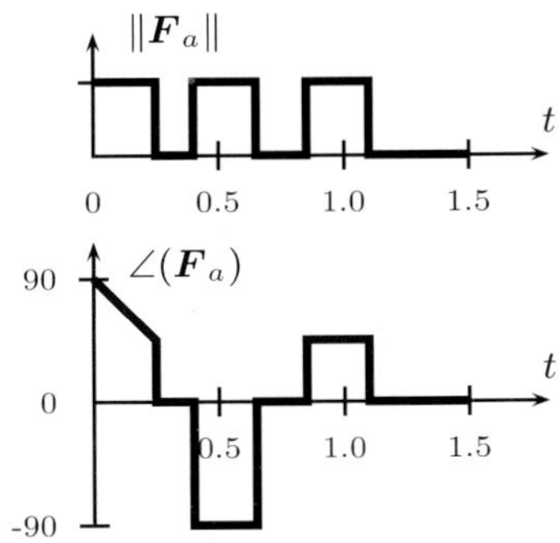

Fig. 5. Applied force profile, absolute value and direction

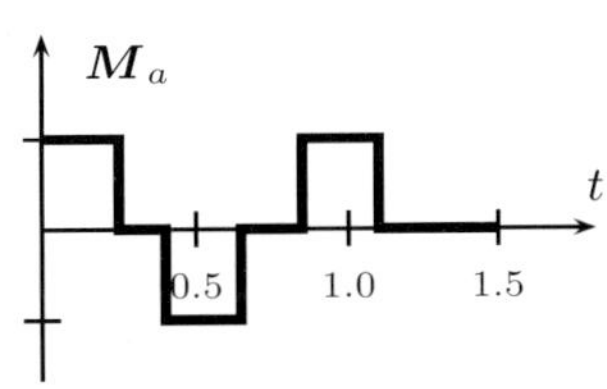

Fig. 6. Applied torque profile

These applied loads were chosen arbitrarily to obtain a random motion of the system. The motion contains alternating phases of sticking and sliding to cover the entire model, especially the transitions from sticking to sliding and vice versa.

Three model parameters, namely c_i, d_i and ε_i of each contact point i, are still unknown. The stiffness parameter c_i was estimated with the help of a regularization parameter as outlined in [10]. Damping d_i was set to a critical value $d_i = 2m_i\sqrt{c_i/m_i}$ in order to prevent oscillations caused by the modeling. Vanishing relative velocities were detected with the small numeric threshold ε_i, which was set to one thousandth of the average relative velocity.

Direct Solution

To solve (24), an event-based integration scheme was used. The two cases of (23) were involved as event functions. Alternatively, one could approximate the discontinuity function $\alpha(\cdot)$ by a piecewise continuous function, as suggested in [1].

The resulting system of differential equations was integrated using a Rosenbrock formula for stiff equations. Figure 7 shows the result obtained with 12 discrete contact points. Then one of the contact points was modified to examine the sensitivity of the result to the distribution of frictional forces. The local value of the coefficient of friction was increased by a factor of three, which corresponds to a change in the total amount of friction of approximately 16%, and the obtained motion deviates quite significantly from the original result, as can be seen in Figs. 8 and 9.

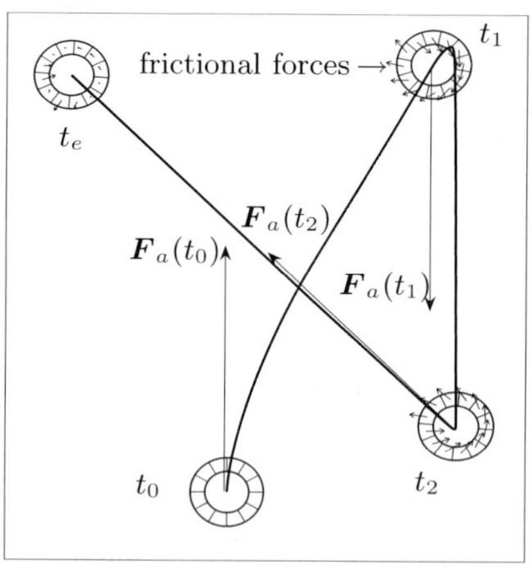

Fig. 7. Trajectory of ring's center of mass. Positions and forces are shown at $t_0 = 0$ s, $t_1 = 0.4$ s, $t_2 = 0.85$ s and $t_e = 1.5$ s

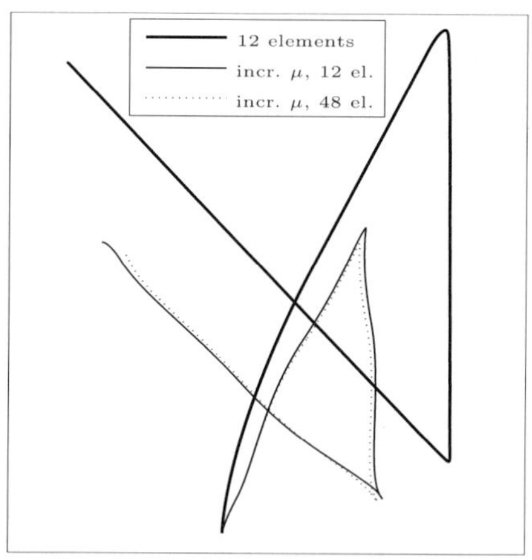

Fig. 8. Trajectories of ring's center of mass for different discretizations and local μ

The same simulation was repeated, but this time with 48 surface elements in the contact interface. The region of increased friction covered four elements which had the same size as a single element in the previous simulation. As can be seen in Figs. 8 and 9, the results for the two levels of discretization are in good agreement. Note that a discretization with just one element would lead to no motion at all, because there is no resistance to the applied torque in a single contact point. The applied force $\boldsymbol{F}_a$ alone is not sufficient to start motion of the disk.

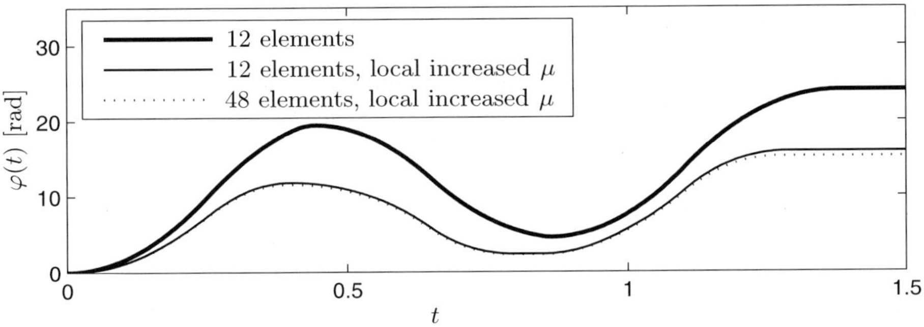

Fig. 9. Angular position of ring moving on a plane

Simplified Solution

For macroscopic sliding, when $\dot{\boldsymbol{z}}_i \approx 0$ and $\dot{\boldsymbol{w}}_i \approx \dot{\boldsymbol{x}}_i$, computation time and memory can be saved by setting $\dot{\boldsymbol{z}}_i = 0$. The corresponding differential equations for $\boldsymbol{z}_i$ can be removed from the system during that time. This leads to a time-variant set of equations. When all contact points are sliding, the system's degrees of freedoms reduce to six, no matter the number of contact points. Two additional criteria are needed to find the instants of removal and insertion of the inner variables $\boldsymbol{z}_i$. The first criterium is: one can remove an inner state variable $\boldsymbol{z}_i$ whenever plastic deformation takes place ($\alpha = 1$) *and* the angle between regularized and COULOMB sliding friction force becomes very small:

$$\angle(\boldsymbol{F}_{pl,i}, \dot{\boldsymbol{x}}_i / \|\dot{\boldsymbol{x}}_i\|) < \psi, \quad \psi \ll 1 \tag{26}$$

The second criterum is: the inner state variable $\boldsymbol{z}_i$ must be introduced again when sticking is detected. The detection of sticking works in the same way as before, but this time total relative velocity $\dot{\boldsymbol{x}}$ must be used instead of plastic deformation rate $\dot{\boldsymbol{w}}$ to measure the sliding velocity:

$$\|\dot{\boldsymbol{x}}_i\| > \varepsilon_i \tag{27}$$

When new inner variables are inserted, the springs get a pre-stressing $\boldsymbol{z}_p$ in such a way that the spring reaction force immediately after activating the new variable equals the Stribeck frictional force at relative velocity ε_i. Of course, the additional event function (26) needs to be evaluated after each integration timestep and hence the computation time increases. But on the other hand, the time is dramatically decreased by the reduced degrees of freedom, at least for motions with long periods of sliding.

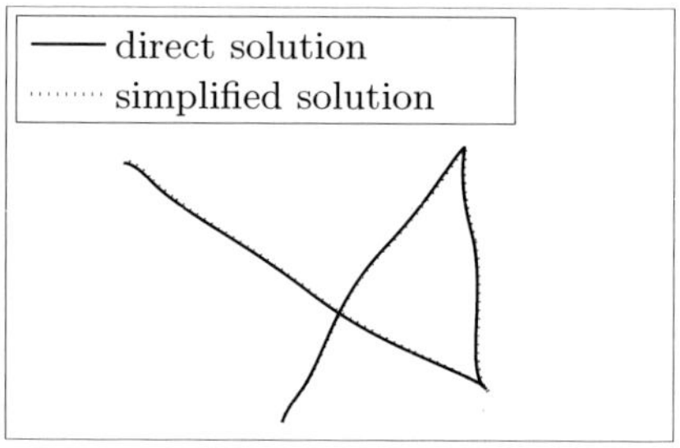

Fig. 10. Comparison of simplified and original solutions

4 Prospects

Future study should concern the extension to contact dynamics in the normal direction. Normal forces and displacements are considered constant in the current model and contact is maintained throughout simulation. But this will not hold for most applications. Such an extension to normal dynamics should include partial or full detachment of the contact interface. However, impacts would be treated as fully plastic impacts only.

References

1. Dupont P, Hayward V, Armstrong B, Altpeter F (2002) IEEE Transactions on Automatic Control 47:787–792
2. Englisch A, Faust H, Friedmann O (2001) Proceedings of the Global Powertrain Congress, Detroit, Michigan
3. Glocker C (2001) Set-Valued Force Laws, Springer, Berlin Heidelberg New York
4. Kachanov L M (1971) Foundations of the Theory of Plasticity, North-Holland
5. Krueger A (2003) PhD Thesis, University of Karlsruhe, Germany
6. Lemaitre J, Chaboche J-L (2000) Mechanics of Solid Materials, Cambridge Univ. Press
7. McNamara S, García-Rojo R, Herrmann H (2005) Physical Review E 72:021304 1–11
8. Palmov V A (1998) Vibrations of Elasto-Plastic Bodies, Springer, Berlin Heidelberg New York
9. Rossmann T, Pfeiffer F, Glocker C (1997) Proc. of the ASME Int. Mech. Eng. Congress and Exposition, Dallas, Texas
10. Vielsack P (1996) ZAMM 76:439–446

Micromechanics of Particle Adhesion

Jürgen Tomas

Mechanical Process Engineering, The Otto-von-Guericke-University,
Universitaetsplatz 2, 39106 Magdeburg, Germany
juergen.tomas@vst.uni-magdeburg.de

Summary. The rapid increasing production of cohesive to very cohesive ultrafine powders ($d < 10$ μm), e.g. very adhering pigment particles, micro-carriers in medicine, auxiliary materials in catalysis make technical problems much serious like undesired adhesion in particle processing, powder handling, and desired, in agglomeration or coating. Thus, it is very essential to understand the fundamentals of particle adhesion with respect to product quality assessment and process performance in powder technology. The state of arts in modelling of elastic, elastic-adhesion, elastic-dissipative, plastic-adhesion and plastic-dissipative contact deformation response of a single, normal loaded, isotropic contact of two smooth spheres is briefly discussed. Then the force-displacement behaviour of elastic-plastic and adhesive contacts is shown. Using the model "stiff particles with soft contacts", the combined influence of elastic and elastic-plastic repulsions in a characteristic particle contact is demonstrated. A sphere-sphere model for van der Waals forces F_{H0} without any contact deformation describes the "stiff" attractive particle adhesion term. A plate-plate model is used to calculate the micro-contact flattening or overlap. Various contact deformation paths for loading, unloading, reloading and contact detachment are discussed. Thus, the varying adhesion forces between particles depend directly on this "frozen" irreversible deformation, the so-called contact pre-consolidation history. The adhesion force is found to be load dependent $F_H(F_N)$. The contribution of this history dependent adhesion on the tangential force in an elastic-plastic frictional contact $F_T(F_N, F_H(F_N))$, the rolling resistance $F_R(F_N, F_H(F_N))$ and the torque of mobilized frictional contact rotation $M_{to}(F_N, F_H(F_N))$ are shown. With this increasing load, normal and tangential contact stiffness, energy absorption, Coulomb friction limit and friction work increase.These constitutive models are generally applicable for solid micro- or nanocontacts but have been shown here for an ultrafine limestone powder ($d_{50} = 1.2 \mu m$).

1 Introduction

The rapid increasing production of cohesive ultrafine powders ($d < 10$ μm), e.g. adhering pigment particles, micro-carriers in biotechnology, auxiliary materials in chromatography, make technical problems much more serious like

Peter Eberhard (ed.), Multiscale Problems in Multibody System Contacts, 301–316.

undesired adhesion in particle conversion or powder handling, and desired in agglomeration or coating. Thus, understanding the fundamentals of particle adhesion with respect to product quality assessment and process performance is very essential in powder technology. First comprehensive physical models for micromechanics of ultrafine particle adhesion were published 1976/78 by Molerus [1, 2] and continued 1983/91 by the author [3, 4] independently of development of the discrete element method (DEM) in rock mechanics [5].

2 State of Arts

It is worth to note here that van der Waals forces discussed in this paper are dominant and approximately $10^4 - 10^6$ times the gravitational force acting on ultrafine particles, see calculations of Rumpf [6] and Schubert [7]. The fundamentals of molecular attraction potentials and the mechanics of adhesion are treated for example by Krupp [8], Israelachvili [9], Maugis [10] and Kendall [11]. Contact and impact mechanics are described by Johnson [12] and Stronge [13].

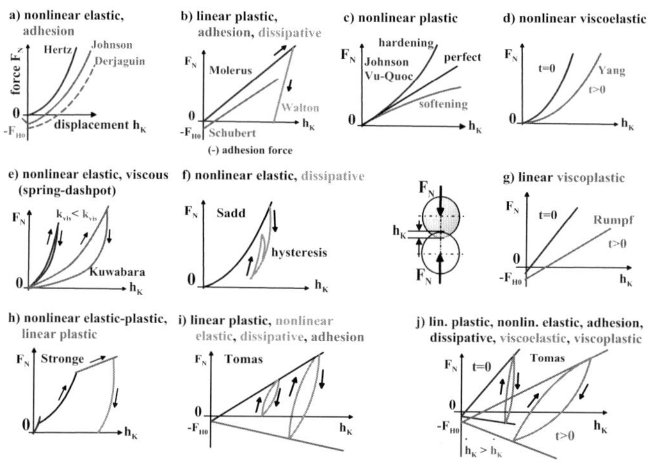

Fig. 1. Constitutive models of contact deformation of smooth spherical particles in normal direction without adhesion (only compression, +sign) and with adhesion (tension, -sign).

The basic models for elastic behaviour were derived by Hertz [14], Fig. 1a), for viscoelasticity by Yang [15], Fig. 1c), and for constant adhesion by Johnson et al. [16] and Derjaguin et al. [17], Fig. 1a). Plastic behaviour was described by Stieß [18], Thornton and Ning [20] and Walton and Braun [19]. But the increase of adhesion force due to plastic contact deformation was introduced by Molerus [1] and Schubert et al. [21], Fig. 1b). Nonlinear plastic, displacement-driven contact hardening was investigated by Johnson [12] and Vu-Quoc [22],

Fig. 1c). Additionally, contact softening could be included [29, 30], Fig. 1c). Viscoelasticity and relaxation was considered by Yang [15], Fig. 1d). Energy dissipation of the nonlinear elastic contact with viscous spring-dashpot behaviour was modelled by Kuwabara [23], Fig. 1e), during one unload/reload cycle by Sadd et al. [24], Fig. 1f). Different elastic, elastic-plastic and fully plastic behaviour were recently described by Stronge [13], Fig. 1h). Time dependent viscoplasticity was modelled by Rumpf et al. [25] and Luding [26], Fig. 1g). Considering constitutive models, one obtains a general contact model for load, time and rate dependent viscoelastic, plastic, viscoplastic, adhesion and dissipative behaviour, Tomas [27 - 30], Fig. 1i) and j), see section 3.1. This inelastic contact flattening leads to an increase of adhesion force $F_H(F_N)$ depending on the applied normal load F_N- the load or pre-consolidation history. This essential effect of a soft particle contact was modelled by Krupp [8], Dahneke [31], Rumpf [25], Molerus [1], Maugis [32], Castellanos [33] and Tomas [4, 27 - 30], section 3.2.

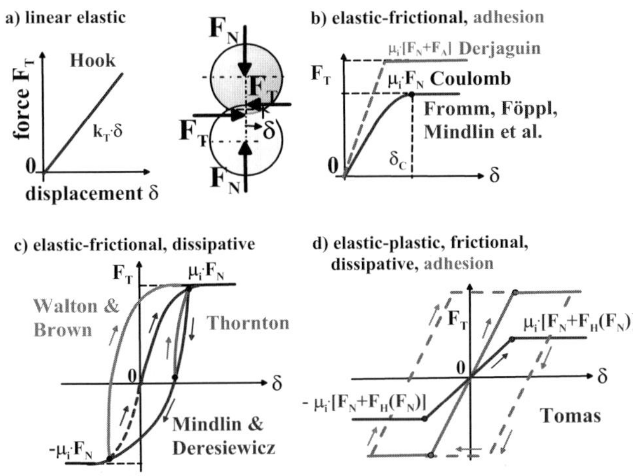

Fig. 2. Constitutive models of contact displacement of smooth spherical in tangential direction (-sign means reverse shear and displacment directions).

Besides the linear elastic tangential force-displacement relation, Hook's law Fig. 2 panel a), Fromm [34], Cattaneo [35], Föppl [36], Mindlin [38, 39] and Sonntag [37] modelled the non-linear contact loading path up to Coulomb friction as the limit, Fig. 2 panel b). The contribution of adhesion forces in Coulomb friction was considered by Derjaguin [17] and Thornton [40], Fig. 2 panel b). Mindlin and Deresiewicz [39], Walton and Braun [19], Thornton [40], Di Renzo and Di Maio [41] modelled different non-linear paths for load, unload, reload, reverse shear load, unload and load Fig. 2 panel c). The effect of load dependent adhesion $F_H(F_N)$ in Coulomb friction, see Tomas [27 - 29], is demonstrated in Fig. 2 panel d) and explained in section 3.3. The sources of an additional rolling resistance F_R should be considered by a so-called "microslip", micro-roughness of contact surfaces, and especially, by contact deformations [12, 34, 37]. Another effect is that the sphere can rotate (twist or spin) around its principal axis within the contact plane. The torque M_{to} as radial distribution versus radius coordinate of circular elastic contact area as function of the rotation angle ϕ was calculated by Mindlin [38], Cattaneo [42], Deresiewicz [43] and Johnson [12], section 3.4 - 3.6.

3 Constitutive Models for Elastic-Plastic, Dissipative Behaviour and Load Dependent Adhesion

3.1 Particle Contact Constitutive Model for Normal Loading

This paper is intended to focus on the model of isotropic, stiff, and linear elastic, spherical particles that are approaching to soft contacts by attractive adhesion forces of smooth surfaces, Fig. 3. Thus, this soft or compliant contact displacement is assumed to be small $h_K/d << 1$ compared to the size (diameter) of the stiff particle. The particles may have a certain material stiffness so that the volume deformation is negligible. During surface stressing the rigid particle is not so much deformed that it undergoes a certain change of the particle shape. This is equivalent to a model of "healing contacts" after stressing and deformation.

A typical normal force-displacement diagram is demonstrated in Fig. 3. The zero-point of this diagram $h_K = 0$ is equivalent to the characteristic adhesion separation of a direct contact $a_{F=0}$. After approaching $F_N \propto a^{-2}$ (with $a = a_{F=0} - h_K$) from an infinite distance $-\infty$ to a minimum separation $a = a_{F=0}$, Fig. 3a), the sphere-sphere contact without any contact deformation is formed by the short-range attractive adhesion force F_{H0} (the so-called jump in). Then the contact may be loaded from point F_{H0} to Y and, as the response, is elastically deformed with an approximated circular contact area, Fig. 3b). When this Hertz curve intersects the abscissa the total force equilibrium $F_N = 0$ is obtained within the self-equilibrating contact. With increasing external normal load this soft contact starts at a pressure pf with plastic yielding at the point Y, Fig. 3c). This yield point Y is located here below the abscissa, i.e. the contact force equilibrium $F_N = 0$ includes a certain elastic-plastic deformation as response of adhesion force $(1+\kappa).F_{H0}$. The micro-yield surface is reached and this maximum pressure p_f cannot be exceeded and results in a combined elastic-plastic yield limit of the plate-plate contact with an annular elastic zone and a circular centre. A confined plastic field is formed inside of the contact circle. Constant mechanical properties provided, the finer the particles the smaller is the force necessary for yielding [27 - 30]. An initial and exclusive elastic contact deformation at loading, see the position of yield point Y in Fig. 3, has no relevance for irreversible compression of fine to ultrafine powders and can be excluded here. Thus, one has to consider an elastic-plastic contact behaviour, which is characterized by the elastic-plastic yield

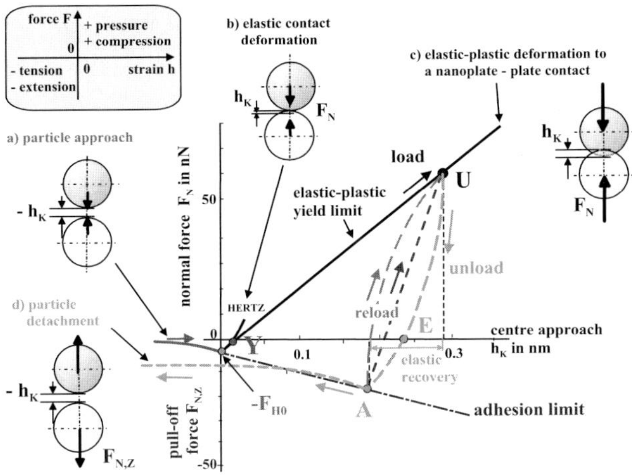

Fig. 3. Normal force - displacement diagram of characteristic contact flattening of limestone particles, modelled as smooth mono-disperse spheres, median diameter $d_{50} = 1.2\ \mu m$. For convenience, pressure and compression are defined as positive but tension and extension are negative. The material data are: modulus of elasticity $E = 150\ kN/mm^2$, plastic micro-yield strength $p_f = 350\ N/mm^2$, Poisson ratio $\nu = 0.28$, Hamaker constant $C_{H,sls} = 3.8 \cdot 10^{-20}$ J, equilibrium separation for dipole interaction $a_{F=0}$= 0.336 nm, elastic-plastic contact area coefficient $\kappa_A = 5/6$ and the plastic repulsion coefficient $\kappa_p = 0.15$.

limit. This line cannot be crossed. Between the elastic-plastic yield limit and the adhesion limit the elastic domain is located. Any load F_N yields an increasing displacement h_K. But, if one would unload, beginning at arbitrary point U, the elastically deformed, annular contact zone recovers along a parabolic curve U - A and the plate-plate contact remains with a "frozen" radius $r_{K,A}$ or reduced plastic displacement $h_{K,A}$. The adhesion or contact failure limit at point A in the diagram of Fig. 3 is reached and the contact plates detach with the increasing distance $a = a_{F=0} + h_{K,A} - h_K$ [27, 28], see Fig. 3d). This actual particle separation a can be considered for the calculation by means of a long-range hyperbolic adhesion force curve $F_{N,Z} \propto p_{VdW}(a) \propto a^{-3}$ [9]. These theoretical predictions of normal force-displacement behaviour are very useful to describe the compression of primary elastic, elastic-plastic particle compounds or elastic-plastic granules at conveying and handling operations. The elastic-plastic load and Hertzian unload curves and the unload/reload hysteresis were experimentally confirmed for coarse granules [44].

3.2 Load Dependent Adhesion Force

The slopes of elastic-plastic yield and adhesion limits in Fig. 3 are characteristics of irreversible particle contact stiffness or compliance. Consequently, if one eliminates the centre approach h_K of the loading and unloading functions an implied positive function for the contact pull-off force $F_H = -F_{N,Z}$ at the detachment point A is obtained

$$F_{H,(1)} = F_{H0} + \kappa \cdot (F_N + F_{H0}) - \pi \cdot r_{1,2}^2 \cdot p_{VdW} \cdot \left[\frac{3 \cdot (F_N + F_{H0})}{2 \cdot r_{1,2}^2 \cdot E^*} \cdot \left(1 + \frac{F_{H,(0)} - F_{H0}}{F_N + F_{H0}} \right) \right]^{2/3} \quad (1)$$

with the median particle radius $r_{1,2}$ or characteristic radius of contact surface curvature (r_1 and r_2 are the principal radii of surface curvatures before flattening),

$$r_{1,2} = \left(\frac{1}{r_1} + \frac{1}{r_2} \right)^{-1} \quad (2)$$

and with the averaged material stiffness as series of elastic elements 1 and 2 (which is equivalent to the sum of element compliances, E modulus of elasticity, ν Poisson ratio):

$$E^* = 2 \cdot \left(\frac{1 - \nu_1^2}{E_1} + \frac{1 - \nu_2^2}{E_2} \right)^{-1} \quad (3)$$

The adhesion force per unit planar surface area or attractive pressure p_{VdW} which is used here to describe the van der Waals interactions at contact, is equivalent to a theoretical bond strength and can be simply calculated as [9] (e.g. $p_{VdW} \approx 3$ - 600 MPa, σ_{sls} 0.25 - 50 mJ/m^2 surface tension):

$$p_{VdW} = \frac{C_{H,sls}}{6 \cdot \pi \cdot a_{F=0}^3} = \frac{4 \cdot \sigma_{sls}}{a_{F=0}} \quad (4)$$

The plastic repulsion coefficient κ_p describes a dimensionless ratio of attractive van der Waals pressure p_{VdW} for a plate-plate model, Eq. (4), to repulsive micro-yield strength p_f (particle micro-hardness), which is temperature sensitive ($C_{H,sls} = (0.2 - 40) \cdot 10^{-20}$ J Hamaker constant):

$$\kappa_p = \frac{p_{VdW}}{p_f} = \frac{C_{H,sls}}{6 \cdot \pi \cdot a_{F=0}^3 \cdot p_f} = \frac{4 \cdot \sigma_{sls}}{a_{F=0} \cdot p_f} \quad (5)$$

For a rigid contact this coefficient is infinitely small, i.e. $\kappa_p \approx 0$ and for a soft contact $\kappa_p \to 1$.

The unloading point U is stored in the memory of the contact as preconsolidation history. This general non-linear adhesion force model, dashed curve in Fig. 4, implies the dimensionless, elastic-plastic contact consolidation coefficient κ, the "stiff" contribution of adhesion force $F_{H0}(r_{1,2})$ without any contact flattening and, additionally, the influence of adhesion, stiffness, averaged particle radius $r_{1,2}$ and the averaged modulus of elasticity E* in the last term of the equation. This non-linear adhesion-normal force function $F_H = f(F_N)$ can be linearised as shown in Fig. 4:

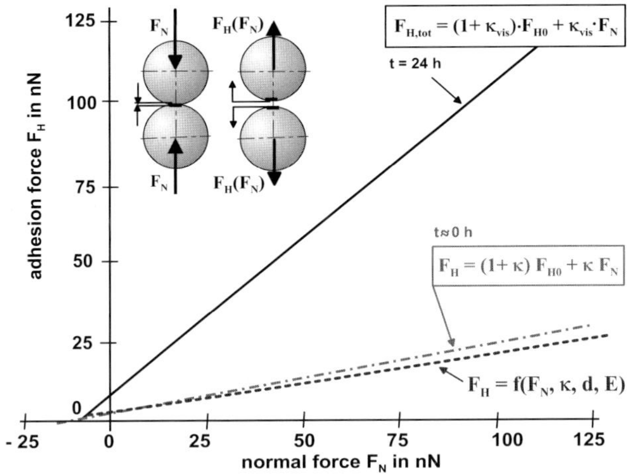

Fig. 4. Adhesion force - normal force diagram of limestone according to Eqs. (1) and (6), surface moisture $X_W = 0.5\%$. Adhesion force parameters are the elastic-plastic contact consolidation coefficient $\kappa = 0.224$ and the adhesion force without any contact flattening $F_{H0} = 2.64$ nN. For comparison, the total adhesion force $F_{H,tot}$ after a time consolidation of 24 h is also shown in Fig. 4 by a full line. This time variable slope is given by the elastic-plastic and viscoplastic contact consolidation coefficient $\kappa_{vis} = 1.014$ which is obtained by the sum of rheological models [29].

$$F_H = \frac{\kappa_A}{\kappa_A - \kappa_p} \cdot F_{H0} + \frac{\kappa_p}{\kappa_A - \kappa_p} \cdot F_N = (1 + \kappa) \cdot F_{H0} + \kappa \cdot F_N \tag{6}$$

The dimensionless elastic-plastic contact consolidation coefficient κ is given by the slope of adhesion force F_H influenced by predominant plastic contact failure. This displacement coefficient κ characterizes the irreversible particle contact stiffness or softness as well. A shallow slope implies low adhesion level $F_H \approx F_{H0}$ because of stiff particle contacts, but a large slope means soft contacts, or i.e., a cohesive powder flow behaviour. This model considers, additionally, the flattening of soft particle contacts caused by the adhesion force $\kappa \cdot F_{H0}$. Thus, the total adhesion force F_H consists of a stiff contribution F_{H0} and a soft, displacement influenced term $\kappa(F_{H0} + F_N)$, Fig. 4. Thus, Eq. (6) can be interpreted as a general linear constitutive model, i.e. linear in forces, but non-linear in material characteristics [27].

3.3 Elastic-Plastic, Frictional Tangential Force of Load Dependent Adhesive Contact

The tangential stiffness according to Mindlin [38] can be written for the elastic-plastic contact flattening caused by a sufficiently large normal force F_N and, consequently, partially sticking by load dependent adhesion force $F_H(F_N)$, Eq. (6), as function of tangential displacement δ (index H for adhesive contact):

$$k_{T,H} = \frac{dF_T}{d\delta} = 4 \cdot G^* \cdot r_K \cdot \left(1 - \frac{\delta}{\delta_{C,H}}\right)^{1/2} \tag{7}$$

$G_i = E_i/2(1+\nu_i)$ i = 1,2 is the shear modulus and the averaged shear modulus G* is given as:

$$G^* = 2 \cdot \left(\frac{2-\nu_1}{G_1} + \frac{2-\nu_2}{G_2}\right)^{-1} \tag{8}$$

The initial stiffness $k_{T,H0}$ is found at tangential displacement $\delta = 0$:

$$k_{T,H0} = \left.\frac{dF_T}{d\delta}\right|_{\delta=0} = 4 \cdot G^* \cdot r_K = 4 \cdot G^* \cdot \sqrt{r_{1,2} \cdot h_K} \tag{9}$$

Thus, the tangential force-displacement relation can be expressed by a linear elastic contribution for the no-slip region within the contact area according to Hook's law, Fig. 2 panel a):

$$F_T = 4 \cdot G^* \cdot r_K \cdot \delta \tag{10}$$

The contact radius r_K depends on the applied normal force F_N of the elastic-plastic yield limit, Eq.(11),

$$r_K^2 = \frac{F_N + F_{H0}}{\pi \cdot p_f \cdot (\kappa_A - \kappa_p)} \tag{11}$$

and on the effective adhesion force $F_H(F_N)$, Eq. (6),

$$F_T = 4 \cdot G^* \cdot \sqrt{\frac{(1+\kappa) \cdot (F_{H0} + F_N)}{\pi \cdot \kappa_A \cdot p_f}} \cdot \delta \tag{12}$$

so that F_T and the initial tangential stiffness $k_{T,H0}(F_N)$, Eq. (9), increase with increasing pre-consolidation $F_H(F_N)$, see Fig. 5. This stiffness amounts to $k_{T,H0}$ = 800 - 2000 N/m for limestone particles. The Coulomb friction limit of the tangential force is described by the coefficient of internal friction μ_i and depends on the elastic-plastic flattening by the normal force F_N and the variable adhesion force $F_H(F_N)$ as well, Eq. (6). Using Eq. (10) the transition between the linear elastic range and incipient contact failure by Coulomb friction is calculated by the friction limit of tangential displacement $\delta_{0,H}$, see the red points for yield in Fig. 5:

$$\delta_{0,H} = \frac{\mu_i \cdot [F_N + F_H(F_N)]}{4 \cdot G^* \cdot r_K(F_N)} = \frac{\mu_i}{4 \cdot G^*} \cdot \sqrt{\pi \cdot \kappa_A \cdot p_f \cdot (1+\kappa) \cdot (F_{H0} + F_N)} \tag{13}$$

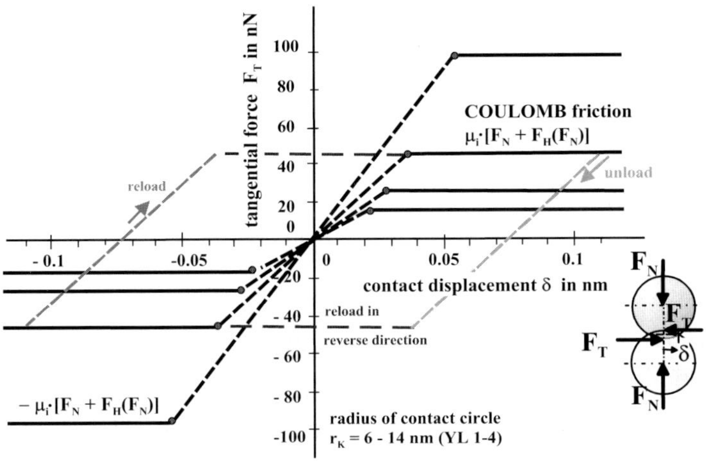

Fig. 5. Representative tangential force-displacement diagram for limestone particles modelled as smooth mono-disperse spheres (median $d_{50,3} = 1.2\ \mu m$, shear modulus $G^* = 34\ kN/mm^2$, friction coefficient $\mu_i = 0.76$). Compared to the contact radius $r_K = 6 - 14\ nm$, the elastic range is very small and limited by the tangential displacement $\delta_{0,H} < 0.05\ nm$. At this yield point the elastic behaviour is transmitted into the frictional behaviour of contact sliding shown by constant tangential force. The elastic-plastic contact flattening in normal direction and resulting load dependent adhesion force $F_H(F_N)$ is a parameter of this friction limit. Contact shear, unload and reverse shear generate large hysteresis loops and, as consequence, energy dissipation by friction work.

Moreover, the contact looses his elastic tangential stiffness at $k_{T,H} = 0$ and completely mobilized contact sliding is obtained for the new friction limit of displacement $\delta = \delta_{C,H}$ under load dependent adhesion force $F_H(F_N)$, Eq. (6):

$$\delta_{C,H} = \frac{3 \cdot \mu_i \cdot (1+\kappa) \cdot (F_{H0} + F_N)}{8 \cdot G^* \cdot r_K} = \frac{3 \cdot \mu_i}{8 \cdot G^*} \cdot \sqrt{\pi \cdot \kappa_A \cdot p_f \cdot (1+\kappa) \cdot (F_{H0} + F_N)} \tag{14}$$

Both friction limits, Eqs. (13) and (14) are geometrically similar $\delta_{CH} = 1.5 \cdot \delta_{O,H}$ and increase with increasing load F_N. Mindlin's nonlinear model $F_T = f(\delta)$ is supplemented here for first loading:

$$F_T = F_{T,C,H} \cdot \left[1 - \left(1 - \frac{\delta}{\delta_{C,H}}\right)^{3/2}\right] \tag{15}$$

Thus, the extended functions for unloading and shear in reverse direction (- sign) [39] are

$$F_{T,U} = F^*_{T,U} - 2 \cdot F_{T,C,H} \cdot \left[1 - \left(1 - \frac{\delta_U - \delta}{2 \cdot \delta_{C,H}}\right)^{3/2}\right] \quad (16)$$

and for reloading in the previous (+) shear direction for $F_{T,reload} = -F_{T,U}$ *and* $\delta_{reload} = -\delta_U$:

$$F_{T,reload} = -F^*_{T,reload} + 2 \cdot F_{T,C,H} \cdot \left[1 - \left(1 - \frac{\delta + \delta_{reload}}{2 \cdot \delta_{C,H}}\right)^{3/2}\right] \quad (17)$$

3.4 Elastic-Plastic, Frictional Rolling Resistance of Load Dependent Adhesive Contact

Using Johnson's creep model [12], supplemented, rearranged and differentiated, and the stiffness $k_{R,H}$ of rolling resistance F_R by elastic-plastic contact flattening and consequently, partially sticking, follows as function of relative rolling angle γ of both spheres (index H for adhesion):

$$k_{R,H} = \frac{dF_R}{d\gamma} = \frac{16 \cdot G}{\pi \cdot (4 - 3 \cdot \nu)} \cdot \frac{(1 + \kappa) \cdot (F_{H0} + F_N)}{\kappa_A \cdot p_f} \cdot \left(1 - \frac{\gamma}{\gamma_{C,H}}\right)^2 \quad (18)$$

$$\gamma = \tan\left(\frac{\delta_1}{r_1}\right) - \tan\left(\frac{\delta_2}{r_2}\right) \approx \frac{\delta_1}{r_1} - \frac{\delta_2}{r_2} \quad (19)$$

The initial stiffness $k_{R,H0}$ is found at zero of rolling angle $\gamma = 0$:

$$k_{R,H0} = \left.\frac{dF_R}{d\gamma}\right|_{\gamma=0} = \frac{16 \cdot G}{\pi \cdot (4 - 3 \cdot \nu)} \cdot \frac{(1 + \kappa) \cdot (F_{H0} + F_N)}{\kappa_A \cdot p_f} \quad (20)$$

The contact looses his elastic rolling stiffness at $k_{R,H} = 0$ and the rolling friction limit with respect to a critical rolling angle $\gamma = \gamma_{C,H}$ is obtained:

$$\gamma_{C,H} = \frac{3 \cdot (4 - 3 \cdot \nu)}{16 \cdot G \cdot r_{1,2}} \cdot \sqrt{\pi \cdot \kappa_A \cdot p_f \cdot (1 + \kappa) \cdot (F_N + F_{H0})} \quad (21)$$

This friction limit $\gamma_{C,H}$ is geometrically similar to a tangent limit $\gamma_{0,H}$ calculated by ratio of force limit $F_{R,C.H}$ and initial stiffness $k_{R,H0}$, i.e. $\gamma_{C,H} = 3 \cdot \gamma_{0,H} = 3 \cdot F_{R,C,H}/k_{R,H0}$. It increases with increasing load F_N like the friction limit of tangential displacement $\delta_{C,H}$, Eq. (14). With Eq. (6) for elastic-plastic contact displacement in normal direction and load dependent adhesion force, by a tilting moment relation at the perimeter of flattened contact, Fig. 6 panel right, the critical rolling resistance $F_{R,C.H}$ results in:

$$F_{R,C.H} = \frac{r_K}{r} \cdot (1 + \kappa) \cdot (F_{H0} + F_N) \quad (22)$$

Substitution of the contact radius r_K of the elastic-plastic yield limit, Eq. (11), results nearly in a proportionality as $F_{R,C,H}(F_N) \propto F_N^{3/2}$:

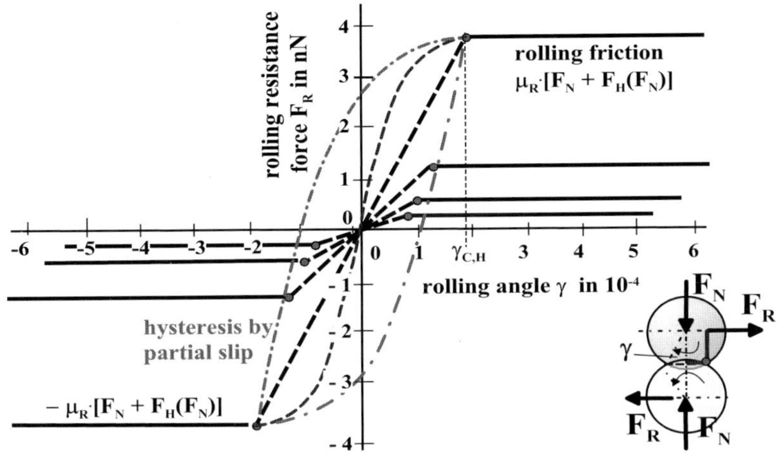

Fig. 6. Representative rolling resistance force - rolling angle diagram for limestone particles modelled as smooth mono-disperse spheres (median $d_{50,3} = 1.2\ \mu$m, shear modulus $G^* = 34\ kN/mm^2$, rolling friction coefficient $\mu_R \approx 0.01 - 0.024$). This frictional rolling resistance of smooth soft spheres can be considered by a tilting moment relation of the force pair around the red pivot at the perimeter of contact circle. As the response of contact flattening a lever arm of contact radius r_K with respect to F_N is generated which is equilibrated by rolling resistance F_R acting perpendicular to direction of F_N with the lever arm $r - h_K/2$, see the blue lines of panel right. The linear elastic range is very small and limited by the critical rolling angle $\gamma_{C,H} = 8 \cdot 10^{-5} - 1.9 \cdot 10^{-4}$. At this yield point the elastic behaviour is transmitted into the frictional behaviour of contact rolling shown by constant rolling resistance force, Eq. (23). The elastic-plastic contact flattening in normal direction and resulting load dependent adhesion force $F_H(F_N)$ is also parameter of this rolling friction limit.

$$F_{R,C,H} = \mu_R \cdot (1+\kappa) \cdot (F_{H0} + F_N) = \sqrt{\frac{(1+\kappa)^3 \cdot (F_{H0} + F_N)^3}{\pi \cdot r_{1,2}^2 \cdot p_f \cdot \kappa_A}} \tag{23}$$

By integration of Eq. (18) the force - rolling angle function for the first loading is obtained:

$$F_R = F_{R,C,H} \cdot \left[1 - \left(1 - \frac{\gamma}{\gamma_{C,H}}\right)^3\right] \tag{24}$$

Similar to the constitutive tangential force - displacement model, Eqs. (16) and (17), the function for unloading and rolling in reverse direction (- sign) is obtained as

$$F_R = F_{R,U} - 2 \cdot F_{R,C.H} \cdot \left[1 - \left(1 - \frac{\gamma_U - \gamma}{2 \cdot \gamma_{C,H}}\right)^3\right] \tag{25}$$

and for reloading in the previous (+) rolling direction, i.e. $F_{R,reload} = -F_{R,U}$ and $\gamma_{reload} = -\gamma_U$:

$$F_R = -F_{R,reload} + 2 \cdot F_{R,C.H} \cdot \left[1 - \left(1 - \frac{\gamma_{reload} + \gamma}{2 \cdot \gamma_{C,H}}\right)^3\right] \tag{26}$$

3.5 Elastic-Plastic, Frictional Twisting of Load Dependent Adhesive Contact

The torsional stiffness according to Deresiewicz [43] can be written for elastic-plastic contact flattening and consequently, partially sticking by load dependent adhesion force $F_H(F_N)$, Eq. (6), as function of torsional moment M_{to}:

$$k_{to,H} = \frac{dM_{to}}{d\phi} = \frac{8 \cdot G \cdot r_K^3}{3} \cdot \left[2 \cdot \left(1 - \frac{M_{to}}{M_{to,C,H}}\right)^{-1/2} - 1\right]^{-1} \tag{27}$$

The initial stiffness $k_{to,H0}$ is found at vanishing torsional moment $M_{to} = 0$:

$$k_{to,H0} = 2 \cdot (1 - \upsilon) \cdot r_{1,2} \cdot (1 + \kappa) \cdot (F_{H0} + F_N) \tag{28}$$

This Eq. (28) corresponds to the initial torsional stiffness of Mindlin [38] supplemented by the load dependent adhesion force $F_H(F_N)$. For the limestone powder ($d_{50} = 1.2\,\mu$m) it gives very small values $k_{to,H} = (35–500)\cdot 10^{-15}$ Nm. Thus for simplicity, the elastic moment - rotation angle function can be approached as linear relation, Fig. 7.

The contact looses his elastic torsional stiffness at $k_{to,H} = 0$ and completely mobilized "frictional" contact rotation is obtained for the friction limit of moment $M_{to} = M_{to,C,H}$ under load dependent adhesion force $F_H(F_N)$:

$$M_{to,C,H} = \frac{2 \cdot \mu_i \cdot (1 + \kappa) \cdot (F_{H0} + F_N) \cdot r_K}{3} \tag{29}$$

If one integrates Eq. (27) the torsional moment M_{to} as function of relative rotation angle $\phi = \phi_1 - \phi_2$ is obtained with Eq. (6) for loading:

$$M_{to} = 4 \cdot M_{to,C,H} \cdot \left[\sqrt{\frac{3 \cdot \phi}{\phi_{C,H}} + 1} - \frac{3 \cdot \phi}{4 \cdot \phi_{C,H}} - 1\right] \tag{30}$$

With the moment of completely mobilized frictional contact rotation $M_{to,C,H}$, Eq. (29), results in the critical rotation angle $\phi_{C,H}$ under load dependent adhesion force $F_H(F_N)$:

$$\phi_{C,H} = \frac{3 \cdot \mu_i \cdot (1 + \kappa) \cdot (F_{H0} + F_N)}{4 \cdot G \cdot r_K^2} \tag{31}$$

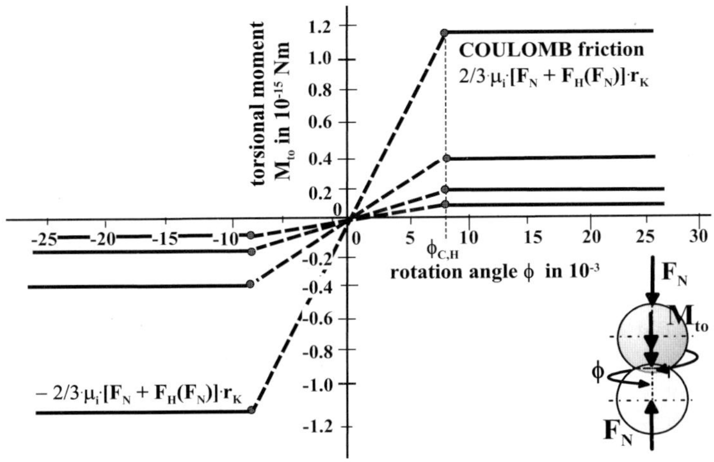

Fig. 7. Representative torsional moment - rotation angle diagram for limestone particles modelled as smooth mono-disperse spheres (median $d_{50,3} = 1.2\ \mu$m, shear modulus $G^* = 34\ kN/mm^2$, friction coefficient $\mu_i \approx tan\phi_i = 0.76$). The linear elastic range is very small and limited by the critical rotation angle $\phi_{C,H} = 0.0076$. At this yield point the elastic behaviour is transmitted into the frictional behaviour of contact sliding shown by constant torque, Eq. (29). The elastic-plastic contact flattening in normal direction and resulting load dependent adhesion force $F_H(F_N)$ is the parameter of this friction limit as well.

Substituting the contact radius r_K by Eq. (11) yields the simple constitutive relation for the frictional limit of rotation of an elastic-plastic contact with adhesion:

$$\phi_{C,H} = \frac{3 \cdot \pi \cdot \mu_i \cdot \kappa_A \cdot p_f}{4 \cdot G} \tag{32}$$

This friction limit $\phi_{C,H}$ is geometrically similar to the tangent limit $\phi_{0,H}$ calculated by ratio of moment limit $M_{to,C.H}$ and initial stiffness $k_{to,H0}$:

$$\phi_{C,H} = 3 \cdot \phi_{0,H} = 3 \cdot \frac{M_{to,C,H}}{k_{to,H0}} \tag{33}$$

It is independent of the load F_N (provided that $\kappa_A =$ const.) and depends only on the particle properties. This critical value amounts to $\phi_{C,H} \approx 0.0076$ for limestone particles, Fig. 7. This new model for loading of an elastic-plastic frictional contact Eq. (30) may be supplemented by the models for unloading, i.e. twisting in the reverse direction,

$$M_{to,unload} = M_{to,U} - 8 \cdot M_{to,C,H} \cdot \left[\sqrt{\frac{3 \cdot (\phi_U - \phi)}{2 \cdot \phi_{C,H}} + 1} - \frac{3 \cdot (\phi_U - \phi)}{8 \cdot \phi_{C,H}} - 1\right] \tag{34}$$

and for reloading, i.e. twisting in the previous direction:

$$M_{to,reload} = -M_{to,U} + 8 \cdot M_{to,C,H} \cdot \left[\sqrt{\frac{3 \cdot (\phi_U + \phi)}{2 \cdot \phi_{C,H}} + 1} - \frac{3 \cdot (\phi_U + \phi)}{8 \cdot \phi_{C,H}} - 1 \right] \tag{35}$$

3.6 Essential Constitutive Particle Parameters

From these elastic-plastic and frictional force - displacement laws $F_N(h_K)$, $F_T(\delta)$, force - force equations $F_H(F_N)$ and moment - angle models $M_R(\gamma)$, $M_{to}(\phi)$ one can conclude that seven independent mechanical, i.e. physically based, material parameters are necessary to know for the micromechanics of particle adhesion:

1. The characteristic particle size d_{50} or, more precise, the median radius of contact surface curvature before flattening $r_{1,2}$, Eq. (2), is necessary to know.
2. The Hamaker constant $C_{H,sls}$ is the characteristic material parameter of the van der Waals adhesion between practically dry particles for both types of bonds the sphere - sphere interaction of contact points F_{H0} and the plate-plate interaction of flattened contact by an external normal load F_N (macroscopic normal stress σ) and the internal adhesion effect $F_H(F_N)$. From this value the attractive van der Waals pressure p_{VdW} and the surface tension σ_{sls}, Eq.(4), can be obtained.
3. The plastic micro-yield strength p_f characterizes the plastic repulsion behaviour of the sphere contact. The ratio of the attractive van der Waals force per unit planar surface p_{VdW} and this repulsive plastic micro-yield strength p_f results in the dimensionless plastic contact repulsion coefficient κ_p, Eq. (5), and from that the elastic-plastic contact consolidation coefficient κ is obtained.
4. The contact viscosity η_K characterizes the viscoplastic repulsion during time consolidation at rest. Hence, the total viscoplastic contact consolidation coefficient κ_{vis} includes both the elastic-plastic κ_p and the viscoplastic contributions $\kappa_{p,t} = p_{vdW}/\eta_K t$ of contact flattening.
5. The elastic contact repulsion is determined by the modulus of elasticity E and by the
6. Poisson ratio ν. From these material data the shear modulus G is calculated.
7. The frictional contact failure, i.e. the loss of elastic tangential or torsional stiffness, is characterized by the coefficient of internal friction μ_i between the adhesive particle surfaces.

This information is essential to model the plastic yield limits of cohesive powders when the stresses are approaching the maximum shear stress - the

Coulomb-Mohr yield conditions for incipient and steady-state flow and consolidation.

4 Summary and Conclusions

By the model "stiff particles with soft contacts" and the contact force equilibrium, universal equations were derived which include the elastic-plastic particle contact behaviour with adhesion, load-unload hysteresis and a history dependent adhesion force function. This physical approach is used for the advanced data evaluation of various powder product properties concerning particle size distribution (nanoparticles to granules), moisture content (dry, moist and wet), material properties (minerals, chemicals, pigments, waste, plastics, food etc.), flow, compression and compaction behaviour which have been tested and evaluated [3, 4, 27].

5 Acknowledgements

The author would like to thank the colleagues from Mechanics, Physics, Material Sciences, Mathematics and Process Engineering for the fruitful collaboration within the DFG graduate school 828 "Micro-Macro Interactions in Structured Media and Particle Systems".

References

1. Molerus O (1975) Powder Technol 12:259–275
2. Molerus O (1978) Powder Technol 20:161–175
3. Tomas J (1983) Ph.D. Thesis, Freiberger Forschungshefte A677:1–133
4. Tomas J (1991) Habilitation. Sc.D. Thesis, Bergakademie Freiberg
5. Cundall PA, Strack OLD (1979) Geotechnique 29:47–65
6. Rumpf H (1974) Chem Ing Tech 46:1–11
7. Schubert H (1979) Chem Ing Tech 51:266–277
8. Krupp H (1967) Adv Colloid Interface Sci 1:111–239
9. Israelachvili JN (1992)Intermolecular and Surface Forces. Academic Press, London
10. Maugis D (1999) Contact, Adhesion and Rupture of Elastic Solids. Springer, Berlin
11. Kendall K (2001) Molecular Adhesion and its Application. Kluwer Academic/Plenum Publ, New York
12. Johnson KL (1985) Contact Mechanics. Cambridge University Press, Cambridge
13. Stronge WJ (2000) Impact mechanics. Cambridge University Press, Cambridge
14. Hertz H (1882) J reine u angewandte Math 92:156–171
15. Yang WH (1966) ASME J Appl Mech 33:395–401

16. Johnson KL, Kendall K, Roberts AD (1971) Proc Roy Soc Lond A324:301–313
17. Derjaguin BV, Muller VM, Toporov YP (1975) J Colloid Interface Sci 53:314–326
18. Stieß M (1976) Ph.D.Thesis. TH Karlsruhe
19. Walton OR, Braun R.L. (1986) J Rheology 30:949–980
20. Thorton C, Ning Z (1998) Powder Technol 99:154–162
21. Schubert H, Sommer K, Rumpf H (1976) Chem Ing Techn 716
22. Vu-Quoc L, Zhang X (1999) Proc R Soc Lond A 455:4013–4044
23. Kuwabara C, Kono K (1987) Jap J Appl Phys 26:1230–1233
24. Sadd MH, Tai Q, Shukla A (1993) Int J Non-Linear Mechanics 28:251–265
25. Rumpf H, Sommer K, Steier K (1976) Chem Ing Techn 48:300–307
26. Luding S, Manetsberger K, Müller J (2005) J Mech Physics of Solids 53:455–491
27. Tomas J (2003) Particulate Sci Technol 19:95–110, 111–129
28. Tomas J (2003) Mechanics of nanoparticle adhesion - a continuum approach, In: Mittal KL (ed) Particles on Surfaces 8: Detection, Adhesion and Removal, VSB Utrecht
29. Tomas J (2004) Chem Eng Technol 27:605–618
30. Tomas J (2004) Granular Matter 5:75–86
31. Dahnecke B (1972) J Colloid Interface Sci 40:1–13
32. Maugis D, Pollock, HM (1984) Acta Metall 32:1323-1334
33. Castellanos A (2006) Advances in Physics 1–143 (manuscript)
34. Fromm G (1927) Zeitschr Angew Math Mech 7:27–58
35. Cattaneo C (1938) Academia Nationale dei Lincei. Rendiconti. Serie 6. 27:342–348 434–436 474–478
36. Föppl L (1947) Die strenge Lösung für die rollende Reibung. Leibnitz Verlag. München
37. Sonntag G (1950) Zeitschr Angew Math Mech 30:73–83
38. Mindlin RD (1949) Trans ASME J Appl Mech 16:259–267
39. Mindlin RD, Deresiewicz H (1953) Trans ASME J Appl Mech 20:327–344
40. Thornton C (1991) J Phys D: Appl Phys 24:1942–1946
41. Renzo Di A, Maio Di FP (2004) Chem Engng Sci 59:525–541
42. Cattaneo C (1952) Ann Scuola Norm Sup Pisa (Serie III) 6:1–16
43. Deresiewicz H (1954) ASME J Appl Mech 21:52–56
44. Antonyuk S, Tomas J, Heinrich S, Mörl L (2005) Chem Eng Sci 60:4031–4044

Experience and Simulation in Dynamic Systems with Discontinuities

Hans Weber[1], Sandor Divenyi[2], Marcelo Savi[2], and Luiz Franca[3]

[1] PUC/Rio, Department of Mechanical Engineering hans@mec.puc-rio.br
[2] UFRJ/COPPE, Department of Mechanical Engineering savi@mecanica.ufrj.br
[3] CSIRO Petroleum – Drilling Mechanics Luiz.Franca@csiro.au

1 Introduction

Non-smooth nonlinearity is abundant in nature being usually related to either the friction phenomenon or the discontinuous characteristics as intermittent contacts of some system components. Non-smooth systems appear in many kinds of engineering systems and also in everyday life. Examples may be mentioned by the stick-slip oscillations of a violin string or grating brakes [1]. Some related phenomena as chatter and squeal causes serious problems in many industrial applications [2].

The mathematical modeling and numerical simulations of non-smooth systems present many difficulties, which turns their description unusually complex. Moreover, the dynamical behavior of these systems is complex, presenting a rich response. The idea that non-smooth systems can be considered as continuous in a finite number of continuous subspaces and also that the system parameters do not change in an abrupt manner, inspired some authors to try to describe non-smooth systems by a smoothed form [3,4,5,6]. Since non-smooth systems description involves many mathematical and numerical difficulties, experimental studies are of great importance. Literature presents references using experimental approaches to verify the proposed numerical methods and others discussing just the experimental point of view [7,8,9,10,11].

The objective of this research effort is the nonlinear analysis of dynamical systems with discontinuities. A single-degree of freedom system with discontinuous support is considered and, despite the deceiving simplicity of this problem, its nonlinear dynamics is very rich. Numerical and experimental approaches are employed. The mathematical model uses a smoothened switch model [3,4,6] that treats non-smooth systems defining different sets of ordinary differential equations. On the other hand, experimental approach considers an apparatus developed in order to verify numerical results. This apparatus is instrumented to obtain all system state variables, making possible the comparison of experimental results with those obtained by numerical simulations. In general, it is possible to say that numerical and experimental results are in close agreement. Besides that, the experimental apparatus is changed in order to treat different aspects related to the influence of internal impacts within the car and also support characteristics, showing how these phenomena alter the system dynamics.

Peter Eberhard (ed.), Multiscale Problems in Multibody System Contacts, 317–326.

2 Discontinuous Support Model

This contribution analyzes the dynamical response of a single-degree of freedom system with discontinuous support, shown in Figure 1. The oscillator is composed by a mass *m*, two linear springs with stiffness *k*. The dissipation process can be modeled by a linear viscous damping with coefficient *c*. Moreover the support is massless, but modeled as a linear spring with stiffness k_s and, again, a dissipation process represented by a linear damping with coefficient c_s. The displacement of the mass is denoted by *x*, relative to the equilibrium position. A gap *g* defines the distance between the mass and the support. Therefore, the system has two possible modes, represented by a situation where the mass presents contact with the support and another situation when there is no contact.

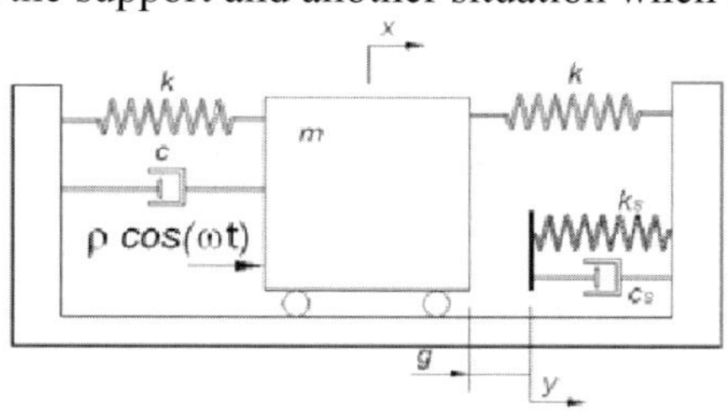

Fig. 1. Non-smooth system with discontinuous support.

By assuming that the relaxation time of the support is much smaller than the time between two contact events, the support dynamics can be neglected. Under this assumption, the system governing equations may be defined by two differential equations. One for situations without contact and the other with contact:

$$\begin{cases} m\ddot{x} + 2kx + c\dot{x} = \rho\cos(\omega t)\,, & \text{without contact.} \\ m\ddot{x} + 2kx + k_s(x-g) + (c+c_s)\dot{x} = \rho\cos(\omega t)\,, & \text{with contact.} \end{cases} \tag{1}$$

The state space $\tilde{\Gamma}$ of a system $\dot{u} = \tilde{f}(u)$, $u = (x, \dot{x})$ may be split into two subspaces $\tilde{\Gamma}_-$ and $\tilde{\Gamma}_+$, separated by a hyper-surface $\tilde{\Sigma}$, which is defined by a scalar indicator function $h(u)$ [3,6]. The description of the oscillator using this approach implies to perform an adequate definition of the transitions. The contact between the mass and the support occurs if the displacement becomes equal to the gap *g*. On the other hand, the mass loses contact with the support when the contact force vanishes, *i.e.*, if $f_s = -[k_s(x-g) + c_s\dot{x}] = 0$. Therefore, there are two subspaces with two different transition conditions. The subspaces are defined by two indicator functions:

$$h_\alpha(x,\dot{x}) = x - g \qquad h_\beta(x,\dot{x}) = -(x-g) - \frac{c_s}{k_s}\dot{x} \tag{2}$$

Now, the mass motion is described considering that there is no contact with the support if the state vector $u = (x, \dot{x}) \in \tilde{\Gamma}_-$. On the other hand, there is contact between the mass and the support if $u = (x, \dot{x}) \in \tilde{\Gamma}_+$. The transitions are related to the hyper-surface $\tilde{\Sigma}$, which consists of the conjunction of two hyper-surfaces $\tilde{\Sigma}_\alpha$ and $\tilde{\Sigma}_\beta$. The hyper-surface $\tilde{\Sigma}_\alpha$ defines the transition from $\tilde{\Gamma}_-$ to $\tilde{\Gamma}_+$, representing situations where the contact is caused when x becomes greater than g. The hyper-surface $\tilde{\Sigma}_\beta$ defines the transition from $\tilde{\Gamma}_+$ to $\tilde{\Gamma}_-$ as the contact is lost when the force of the support vanishes.

This non-smooth system may be smoothened redefining the subspaces and the transition hyper-surfaces, assuming that the transition has a linear variation from $\tilde{\Gamma}_-$ to $\tilde{\Gamma}_+$ (or from $\tilde{\Gamma}_+$ to $\tilde{\Gamma}_-$) in a thin space defined by a narrow band η around the hyper-surface of discontinuity. Figure 2 shows either the subspaces related to both non-smooth and smoothened systems.

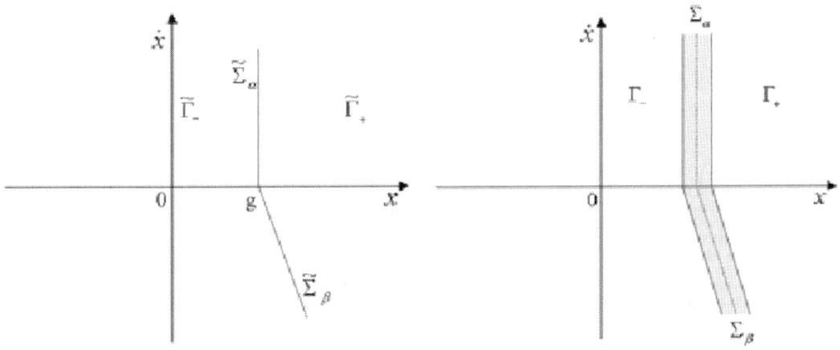

Fig. 2. Subspaces related to the system dynamics.

By assuming a transition through a narrow band with thickness η, made considering $f_\alpha = (1-q)\, f_- + q\, f_+$ with $q = (x - g + \eta)/2\eta$ and $f_\beta = (1-q)\, f_+ + q\, f_-$ with $q = [-x + g + \eta - (c_s / k_s)\dot{x}]/2\eta$ (notice that $0 \le q \le 1$ in both cases), the system $\dot{u} = f(u,t)$ can be written by four differential equations: $f_-, u \in \Gamma_-$, $f_\alpha, u \in \Sigma_\alpha$, $f_\beta, u \in \Sigma_\beta$, $f_+, u \in \Gamma_+$, where the equations are defined as follows [3,6]:

$$f_-(u,t) = \begin{Bmatrix} \dot{x} \\ -\frac{2k}{m}x - \frac{c}{m}\dot{x} + \frac{\rho}{m}\cos(\omega t) \end{Bmatrix} \tag{3a}$$

$$f_+(u,t) = \begin{Bmatrix} \dot{x} \\ -\frac{2k}{m}x - \frac{k_s}{m}(x-g) - \frac{c + c_s}{m}\dot{x} + \frac{\rho}{m}\cos(\omega t) \end{Bmatrix} \tag{3b}$$

$$f_\alpha = \begin{Bmatrix} \dot{x} \\ -\frac{2k}{m}x - \frac{c + c_s}{m}\dot{x} + \frac{\rho}{m}\cos(\omega t) - \frac{k_s}{2m}(x - g + \eta) + \frac{c_s}{m}\dot{x}_a\left(1 - \frac{x - g + \eta}{2\eta}\right) \end{Bmatrix} \tag{3c}$$

$$f_\beta = \begin{Bmatrix} \dot{x} \\ -\dfrac{2k}{m}x - \dfrac{c}{m}\dot{x} + \dfrac{\rho}{m}\cos(\omega t) - \dfrac{k_s}{2m}(x - g + \eta) - \dfrac{c_s}{2m}\dot{x} \end{Bmatrix} \tag{3d}$$

The subspaces and transitions hyper-surfaces are now represented by the following sets meaning that the calculation in the transition region may be started once the value $\dot{x}_a$ is known, *i.e.*, the velocity of the first point after the crossing of the border. With this starting, the whole transition is calculated and the mathematical process reproduces the experiments when all subspaces are crossed by the trajectory of the system. This means that following subspaces and transitions will be in the way of the trajectory:

$$\Gamma_- = \{u \in R^2 \,/\, h_\alpha(u) \le -\eta \text{ or } h_\beta(u) \ge \eta\} \tag{4a}$$

$$\Gamma_+ = \{u \in R^2 \,/\, h_\alpha(u) \ge \eta \text{ and } h_\beta(u) \le -\eta\} \tag{4b}$$

$$\Sigma_\alpha = \{u \in R^2 \,/\, -\eta < h_\alpha(u) < +\eta \text{ and } h_\beta(u) \le -h_\alpha(u)\} \tag{4c}$$

$$\Sigma_\beta = \{u \in R^2 \,/\, h_\alpha(u) > -h_\beta(u) \text{ and } -\eta < h_\beta(u) < +\eta\} \tag{4d}$$

The great deal in this point is the adequate definition of η to guarantee the good results of the method. This approach allows one to deal with non-smooth systems employing a smoothened system [6].

3 Experimental Apparatus

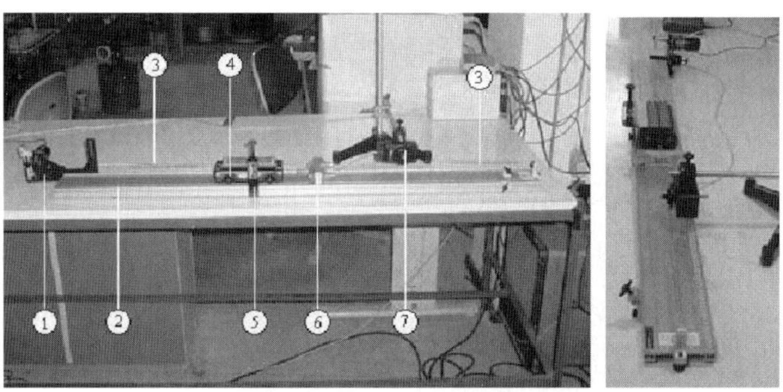

Fig. 3. Experimental apparatus.

In order to perform an experimental analysis of non-smooth systems, it is developed an apparatus of the one-degree of freedom oscillator discussed in the previous section (Figure 3). Basically, the apparatus is composed by an oscillator designed as a car (4), free to move on a rail (2), connected to an excitation system composed by springs (3), strings and a DC motor (1) (PASCO ME-8750 with 0-12 V e 0-0.3 A). Moreover, the apparatus has an adjustable magnetic damping device (5). The discontinuous support (6) is constructed considering a spring with a gap related to the car position. The movement is measured with the aid of a rotary sensor (7), PASCO CI-6538, which has a precision of ±0.25 degrees, maximum velocity of 30 rev/s and maximum sampling frequency of 1000 Hz. Parameters identification is done by different procedures. For a detailed description of the procedures, see reference [12].

4 Numerical and Experimental Results

This section considers the analysis of the system with discontinuous support considering both numerical and experimental approaches. The mass of the oscillator is identified as m = 0.838kg, while the stiffness and damping are, respectively, $k = 8.47$ N/m and $c = 0.87$ Ns/m. With respect to the support, it has a stiffness k_s = 1210 N/m obtained with the procedure discussed in the previous section. By analyzing an impulsive response of the system it is possible to employ classical expressions of logarithmic decrement in order to define $c_s = 0.60$ Ns/m, or $\xi_s = 0.0075$.

The proposed mathematical formulation is integrated using the Runge-Kutta-Fehlberg method, and numerical parameters need to be properly chosen. Divenyi *et al.* [6] show the importance of the choice of the narrow band thickness η, which is related to the proper transition of the system response in all subspaces. On this basis, it is assumed $\eta = 10^{-5}$ for all simulations.

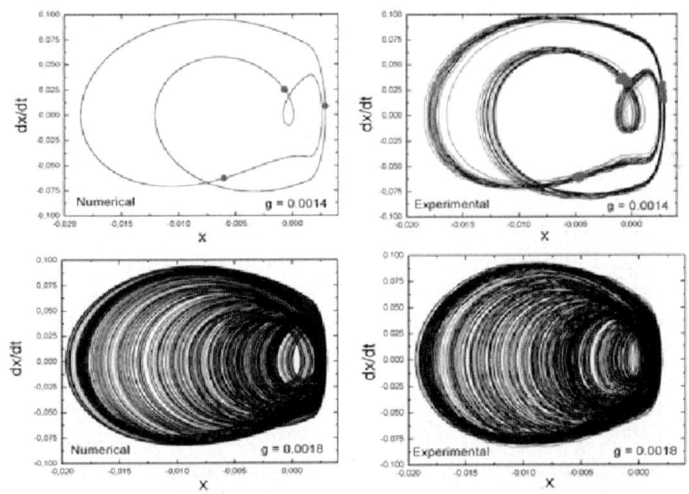

Fig. 4. State space comparing numerical and experimental results.

The gap influence is now focused on by assuming that $\omega = 11.15$ rad/s and ρ = 0.33. Figure 4 presents the state space for different values of the gap, g (0.0014 and 0.0018) showing, respectively, period-3 and chaotic responses. It should be pointed out the close agreement between numerical and experimental results. Strange attractors related to this chaotic response are presented in Figure 5, considering different positions of the Poincaré section. The left side presents numerical simulations while the right side presents those results obtained by the experimental apparatus. Once again, notice the agreement between them.

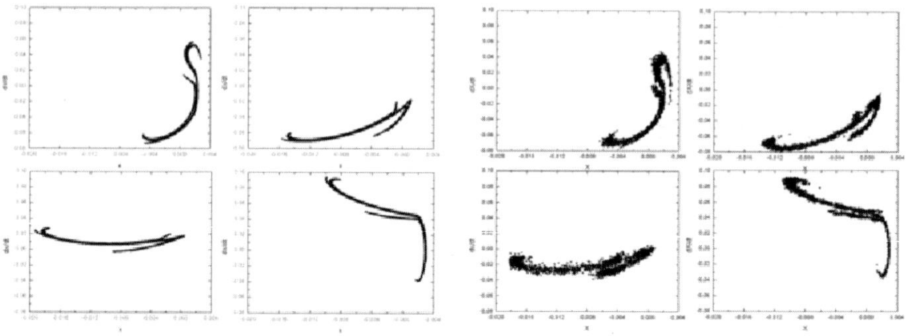

Fig. 5. Strange attractors related to numerical simulations (left side) and experimental data (rigth side), $\rho = 0.33$ N, $\omega = 11.15$ rad/s and $g = 0.0018$ m.

The forthcoming topics analyze the influence of different system characteristics on the dynamical response. Basically, two situations are trated: car characteristics and support characteristics. With respect to the car characteristics, there is considered a further impact system within the car. Concerning the support characteristics, different properties as stiffness and damping are considered, and also the inclusion of inertia aspects, connecting a mass to the support.

5 Influence of the Car Characteristics

At this point, the influence of the car characteristics is investigated. The goal here is to consider an impact mass, which may be free to move through a guide within the car, impacting at its ends. Since this impact mass may be fixed to the car, it is possible to establish a comparison between the dynamics with or without internal impacts. Figure 6 presents the car photo, showing the device used to promote internal impacts. Different parameters configurations are carried out changing the initial position of the impact mass and also the lubrication of the gide. Under this new configuration, the car has a total mass $m = 0.471$ kg.

Fig. 6. Different configurations related to the initial condition of the impact mass.

A comparison of the system dynamics considering the car with and without the internal impact is presented in Figure 7. The situation without impact is generated restricting the internal mass movement. The result without impacts presents large amplitudes and the car is in the imminence to jump out of the rail. On the other hand, internal impacts tend to dissipate system energy and, therefore, small amplitudes are expected.

After this, it is analyzed the influence of the initial position of the impact mass. Two different situations are considered. Figure 6 presents these two initial configurations. In the first one, the impact mass starts at the end near the support and therefore, is associated with a smaller first impact. The second one, on the other hand, the impact mass starts the movement in the opposite end of support being related to a greater first impact.

Figure 8 presents the comparison between the responses of these two configurations for $g = 62.1$ mm, $\rho = 0.76$ N, $\omega = 1.03$ Hz. When the system has a smaller first internal impact, the system response is related to a period-2 orbit. By considering the other situation, the first internal impact changes the orbit and the system response is related to a period-1 orbit.

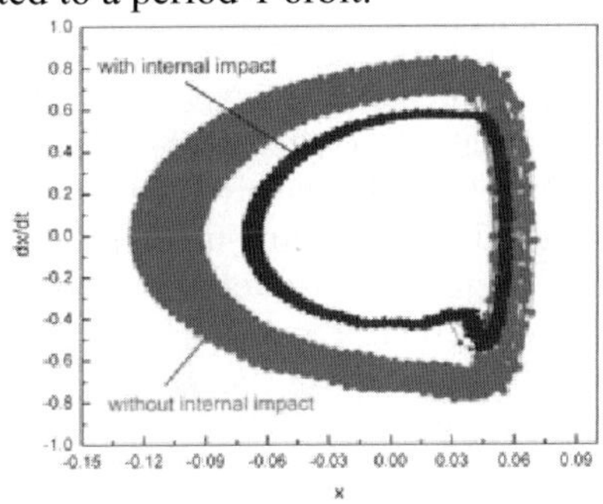

Fig. 7. Effect of internal impact. $g = 22.65$ mm, $\rho = 0.76$ N, $\omega = 1.51$ Hz.

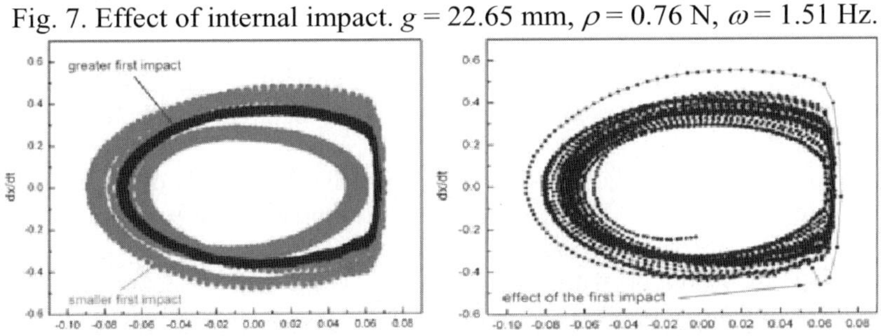

Fig. 8. Effect of internal impact: initial position of the internal mass. $g = 62.1$ mm, $\rho = 0.76$ N, $\omega = 1.03$ Hz

Now, the mass guide is lubricated causing differences in the system response. Figure 9 presents a comparison between the system with and without a special lubrication assuming $g = 50.4$ mm, $\rho = 0.76$ N, $\omega = 1.41$ Hz. Notice that the internal impacts causes perturbation in the system response, and the original orbit presents oscillations during a cycle. These oscillations are probably due to the force increase during internal impacts. Other situations may also cause the increase of the number of impacts for the same energy level.

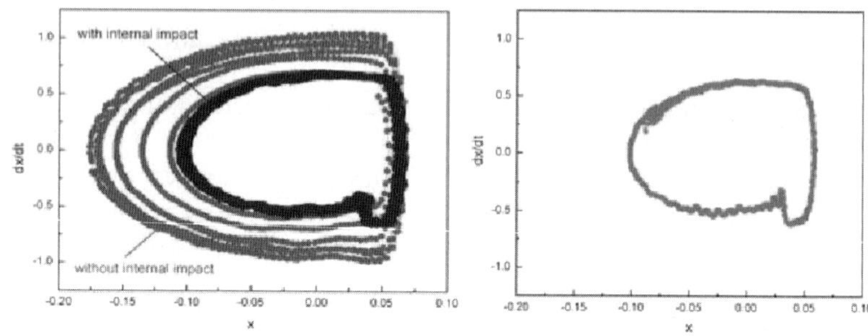

Fig. 9. Effect of internal impact: lubricated guide. g= 50.4 mm, ρ = 0.76 N, ω = 1.41 Hz

6 Influence of the Suppport Characteristics

In order to analyze the influence of the support characteristics in the system dynamics, the experimental set up is altered. The first consideration is just the change of the support elastic element to a spring with higher stiffness and also to a rigid rubber. It is introduced a constraint in order to avoid the impact mass movement and, therefore, there are no internal impacts. The different supports employed in the analysis represents three situations: low stiffness spring, high stiffness spring and rubber. Figure 10 shows the response for each support considering an oscillator with a mass m = 0.838 kg, excitation parameters ρ = 0.75 N, ω = 0.80 Hz and two different gaps: 2.5 and 16.3 mm. For the small gap, g = 2.5 mm it is noticeable that the low stiffness spring impacts just one time in a cycle. The other two supports, high stiffness spring and rubber, on the other hand, impacts twice. Therefore, low stiffness spring presents smaller amplitudes. By changing the gap, the system presents different behavior without the second impact.

The support inertia is now focused on. This investigation is done connecting a new car to the support. Figure 11 shows the apparatus where another car is connected to the support spring. Different support masses are analyzed considering an oscillator with mass m = 0.471 kg. The first noticeable effect associated with the support mass increase is the velocity decrease when the car looses the contact. This behavior may be understood just thinking in terms of momentum conservation. It is also related to this effect the characteristics of the non-contact response, when it is increasing the velocity for the next contact with the support. By observing the state space orbit, it is perceptible the orbit inclination as the mass is increased. Besides, by increasing the support mass, the position where the car looses contact also decreases. The forthcoming results are obtained assuming g = 5.55 mm and ρ = 0.75 N. Figure 12 shows a comparison among systems with different support inertia for ω = 0.88 Hz. The increase of the support inertia causes together with the inclination of the orbit the decrease of the second oscillation amplitude.

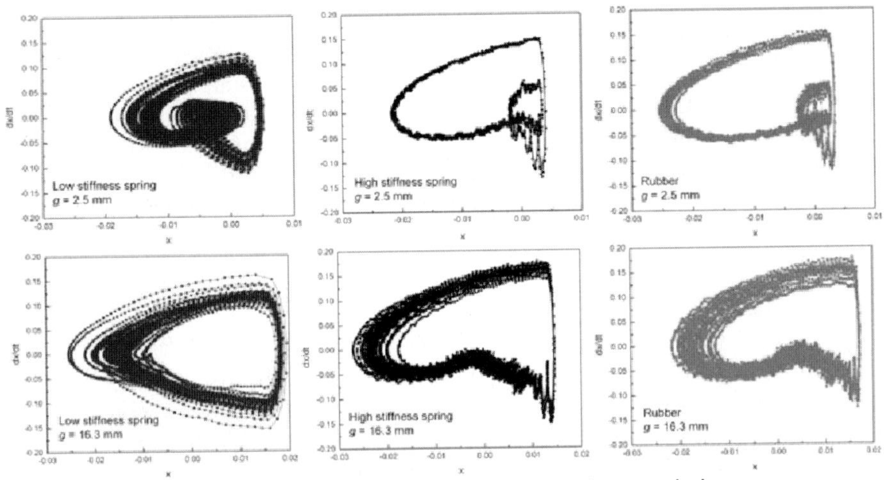

Fig. 10. Influence of different support characteristics.

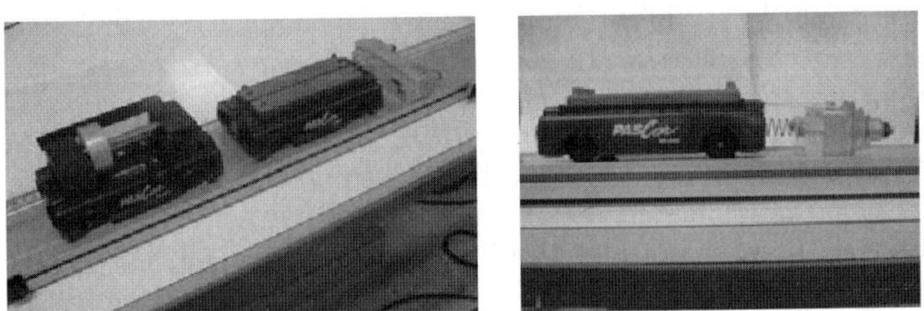

Fig. 11. Experimental apparatus with support inertia.

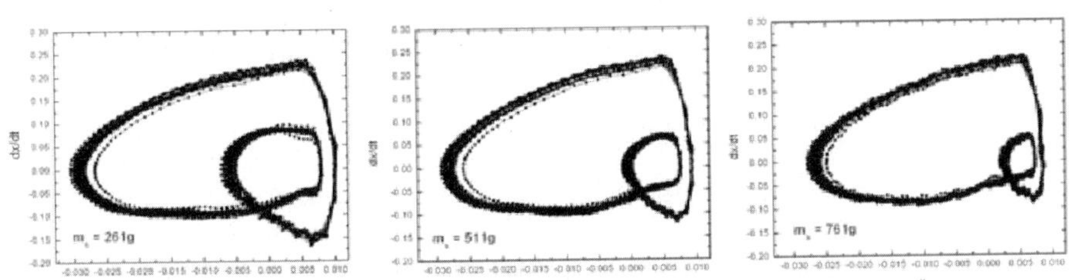

Fig. 12. Comparison of the system dynamics for different values of the support inertia.

7 Conclusions

This article presents the analysis of a non-smooth system with discontinuous support, considering both numerical and experimental approaches. A smoothened switch model is employed splitting the phase space into subspaces, defining finite regions to describe transitions among them. This procedure is useful for numerical simulations representing an effective form to integrate non-smooth equations. Moreover, an experimental apparatus is constructed in order to verify numerical results. Numerical and experimental investigations are carried out allowing the analysis of different aspects related to the system dynamics. In general, numerical and experimental results are in close agreement. Both approaches show a very rich dynamics, presenting dynamical jumps, bifurcations and chaos. In order to analyze some variations of the original experimental apparatus, it is considered an analysis related to the effect of the car characteristics and also the support. With respect to the car characteristics, an internal impact is introduced considering a mass free to move through a guide. Concerning the support, different kinds are considered changing the stiffness and also the influence of support inertia. These aspects can change the system dynamics in a qualitative point view and also may be desirable in order to dissipate energy.

References

1. Hinrichs N., Oestreich M., Popp K. (1998) "On the Modelling of Friction Oscillators", *Journal of Sound and Vibration*, v.216 (3), pp.435-459.
2. Andreaus U., Casini P. (2001) "Dynamics of Friction Oscillators Excited by a Moving Base and/or Driving Force", *Journal of Sound and Vibration*, v.245, n.4, pp.685-699.
3. Wiercigroch M. (2000) "Modelling of Dynamical Systems with Motion Dependent Discontinuities", *Chaos, Solitons and Fractals*, v.11, pp.2429-2442.
4. Leine R.I. (2000) "*Bifurcations in Discontinuous Mechanical Systems of Filippov-Type*", Ph.D. Thesis, Technische Universiteit Eindhoven.
5. Leine R.I., van Campen D.H., van de Vrande B.L. (2000) "Bifurcations in Nonlinear Discontinuous Systems", *Nonlinear Dynamics*, v.23, pp.105-164.
6. Divenyi S., Savi M.A., Franca L.F.P., Weber, H.I. (2006) "Nonlinear Dynamics and Chaos in Systems with Discontinuous Support", *Shock and Vibration*, to appear.
7. Wiercigroch M., Sin V. W. T., Li K. (1998) "Measurement of Chaotic Vibration in a Symmetrically Piecewise Linear Oscillator", *Chaos, Solitons and Fractals*, v.9, n.1/2, pp.209-220.
8. Wiercigroch M., Sin V.W.T. (1998) "Experimental Study of a Symmetrical Piecewise Base-Excited Oscillator", *Journal of Applied Mechanics - ASME*, v.65, pp.657-663.
9. Todd M.D., Virgin L.N. (1997) "An Experimental Impact Oscillator", *Chaos, Solitons & Fractals*, v.8, n.4, pp. 699-714.
10. Begley C.J., Virgin L.N. (1998) "Impact Response and the Influence of Fiction", *Journal of Sound and Vibration*, v.211, n.5, pp.801-818.
11. Slade K.N, Virgin L.N., Bayly P.V. (1997) "Extracting Information from Interimpact Intervals in a Mechanical Oscillator", *Physical Review E*, v.56, n.3, pp.3705-3708.
12. Savi M.A., Divenyi S., Franca L.F.P., Weber H.I. (2006) "Numerical and Experimental Investigations of the Nonlinear Dynamics and Chaos in Non-Smooth Systems", submitted to the *Journal of Sound and Vibration.*

Impacts on Beam Structures: Interactions of Wave Propagation and Global Dynamics

Roland Zander, Martin Förg, and Heinz Ulbrich

Lehrstuhl für Angewandte Mechanik, Technische Universität München,
Boltzmannstraße 15, 85748 Garching, Germany
[zander,foerg,ulbrich] @ amm.mw.tum.de

Summary. The paper investigates the interactions between impacting rigid bodies and the induced elastic waves in planar beam structures. Therefor, the framework for non-smooth dynamics of multibody systems including elastic impacts is applicated to a finite element description for beams in free planar motion. The model is used for two numerical examples: a cantilever beam with multiple impacts to several point masses and an elastic model of the rocking rod, which is used to adapt the restitution coefficient for a rigid model.

1 Introduction

In systems with impacts between elastic and rigid bodies the slow rigid body motions can be significantly influenced by fast effects in the elastic components like wave propagation. Thus, a correct description of the elastic behaviour has to be given in *m*ulti*b*ody *s*ystems (MBS) at least in the sense of the large-scale rigid body motions. To describe elastic effects during impacts in MBS, restitution coefficients can be used for the approximation of instantaneous interactions between different contacts in multiple contact collisions. Nevertheless, effects like wave propagation need to be described separately. The paper investigates a MBS-model for impacts to planar elastic beam structures.

A mathematical description of MBS with impacts is given in section 2. The system motion can be separated into impact free phases and impacts: for the first the contact law in normal direction is described by a unilateral constraint ensuring that the bodies do not penetrate each other. For the latter, the formulation of the impact law is crucial since the spatial discretisation has a strong influence on the local dynamics: for an infinitely fine discretisation of the elastic bodies the elastic impact behaviour is given by the structural elasticity. Due to a finite discretisation, discrete masses are involved in the impact process resulting in an energy loss if only in-penetrability is regarded. In order to control this energy loss, restitution coefficients are introduced using the NEWTON impact law. A theoretical study on the relation between the restitution coefficient and the spatial discretisation can be found e.g. in [3].

Peter Eberhard (ed.), Multiscale Problems in Multibody System Contacts, 327–338.

Section 3 briefly describes the *r*edundant *c*oordinate *m*ethod (RCM), which is used to model the free planar motion of an elastic beam: ideas of flexible MBS are combined with concepts of the *f*inite *e*lement (FE) method. Single elements are treated in a formulation of hybrid MBS giving exact rigid body movements for single elements. To permit the assembly of several elements to one structure, the equations of motion are transformed to a redundant coordinate set motivated by FE-ideas.

Impacts on beam structures are studied for two different numerical examples: Section 4 investigates the interaction between several rigid bodies and a cantilever beam. Multiple impacts occur, leading to changing influences on the elastic wave propagation. In section 5, the rocking rod, a classical example for non-smooth dynamics, is studied for elastic and rigid modelling. An elastic beam model is used to adapt a restitution coefficient for a rigid model.

2 Non-smooth multi-body dynamics

A brief overview is given for the equations describing the dynamics of systems with constraints, friction and frictional impacts. The constraints can be of bilateral as well as of unilateral type. Impacts can occur due to closing unilateral contacts, leading to jumps in the velocities of the system. The related systems are called non-smooth (see e.g. [5]). The general structure of the equations of motion is given in section 2.1. The force laws for bi- and unilateral constraints as well as for friction are formulated in section 2.2. Finally, models for elastic impacts complete the mathematical description.

The overall system dynamics including the contact and impact laws is formulated by an Augmented LAGRANGIAN approach. Within this approach the equations of motion are augmented by projection equations representing the physical constraints. The overall set of non-smooth, nonlinear equations can be solved by a root-finding algorithm [1]. An efficient time-stepping scheme for time integration of non-smooth MBS can be found in [2].

2.1 Dynamical equations for systems with discontinuities

The dynamics of a bi- and unilateral constrained system can be expressed by a finite dimensional measure differential equation [4]

$$\boldsymbol{M}\,\mathrm{d}\boldsymbol{u} = \boldsymbol{h}\,\mathrm{d}t + \boldsymbol{W}\mathrm{d}\boldsymbol{\Lambda} \quad . \tag{1}$$

The matrix $\boldsymbol{M} = \boldsymbol{M}(\boldsymbol{q})$ denotes the symmetric, positive definite mass matrix and depends on the f-dimensional vector of generalised coordinates $\boldsymbol{q} \in \mathbb{R}^f$. The vector $\boldsymbol{u} = \dot{\boldsymbol{q}}$ denotes the velocity vector. The acceleration measure

$$\mathrm{d}\boldsymbol{u} = \dot{\boldsymbol{u}}\,\mathrm{d}t + (\boldsymbol{u}^+ - \boldsymbol{u}^-)\mathrm{d}\eta \tag{2}$$

is the sum of the continuous part $\dot{\boldsymbol{u}}\,\mathrm{d}t$ and the discrete parts $(\boldsymbol{u}^+ - \boldsymbol{u}^-)\mathrm{d}\eta$. The second term is the difference of the left and the right limit of the velocities weighted by the sum of the DIRAC delta functions $\mathrm{d}\delta_i$ at the discontinuities t_i :

$$\mathrm{d}\eta = \sum_i \mathrm{d}\delta_i\,, \quad \mathrm{d}\delta_i = \mathrm{d}\delta(t-t_i) = \begin{cases} \infty & \text{if } t = t_i \\ 0 & \text{if } t \neq t_i \end{cases} \quad . \tag{3}$$

On the right hand side of equation (1) the vector $\boldsymbol{h} = \boldsymbol{h}(\boldsymbol{u}, \boldsymbol{q}, t)$ contains all smooth external, internal and gyroscopical forces. The reaction measure in the contacts $\boldsymbol{W}\mathrm{d}\boldsymbol{\Lambda}$ is decomposed by the generalised force directions $\boldsymbol{W} = \boldsymbol{W}(\boldsymbol{q})$ and the magnitudes $\mathrm{d}\boldsymbol{\Lambda}$. In analogy to the acceleration measure, the reaction measure $\mathrm{d}\boldsymbol{\Lambda}$ contains forces $\boldsymbol{\lambda}$ due to persisting contacts as well as impulses $\boldsymbol{\Lambda}$ due to collisions of bodies at the impact times t_i :

$$\mathrm{d}\boldsymbol{\Lambda} = \boldsymbol{\lambda}\mathrm{d}t + \boldsymbol{\Lambda}\mathrm{d}\eta \quad . \tag{4}$$

Integrating (1) under consideration of the DIRAC delta (3) yields the classical equations of motion for a constrained system and the impact equations.

2.2 Contact and impact laws

The computation of the accelerations $\dot{\boldsymbol{u}}$ as well as the post-impact velocities $\boldsymbol{u}_i^+$ in equation (1) requires the knowledge of the unknown contact reactions $\boldsymbol{\lambda}$ and $\boldsymbol{\Lambda}_i$, respectively. Thus, additional contact laws must be constituted. Contacts between bodies in the system are modelled as discrete point contacts whereby the contact zone is assumed to be totally rigid. Deformations of elastic components are only regarded in form of the overall discretisation, no local deformation e.g. of a beam cross-section is modelled. Consequently, a contact corresponds to a constraint. In this context two different types of contacts are considered, for which different contact laws hold: persisting contacts which are always closed and contacts that may be open or closed.

In the following, the force laws of the two contact types are explained for smooth and non-smooth motion. For this, the contact reactions

$$\boldsymbol{W}\mathrm{d}\boldsymbol{\Lambda} = (\boldsymbol{W}_N\ \boldsymbol{W}_T)\begin{pmatrix}\mathrm{d}\boldsymbol{\Lambda}_N \\ \mathrm{d}\boldsymbol{\Lambda}_T\end{pmatrix} = (\boldsymbol{W}_B\ \boldsymbol{W}_U\ \boldsymbol{W}_T)\begin{pmatrix}\mathrm{d}\boldsymbol{\Lambda}_B \\ \mathrm{d}\boldsymbol{\Lambda}_U \\ \mathrm{d}\boldsymbol{\Lambda}_T\end{pmatrix}$$

are decomposed into components normal (index N) – split up in bilateral (B) and unilateral (U) – and tangential (T) to the contact plane.

2.2.1 Dynamics between impacts

First of all, only smooth motion is considered, i.e. no impacts occur. Then a bilateral contact implies a bilateral constraint of the form

$$g_B = 0, \quad \lambda_B \in \mathbb{R} \quad , \tag{5}$$

where g_B denotes the normal distance of the interacting bodies in the contact

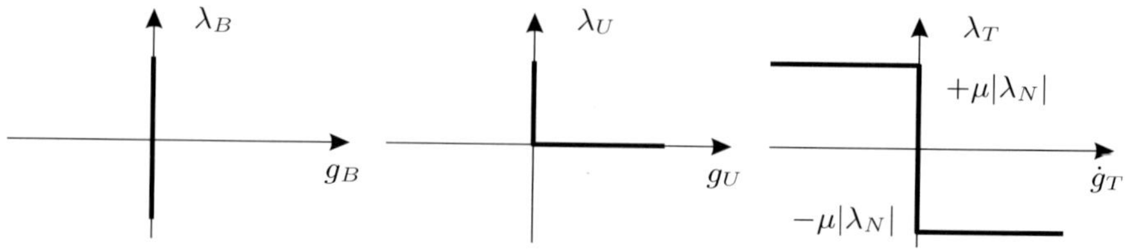

(a) Bilateral constraint (b) Unilateral constraint (c) COULOMB friction

Fig. 1. Force laws for bi- and unilateral contacts and friction.

point. The second type of contact also allows for detachment. The associated unilateral constraint is given by the SIGNORINI-FICHERA-condition

$$g_U \geq 0, \quad \lambda_U \geq 0, \quad g_U \lambda_U = 0 \quad . \tag{6}$$

The respective force laws are shown in figures 1(a) and 1(b).

For both bi- and unilateral constraints dry friction is considered. In order to establish COULOMB's law the force of a single contact is decomposed in a component $\lambda_N \in \{\lambda_B, \lambda_U\}$ normal to the contact plane and – in case of planar dynamics – a tangential component λ_T in friction direction. Using the relative tangential velocity $\dot{g}_T$, COULOMB's friction law is given by

$$\begin{aligned} \dot{g}_T &= 0 \quad \Rightarrow \quad |\lambda_T| \leq \mu_0 |\lambda_N| \\ \dot{g}_T &\neq 0 \quad \Rightarrow \quad \lambda_T = -\frac{\dot{g}_T}{|\dot{g}_T|} \mu |\lambda_N| \end{aligned} \quad . \tag{7}$$

The force law of a tangential frictional contact is plotted in figure 1(c).

2.2.2 Impact dynamics

In contrast to persisting and detaching contacts, a closing contact implies a discontinuity in the relative velocities and must be treated separately. The effect of an impact of a specific contact may concern all other constraints, the bilateral as well as the unilateral ones.

The impact law for a bilateral contact is given by

$$\dot{g}_B^+ = 0, \quad \Lambda_B \in \mathbb{R} \tag{8}$$

and ensures that relation (5) is not violated after collisions. For a unilateral constraint, reversibility within the impact event is considered. Therefor the normal relative velocity is interpolated:

$$\dot{\bar{g}}_U = \frac{\varepsilon_U}{1+\varepsilon_U} \dot{g}_U^- + \frac{1}{1+\varepsilon_U} \dot{g}_U^+$$

The restitution coefficient $0 \leq \varepsilon_U \leq 1$ allows to choose an impact behaviour between fully plastic and elastic for the normal direction. Given on impulsive level, NEWTON's impact law in the formulation of MOREAU [4]

$$\dot{g}_U^+ \geq 0, \quad \Lambda_U \geq 0, \quad \dot{g}_U^+ \Lambda_U = 0 \tag{9}$$

and COULOMB's friction law with the normal reaction $\Lambda_N \in \{\Lambda_B, \Lambda_U\}$

$$\begin{aligned} \dot{g}_T^+ &= 0 \quad \Rightarrow \quad |\Lambda_T| \leq \mu|\Lambda_N| \\ \dot{g}_T^+ &\neq 0 \quad \Rightarrow \quad \Lambda_T = -\frac{\dot{g}_T^+}{|\dot{g}_T^+|}\mu|\Lambda_N| \end{aligned} \tag{10}$$

hold for active contacts with $g_N = 0$ only.

3 Finite Element description in elastic MBS

After a short discussion of discretisation schemes for elastic components in section 3.1, section 3.2 briefly describes the model used for planar beam structures. Detailed descriptions can be found in [6]. The kinematics of a point to structure contact situation is described in section 3.3. Even though the paper focuses on planar beam dynamics, models for three dimensional motion can be derived following the same schemes. For a better readability, all dynamical equations are given for smooth phases only (compare section 2).

3.1 Discretisation strategies for constrained systems

The classical approach for the discretisation of elastic structures within MBS is to use globally defined *s*hape *f*unctions (SF), usually eigenforms. *M*oving *f*rame of *r*eference (MFR) approaches describe deformations of elastic bodies related to one moving reference. This approach is suitable for smooth and especially stationary motion using a compact mathematical formulation. However, the formulation does not provide a correct description for spatially bounded effects. For example, in impact situations local disturbances are induced into the structure which need to be described locally. Figure 2(a) compares a local disturbance and the approximation given by global harmonic SF. For a finite set of SF, a spontaneous global reaction of the discretised structure can

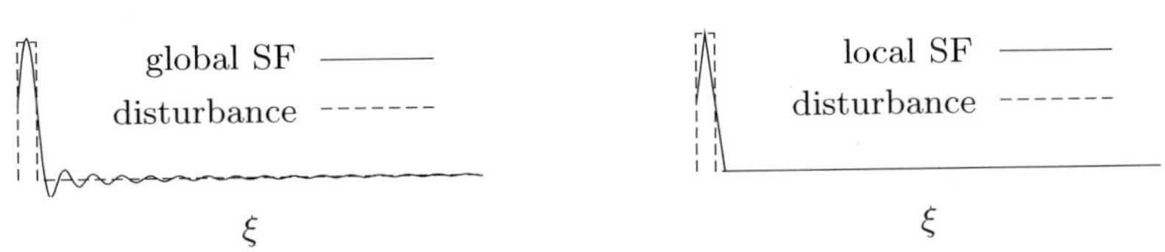

(a) global eigen-forms (b) local shape functions, FE

Fig. 2. Global and local shape functions for local disturbances.

be seen. In a physical interpretation this means an infinite velocity of wave propagation. Moreover, when using constraints for contact description, this leads to an instantaneous coupling of all contacts to one structure independent from their spatial distance, probably leading to a high computational effort in numerical simulation.

In contrast, *f*inite *e*lement (FE) approaches can be developed based on locally defined SF. This local definition restricts the spontaneous influence of a local disturbance to the physical representation of the element: figure 2(b) depicts this situation. The velocity of wave propagation is finite, only contacts to the same element of one structure are coupled within the contact problem.

To maintain the advantages of MFR discretisations, a single FE is described in a separate MFR: a single FE is treated in a formulation of hybrid MBS giving exact rigid body motion. The compact form of equations is maintained, longitudinal and transversal deformations can be approximated independently. To permit the assembly of several elements to one structure, the equations of motion are transformed to a redundant coordinate set. The *r*edundant *c*oordinate *m*ethod allows for large translations and geometrically large deformations of the entire structure.

3.2 Equations of motion and redundant coordinate sets

To describe the motion of the centre of gravity – which can be seen as rigid body or reference motion – and local deformations, an internal coordinate set

$$\boldsymbol{q}_{\mathrm{i}} = (x_S, y_S, \varphi_S, \tilde{\varepsilon}, a_l, \beta_l, a_r, \beta_r)^T \qquad \in \mathbb{R}^8$$

is used. The rigid body movements are described by x_S, y_S, φ_S, the longitudinal strain is approximated by $\tilde{\varepsilon}$ and $a_l, \beta_l, a_r, \beta_r$ describe the bending deflections. Except for $\tilde{\varepsilon}$, these degrees of freedom are illustrated in figure 3(a).

Figure 3(b) depicts the global coordinate set

$$\boldsymbol{q}_{\mathrm{g}} = (x_1, y_1, \varphi_1, a_1, a_2, x_2, y_2, \varphi_2)^T \qquad \in \mathbb{R}^8$$

which is used for coupling different FE to one discretised description of a

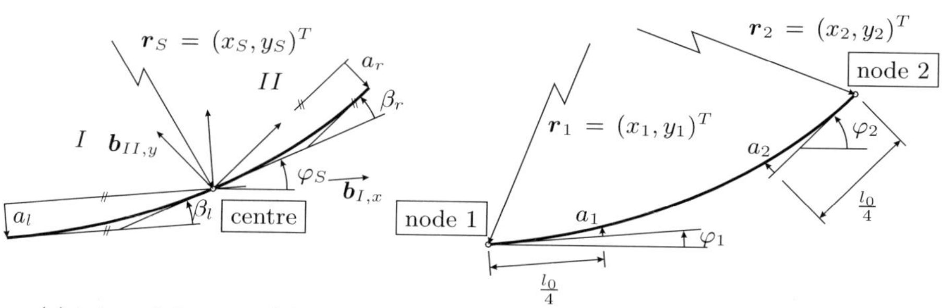

(a) internal degrees of freedom (b) global degrees of freedom

Fig. 3. Coordinate sets of one finite element.

structure without introducing constraints. The absolute coordinates x_1, y_1, φ_1 give the position and orientation of the FE-node 1, x_2, y_2, φ_2 of node 2. They are supplemented by the deflections a_1 and a_2 within the beam element. This FE-inspired set is used for time integration of the entire dynamical system.

Since both coordinate sets describe the same deformed element, $\boldsymbol{q}_\mathrm{i}$ and $\boldsymbol{q}_\mathrm{g}$ of one single FE are subjected to an explicit relation:

$$\boldsymbol{q}_\mathrm{i} = \boldsymbol{Q}(\boldsymbol{q}_\mathrm{g}) \qquad \in \mathrm{I\!R}^8 \tag{11}$$

This relation is used to derive dependencies for velocities and accelerations. Then these equations are used to transform the equations of motion of one single FE into a form in terms of the global coordinates.

Using the velocity $\boldsymbol{u}_\mathrm{i} = \frac{\mathrm{d}\,\boldsymbol{q}_\mathrm{i}}{\mathrm{d}\,t}$, the derivatives of the equality form (11) with respect to the time t are given by:

$$\boldsymbol{u}_\mathrm{i} = \boldsymbol{G}_\mathrm{ig}\boldsymbol{u}_\mathrm{g} \qquad \text{with} \qquad \boldsymbol{G}_\mathrm{ig} = \boldsymbol{G}_\mathrm{ig}(\boldsymbol{q}_\mathrm{g}) = \frac{\partial\,\boldsymbol{q}_\mathrm{i}}{\partial\,\boldsymbol{q}_\mathrm{g}} = \frac{\partial\,\boldsymbol{u}_\mathrm{i}}{\partial\,\boldsymbol{u}_\mathrm{g}} \tag{12}$$

$$\dot{\boldsymbol{u}}_\mathrm{i} = \dot{\boldsymbol{G}}_\mathrm{ig}\boldsymbol{u}_\mathrm{g} + \boldsymbol{G}_\mathrm{ig}\dot{\boldsymbol{u}}_\mathrm{g} \qquad \text{with} \qquad \dot{\boldsymbol{G}}_\mathrm{ig} = \dot{\boldsymbol{G}}_\mathrm{ig}(\boldsymbol{q}_\mathrm{g}, \boldsymbol{u}_\mathrm{g}) = \frac{\mathrm{d}\,\boldsymbol{G}_\mathrm{ig}}{\mathrm{d}\,t} \tag{13}$$

Equation (12) describes the linear dependencies between the internal and global velocities $\boldsymbol{u}_\mathrm{i}$ and $\boldsymbol{u}_\mathrm{g}$ given by the JACOBIAN-matrix $\boldsymbol{G}_\mathrm{ig}$.

The development of the equations of motion for a single FE in terms of the internal parameters $(\boldsymbol{q}_\mathrm{i}, \boldsymbol{u}_\mathrm{i})$ follows common approaches for flexible multibody systems and is described in [6] in detail. The position $\boldsymbol{r} = \boldsymbol{r}(\boldsymbol{q}_\mathrm{i}, s)$ and the velocity $\boldsymbol{v} = \boldsymbol{v}(\boldsymbol{q}_\mathrm{i}, \boldsymbol{u}_\mathrm{i}, s)$ of a cross-section at the material (or LAGRANGE) coordinate s are described in the moving coordinate systems I and II (see figure 3(a)) using the internal coordinates $\boldsymbol{q}_\mathrm{i}$ and velocities $\boldsymbol{u}_\mathrm{i}$. Using the LAGRANGE-method and introducing the contact forces $\boldsymbol{W}_\mathrm{i}\,\boldsymbol{\lambda}$, the equations of motion of one FE are derived in terms of the internal parameters:

$$\boldsymbol{M}_\mathrm{i}\dot{\boldsymbol{u}}_\mathrm{i} = \boldsymbol{h}_\mathrm{i} + \boldsymbol{W}_\mathrm{i}\,\boldsymbol{\lambda} \tag{14}$$

Here, $\boldsymbol{M}_\mathrm{i} = \boldsymbol{M}_\mathrm{i}(\boldsymbol{q}_\mathrm{i})$ is the mass-matrix, the vector $\boldsymbol{h}_\mathrm{i} = \boldsymbol{h}_\mathrm{i}(\boldsymbol{q}_\mathrm{i}, \boldsymbol{u}_\mathrm{i}, t)$ holds gyroscopical and elastic forces as well as smooth external loads. The generalised force directions $\boldsymbol{W}_\mathrm{i} = \boldsymbol{W}_\mathrm{i}(\boldsymbol{q}_\mathrm{i})$ of the contacts are addressed in the following section 3.3. The physical properties of the structure are described within $\boldsymbol{M}_\mathrm{i}$ and $\boldsymbol{h}_\mathrm{i}$ by the modulus E of elasticity, the density ρ and the cross section area A as well as the moment of inertia I .

Using the equations (11) to (13), the equations of motion (14) can be transfered to a form given in terms of the global coordinates $\boldsymbol{q}_\mathrm{g}$:

$$\underbrace{\boldsymbol{G}_\mathrm{ig}^T\boldsymbol{M}_\mathrm{i}\boldsymbol{G}_\mathrm{ig}}_{\boldsymbol{M}_\mathrm{g}}\,\dot{\boldsymbol{u}}_\mathrm{g} = \underbrace{\boldsymbol{G}_\mathrm{ig}^T\left(\boldsymbol{h}_\mathrm{i} - \boldsymbol{M}_\mathrm{i}\dot{\boldsymbol{G}}_\mathrm{ig}\boldsymbol{u}_\mathrm{g}\right)}_{\boldsymbol{h}_\mathrm{g}} + \underbrace{\boldsymbol{G}_\mathrm{ig}^T\,\boldsymbol{W}_\mathrm{i}}_{\boldsymbol{W}_\mathrm{g}} \tag{15}$$

To maintain the compact form (14) of the equations of motion for implementation, this transformation has to be performed during every evaluation of (15).

3.3 Contact kinematics and generalised force directions

Contacts and constraints are not restricted to nodes but can be applied to arbitrary positions on the structure. Furthermore – especially in the case of unilateral contacts – they can act on time-variant positions. If the contact is not at a constant position $s_c = \text{const.}$, its position s_c needs to be calculated in a first step. Assuming a point contact to a partner at $\boldsymbol{a}$, the root function

$$\boldsymbol{t}^T(\boldsymbol{a} - \boldsymbol{r}) = 0 \tag{16}$$

is solved for the FE nearest to $\boldsymbol{a}$ in order to compute s_c. Since the structure can not be said to be convex, multiple solutions of (16) may occur of which the one related to the smallest gap distance g_N is selected.

The geometric setup with the cross-sections position $\boldsymbol{r} = \boldsymbol{r}(\boldsymbol{q}_i, s)$ and tangent $\boldsymbol{t} = \boldsymbol{t}(\boldsymbol{q}_i, s)$ is depicted in figure 4. The contact is described by the relative distance $\boldsymbol{r}_{ac} = (\boldsymbol{a} - \boldsymbol{r}_c)$ and the relative velocity $\boldsymbol{v}_{ac} = (\dot{\boldsymbol{a}} - \boldsymbol{v}_c)$ with the position $\boldsymbol{r}_c = \boldsymbol{r}(\boldsymbol{q}_i, s_c)$ and the velocity $\boldsymbol{v}_c$ of the cross-section in the contact point. With the normal $\boldsymbol{n}_c = \boldsymbol{n}(\boldsymbol{q}_i, s_c)$ on the deformed FE in the contact point, the normal distance g_N as well as the normal and the tangential velocity

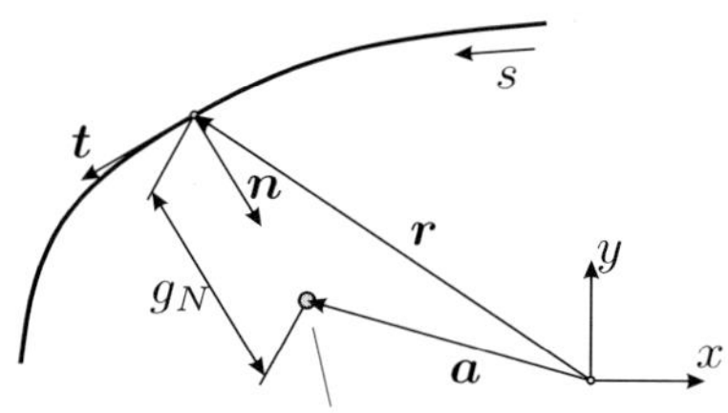

Fig. 4. Contact between beam structure and point obstacle.

$$g_N = \boldsymbol{n}_c^T \boldsymbol{r}_{ac} \quad , \quad \dot{g}_N = \boldsymbol{n}_c^T \boldsymbol{v}_{ac} \quad , \quad \dot{g}_T = \boldsymbol{t}_c^T \boldsymbol{v}_{ac}$$

of the contact points of the interacting bodies are projections in normal respectively tangential directions. Using the position $\boldsymbol{r}_c$ of the the contact point, the JACOBIAN-matrix of translations between the system-space and the generalised coordinates of one FE can be developed:

$$\boldsymbol{J}_i = \left(\frac{\partial \boldsymbol{r}_c}{\partial \boldsymbol{q}_i}\right)^T = \left(\frac{\partial \boldsymbol{v}_c}{\partial \boldsymbol{u}_i}\right)^T \quad . \tag{17}$$

Then the generalised force directions are the products

$$\boldsymbol{W}_{i,B} = \boldsymbol{J}_i \boldsymbol{f} \quad , \quad \boldsymbol{W}_{i,U} = \boldsymbol{J}_i \boldsymbol{n} \quad , \quad \boldsymbol{W}_{i,T} = \boldsymbol{J}_i \boldsymbol{t}$$

of $\boldsymbol{J}_i$ and the directions $\boldsymbol{f}$, $\boldsymbol{n}$ and $\boldsymbol{t}$ of constraint forces in the system-space.

4 Multiple impact situation

To investigate the interaction between point masses and a beam structure, a simple cantilever beam is regarded. One of the point masses excites the structure due to an initial impact, leading to impacts of the structure on the other masses. Especially the convergence towards one common solution for finer spatial discretisations is regarded.

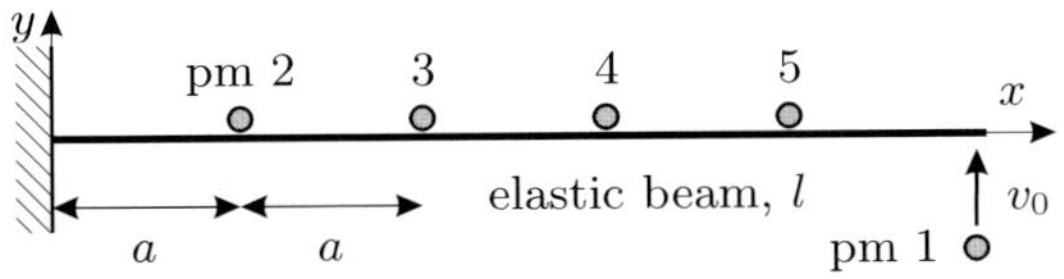

Fig. 5. Multiple impact situation.

4.1 Mechanical setup

The system, consisting of a cantilever beam which is fixed at one end and 5 *point masses* (pm) of mass m, is depicted in figure 5. Four of the point masses are aligned in equal distances a at the top of the beam with closed contacts to the beam. Initially the fifth point mass is detached from the beam and has a velocity v_1 in positive y-direction. The first impact occurs near the free end of the beam. The following parameters describe the system properties:

point masses:	$m = 0.25\,\mathrm{kg}$	
	$a = 0.1\,\mathrm{m}$	$v_1 = 1.0\,\frac{\mathrm{m}}{\mathrm{s}}$
beam:	$l = 0.5\,\mathrm{m}$	
	$E = 14.\cdot 10^9\,\frac{\mathrm{N}}{\mathrm{m}^2}$	$\rho = 8\cdot 10^3\,\frac{\mathrm{kg}}{\mathrm{m}^3}$
	$A = 6\cdot 10^{-5}\,\mathrm{m}^2$	$I = 4.5\cdot 10^{-11}\,\mathrm{m}^4$

Note that no gravity is regarded for the system and all contacts are treated as frictionless. Only fully plastic impacts ($\varepsilon = 0$) are regarded for all contacts.

4.2 Results

As reaction to the initial impact, a shock wave moves through the beam. Due to their contact to the beam, the point masses 2-5 are influenced by this wave propagation and excited to motions upwards. Figure 6(a) shows a good convergence in the velocities $\dot{y}$ of the point masses for finer spatial discretisations of the beam: the number n of FE is given as n FE. Especially the times of impacts between beam and the masses – which are marked as jumps in the velocity of the involved mass – are resolved very accurate.

Figure 6(b) shows a deformed state of the beam shortly after the initial impact. Note that the resulting deformation is determined by the contacts of the point mass in the middle and at the right side of the picture. An animation of the system dynamics can be found at the internet homepage of the institute.

5 Elastic rocking rod

One of the classical examples for a non-smooth dynamical system is the rocking rod. To allow for elastic impacts, the rigid description of the rod was extended by restitution coefficients for the contacts. Here, an elastic model is used as comparison and to define the restitution coefficient for the rigid model in order to approximate – as far as possible – the motion of the elastic model.

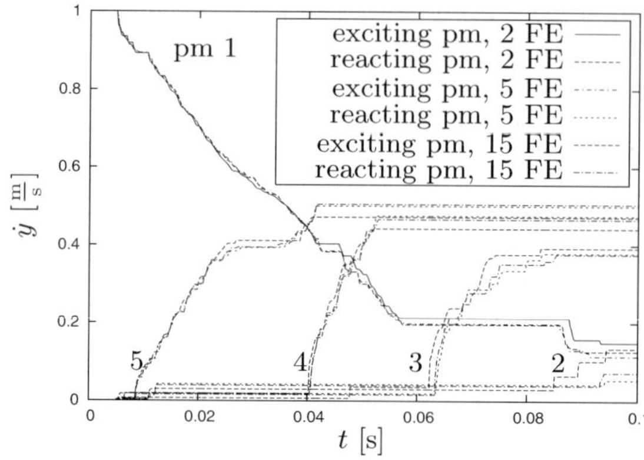

(a) Velocities of the 5 point masses

http://www.amm.mw.tum.de
→Research
→Animations

(b) Deformed state

Fig. 6. Simulation results for the multi-impact on the cantilever beam.

5.1 Mechanical setup

The rod is initialised in a free situation above the two pins 1 and 2 as displayed in figure 7. The height h gives the distance between the level of the pins with distance a and the rods centre of gravity. Except for the modulus E, the parameters for the elastic and the rigid model are identical:

pins:	$a = 0.4\,\text{m}$	$\mu = 0.1$
beam:	$h = 0.2079\,\text{m}$	$\varphi_0 = 12.0°$
	$l = 1.0\,\text{m}$	
	$E = 21. \cdot 10^{11}\,\frac{\text{N}}{\text{m}^2}$	$\rho = 8 \cdot 10^3\,\frac{\text{kg}}{\text{m}^3}$
	$A = 6 \cdot 10^{-5}\,\text{m}^2$	$I = 4.5 \cdot 10^{-11}\,\text{m}^4$
gravity:	$g = 9.81\,\frac{\text{m}}{\text{s}^2}$	

Equal parameters are used to describe both contacts: collisions for the elastic model are considered with $\varepsilon_e = 0$, whereas the restitution coefficient of the rigid model is chosen as $0 < \varepsilon_r < 1$. Friction is considered for both contacts.

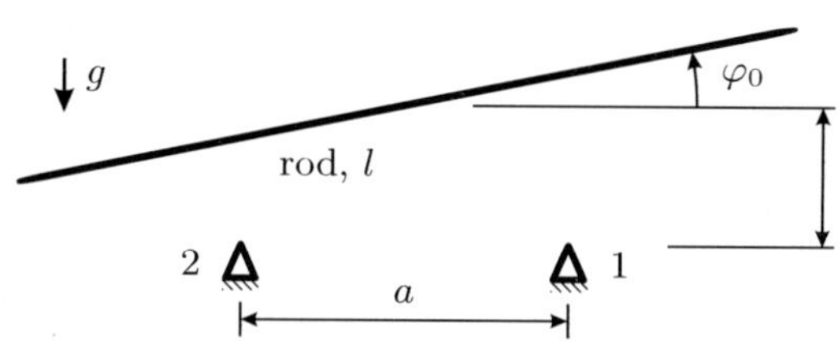

Fig. 7. Elastic rocking rod investigated with both models.

5.2 Optimisation of ε

To gain a reference solution, which can be used for the adaption of the restitution factor ε_{r}, a fine discretisation with 12 elements is used for the elastic model. Based on the positions $\boldsymbol{r}_{\mathrm{a}}$ and $\boldsymbol{r}_{\mathrm{b}}$ of the left and right endpoints of the rod, the movements of the rigid and the elastic model (index r and e) can be compared. Using the cumulative displacement

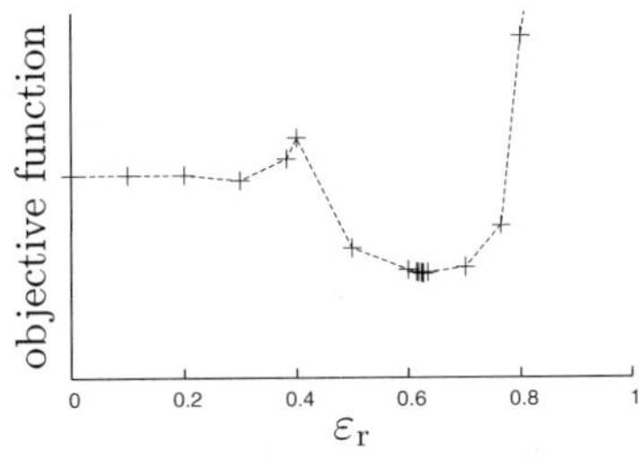

Fig. 8. Objective function for determination of ε_{r} .

$$f = \int_0^{t_{\mathrm{end}}} \left((\boldsymbol{r}_{\mathrm{a,r}} - \boldsymbol{r}_{\mathrm{a,e}})^2 + (\boldsymbol{r}_{\mathrm{b,r}} - \boldsymbol{r}_{\mathrm{b,e}})^2 \right) \mathrm{d}t$$

over simulation time $[0, t_{\mathrm{end}}]$, the restitution coefficient ε is found as solution of the minimisation problem $\min_{\varepsilon_{\mathrm{r}} \in [0,1]} f$. The objective function $f(\varepsilon_{\mathrm{r}})$ is plotted without scaling in figure 8, giving $\varepsilon_{\mathrm{r,opt}} = 0.6262$ as the restitution coefficient with the best description of the elastic model behaviour for rigid modelling.

5.3 Simulation results

The normal distance in both contacts 1 and 2 is plotted in figure 9(a) for the rigid model with $\varepsilon_{\mathrm{r,opt}}$ and the elastic model. Since only gravity acts until the first impact event (at the same time for both models), the results are identical in this time span. Due to $\varepsilon_{\mathrm{r,opt}} > 0$, the contact of the rigid model reopens instantaneously whereas the contact of the elastic model stays closed for a finite time span. This behaviour can be seen for all state changes in the contacts and can be interpreted as time delay between both models.

Figure 9(a) shows that the impact induces elastic waves in the elastic model, which can be seen as high frequency oscillations in the gap distance. For the dynamics of the rigid model, no elastic effects are included. The energy transfered to elastic waves must be seen as lost for the rigid model. Therefor the restitution coefficient is introduced to control both the energy loss and local rebound effects of the contact. Figure 9(b) shows a sequence of the animation for both models which can be found on the homepage of the institute.

6 Conclusions

The results presented in this paper show that non-smooth dynamics of elastic components can be formulated and simulated using the formalisms for MBS with impacts. These schemes allow to handle discontinuities in the state using a compact form. For elastic MBS, jumps in the velocities usually are related

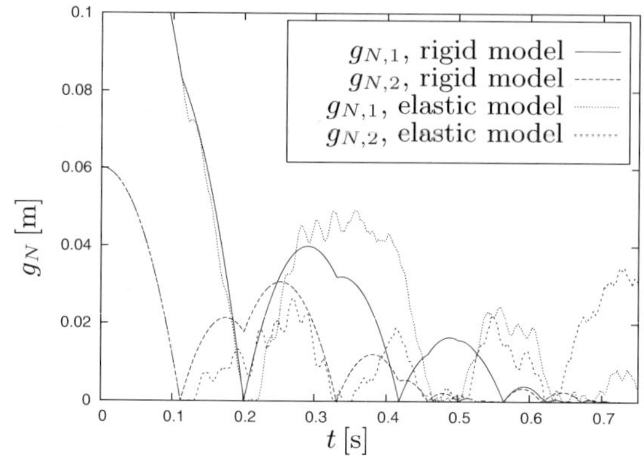

(a) Normal distances in both contacts

`http://www.amm.mw.tum.de`
→Research
→Animations

(b) System movements

Fig. 9. Simulation results for the elastic rocking rod.

to waves that are induced into a structural component. The resolution of these impact waves is defined by the spatial and the time discretisation. The presented model is used for research as well as in an industrial application for the simulation of a push-belt CVT (*c*ontinuously *v*ariabel *t*ransmission) [2].

In further work, enhanced models for contacts of beams to extended contours and other structures should be developed for the planar case. A three dimensional model for non-smooth dynamics of beams will be derived.

References

[1] FÖRG, M. ; NEUMANN, L. ; GEIER, Th. ; ULBRICH, H.: r-factor Strategies for the Augmented Lagrangian Approach in Multi-Body Contact Mechanics. In: *Proceedings of the III European Conference on Computational Mechanics.* Lisbon, Portugal : Springer, 2006

[2] GEIER, T. ; FÖRG, M. ; ZANDER, R. ; ULBRICH, H. ; PFEIFFER, F. ; BRANDSMA, A. ; VELDE, A. van d.: Simulation of a push belt CVT considering uni- and bilateral constraints. In: *ZAMM - Journal of Applied Mathematics and Mechanics.* Weinheim : WILEY-VCH Verlag GmbH, 2006. – to appear

[3] GLOCKER, Ch.: Concepts for modeling impacts without friction. In: *Acta Mechanica* Bd. 168. Wien, New York : Springer, 2004

[4] MOREAU, J. J.: Unilateral Contact and Dry Friction in Finite Freedom Dynamics. In: *International Centre for Mechanical Sciences, Courses and Lectures.* Bd. 302. Montpellier, France : Springer, 1988

[5] PFEIFFER, F. ; GLOCKER, Ch.: *Multibody Dynamics with Unilateral Contacts.* New York : John Wiley Inc., 1996

[6] ZANDER, R. ; ULBRICH, H.: Reference-free mixed FE-MBS approach for beam structures with constraints. In: *Nonlinear Dynamics* (2006). – to appear

IUTAM Bookseries

1. P. Eberhard: *IUTAM Symposium on Multiscale Problems in Multibody System Contacts*. 2007 ISBN 978-1-4020-5980-3